Lecture Notes in Computer Science 6600

Commenced Publication in 1973
Founding and Former Series Editors:
Gerhard Goos, Juris Hartmanis, and Jan van Leeuwen

James F. Peters   Andrzej Skowron
Hiroshi Sakai   Mihir Kumar Chakraborty
Dominik Slezak   Aboul Ella Hassanien
William Zhu (Eds.)

Transactions on Rough Sets XIV

 Springer

Editors-in-Chief

James F. Peters
University of Manitoba, Winnipeg, MB, Canada
E-mail: jfpeters@ee.umanitoba.ca

Andrzej Skowron
University of Warsaw, Poland
E-mail: skowron@mimuw.edu.pl

Guest Editors

Hiroshi Sakai
Kyushu Institute of Technology, Tobata, Kitakyushu, Japan
E-mail: sakai@mns.kyutech.ac.jp

Mihir Kumar Chakraborty
Jadavpur University and Indian Statistical Institute, Calcutta, India
E-mail: mihirc4@gmail.com

Dominik Slezak
University of Warsaw, Poland
E-mail: donimik.slezak@infobright.com

Aboul Ella Hassanien
Cairo University, Orman, Giza, Egypt
E-mail: aboitcairo@gmail.com

William Zhu
Zhangzhou Normal University, Zhangzhou, Fujian, China
E-mail: williamfengzhu@yahoo.com

ISSN 0302-9743 (LNCS)   e-ISSN 1611-3349 (LNCS)
ISSN 0302-9743 (TRS)   e-ISSN 1611-3349 (TRS)
ISBN 978-3-642-21562-9   e-ISBN 978-3-642-21563-6
DOI 10.1007/978-3-642-21563-6
Springer Heidelberg Dordrecht London New York
Library of Congress Control Number: 2011929861
CR Subject Classification (1998): F.4.1, F.1.1, H.2.8, I.5, I.4, I.2

Typesetting: Camera-ready by author, data conversion by Scientific Publishing Services, Chennai, India
Printed on acid-free paper
Springer is part of Springer Science+Business Media (www.springer.com)

Preface

Volume XIV of the *Transactions on Rough Sets* includes extensions of papers presented at the 12th International Conference on Rough Sets, Fuzzy Sets, Data Mining and Granular Computing (RSFDGrC 2009) held in the Indian Institute of Technology, Delhi, India, and published in the *Lecture Notes in Artificial Intelligence* volume 5908. The authors of 14 papers were invited to prepare extended manuscripts. Each submission was peer-reviewed by three reviewers. After two rounds of reviews, 10 articles were accepted for publication.

The contents of this special issue refer to both theory and practice. The topics include various rough set generalizations in combination with formal concept analysis, lattice theory, fuzzy sets and belief functions, rough and fuzzy clustering techniques, as well as applications to gene selection, Web page recommendation systems, facial recognition, and temporal pattern detection. In addition, this publication contains a regular article on rough multiset and its multiset topology.

The editors of this special issue would like to express gratitude to the authors of all submitted papers. Special thanks are due to the following reviewers: Hidenao Abe, Suruchi Chawla, Zied Elouedi, Homa Fashandi, Daryl Hepting, Andrzej Janusz, Na Jiao, Michiro Kondo, Pawan Lingras, Duoqian Miao, E.K.R. Nagarajan, Bidyut Kr. Patra, Sheela Ramanna, and Pulak Samanta. We would also like to acknowledge Sheela Ramanna for technical management of the TRS online system. The editors and authors of this volume extend their gratitude to Alfred Hofmann, Anna Kramer, Ursula Barth, Christine Reiss, and the LNCS staff at Springer for their support in making this volume of the TRS possible.

The Editors-in-Chief were supported by the Ministry of Science and Higher Education of the Republic of Poland research grants N N516 368334, N N516 077837, and the Natural Sciences and Engineering Research Council of Canada (NSERC) research grant 185986, Canadian Network of Excellence (NCE), and a Canadian Arthritis Network (CAN) grant SRI-BIO-05.

February 2011

Hiroshi Sakai
Mihir K. Chakraborty
Dominik Ślęzak
Aboul E. Hassanien
William Zhu
James F. Peters
Andrzej Skowron

LNCS Transactions on Rough Sets

The *Transactions on Rough Sets* series has as its principal aim the fostering of professional exchanges between scientists and practitioners who are interested in the foundations and applications of rough sets. Topics include foundations and applications of rough sets as well as foundations and applications of hybrid methods combining rough sets with other approaches important for the development of intelligent systems. The journal includes high-quality research articles accepted for publication on the basis of thorough peer reviews. Dissertations and monographs up to 250 pages that include new research results can also be considered as regular papers. Extended and revised versions of selected papers from conferences can also be included in regular or special issues of the journal.

Editors-in-Chief: James F. Peters, Andrzej Skowron
Managing Editor: Sheela Ramanna
Technical Editor: Marcin Szczuka

Editorial Board

Table of Contents

Evaluating a Temporal Pattern Detection Method for Finding Research Keys in Bibliographical Data

Hidenao Abe and Shusaku Tsumoto

Department of Medical Informatics, Shimane University,
School of Medicine
89-1 Enya-cho, Izumo, Shimane 693-8501, Japan
abe@med.shimane-u.ac.jp, tsumoto@computer.org

Abstract. According to the accumulation of the electrically stored documents, acquisition of valuable knowledge with remarkable trends of technical terms has drawn the attentions as the topic in text mining. In order to support for discovering key topics appeared as key terms in such temporal textual datasets, we propose a method based on temporal patterns in several data-driven indices for text mining. The method consists of an automatic term extraction method in given documents, three importance indices, temporal pattern extraction by using temporal clustering, and trend detection based on linear trends of their centroids. Empirical studies show that the three importance indices are applied to the titles of two academic conferences about artificial intelligence field as the sets of documents. After extracting the temporal patterns of automatically extracted terms, we discuss the trends of the terms including the recent burst words among the titles of the conferences.

Keywords: Text Mining, Trend Detection, TF-IDF, Jaccard's Matching Coefficient, Temporal Clustering, Linear Regression.

1 Introduction

The development of information systems in every field such as business, academics, and medicine in recent years have been enabled to store huge amount of various type data. Especially, document data Accumulation is advanced to document data by not the exception but various fields. Document data provides valuable findings to not only domain experts but also novice user. For such electrical documents, it becomes more important to sense emergent targets about researchers, decision makers, marketers by using temporal text mining techniques [1]. In order to realize such detection, emergent term detection (ETD) methods have been developed [2,3].

However, because the frequency of words was used in earlier methods, their detections were difficult as long as the word that became an object did not appear. Besides, emergent or new concepts are usually appeared as new combination of multiple words, coinages created by an author, and words with different spellings

J.F. Peters et al. (Eds.): Transactions on Rough Sets XIV, LNCS 6600, pp. 1–17, 2011.

of current words. Most conventional methods did not consider above-mentioned natures of terms and importance indices separately. This causes difficulties in text mining applications, such as limitations on the extensionality of time direction, time consuming post-processing, and generality expansions. For this reason, conventional ETD methods have been developed for detecting the particular state of appeared words or/and phrases[1] without considering any effect nor similarity between the keywords.

After considering these problems, we focus on temporal behaviors of importance indices of terms and their temporal patterns. Temporal behaviors of the importance indices of extracted phrases are paid attention so that a specialist may recognize emergent terms and/or such fields. In order to detect various temporal patterns of behaviors of terms in the sets of documents, we have proposed a method to identify the remarkable terms as continuous changes of multiple metrics of the terms [4]. Furthermore, we improved this method by clustering the temporal behaviors of the index values of each terms for extracting similar terms at the same time.

In this paper, we describe an integrated method for detecting trends of technical terms by combining automatic term extraction methods, importance indices of the terms, and temporal clustering in Section 3. After implementing this framework with the three importance indices, we performed a case study to extract remarkable temporal patterns of technical terms on the titles of AAAI and IJCAI in Section 4. By regarding these results, we evaluate the degree of the clusters of known emergent terms that burst their appearance frequencies in recent years. Finally, in Section 6, we summarize this paper.

2   Related Work

Text mining is a generic framework to find out valuable information from textual data by using some statistical methods. Finding keywords in the set of document is one of the useful approaches in text mining. Based on a corpus, by using clustering methods, the process to find out keywords with similar terms is called term clustering [5]. For this approach, rough set theory can be also applied as vocabulary mining [6]. However, such conventional term clustering methods have not treated temporal information.

There exist some conventional studies on the detection of emergent trend of terms/topics/themes in textual data. As the first step, Lent et. al [2] proposed a method for finding temporal trends of words. Then, by applying various metrics such as frequency [7], n-gram [8], and tf-idf [9], researchers developed some emergent term detection (ETD) methods [3]. The methods developed in [10,11] suggested for finding emergent theme patterns on the basis of a finite state machine by using Hidden Markov Model (HMM) as one of the advanced ETD method. Topic modeling [12] is a related method from the viewpoint of temporal text analysis. In these methods, researchers consider the changes in

---

[1] We call these important words and phrases in a corpus as 'keywords' in this article. In addition, words and phrases including the keywords in the corpus are called 'terms'.

each particular index of the terms rather, and they consider the emergent trend of terms as a discrete status based on appearance of the words.

In conventional studies on the detection of emergent words and/or phrases in documents such as Web pages and particular electronic message boards, researchers did not explicitly treat the trends of the calculated indices of words and/or phrases. However, based on two different techniques, we consider a framework for detecting temporal trends of phrases that consist of from two to nine words. We have focused on short phrases because a considerably long phrase may be a pattern including grammatical structure and anonymous words, as shown in [11].

As for examining the linear trends of the importance indices in temporal set of documents, we detected the two kinds of trends in our previous study [4,13]. In the previous studies, we can find both of emergent and subsiding technical phrases based on the trends of technical phrases. We used the degree based on the linear regression technique and the intercept for y-axis for ranking the two trends, however, some emergent phrases appears different behaviors based on the values of the importance indices compared to the composed indices by using PCA (Principal Component Analysis) [14]. In addition, the ranking lists of the emergent and subsiding remain a difficulty to understand the similarity among the listed terms. In order to overcome the difficulty, we introduced temporal pattern extraction phase to the previous work. This provides an abstracted layer for understanding the meanings of the groups of similar terms on the basis of temporal behaviors of terms, which are calculated as values of an importance index.

3 An Integrated Method for Detecting Remarkable Trends of Technical Terms as Temporal Patterns of Importance Indices

In this section, we describe the difference between conventional ETD methods and our proposal; detecting continuous temporal patterns of terms in temporal sets of documents.

As illustrated in Fig. 1, in order to find remarkable temporal trends of terms, we developed a framework for detecting various temporal trends of technical terms by using multiple importance indices consisting of the following four components:

1. Technical term extraction in a corpus
2. Importance indices calculation
3. Temporal pattern extraction
4. Trend detection

There are some conventional methods for detecting temporal trends of keywords in a corpus on the basis of each particular importance index. Although these methods calculate each index in order to detect important keywords, information

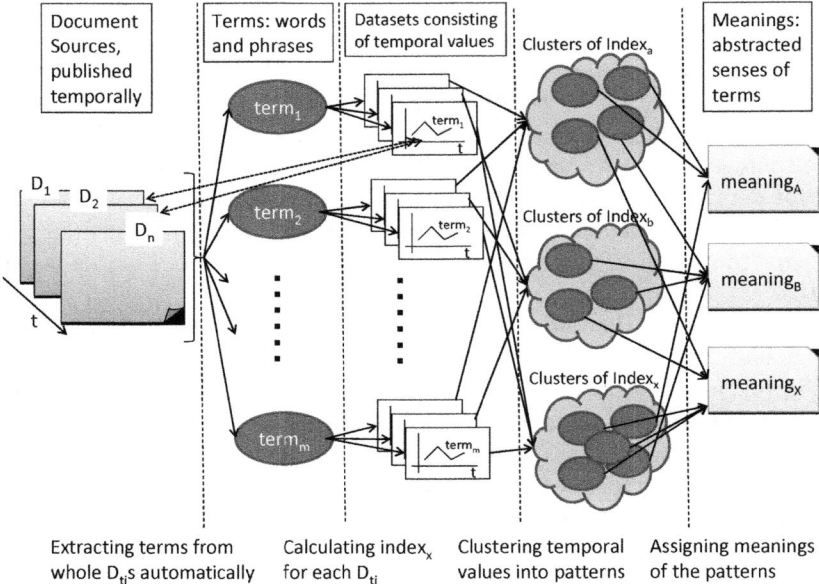

Fig. 1. Overview of the integrated temporal pattern extraction framework for technical terms

about the importance of each word or phrase is lost by cutting off the information with a threshold value. We suggest separating term determination and temporal trend detection based on importance indices. By separating these phases, we can calculate different types of importance indices in order to obtain a dataset consisting of the values of these indices for each term. Subsequently, we can apply many types of temporal analysis methods to the dataset based on statistical analysis, clustering, and machine learning algorithms.

3.1   Extracting Technical Terms in a Given Corpus

First, the system determines terms in the given corpus. There are two reasons why we introduce term extraction methods before calculating importance indices. One is that the cost of building a dictionary for each particular domain is very expensive task. The other is that new concepts need to be detected in a given temporal corpus. Especially, a new concept is often described in the document for which the character is needed at the right time in using the combination of existing words.

Considering the difficulties of the term extraction without any dictionary, we apply a term extraction method that is based on the adjacent frequency of compound nouns. This method involves the detection of technical terms by using the following values for each candidate continued noun CN:

$$FLR(CN) = f(CN) \times (\prod_{l=1}^{L}(FL(N_l) + 1)(FR(N_l) + 1))^{\frac{1}{2L}} \qquad (1)$$

where $f(CN)$ means frequency of the candidates CN, and $FL(N_l)$ and $FR(N_l)$ indicate the frequencies of different words on the right and the left of each noun N_l by taking bi-grams of N_l. In this scoring method, the continued nouns CN are assumed that they consist of $l(1 \leq l \leq L)$ nouns. In the experiments, we selected technical terms with this FLR score as $FLR(CN) > 1.0$. This threshold is important to select adequate terms at the first iteration. However, since our framework assumes the whole process as iterative search process for finding required trends of terms by a user, the user can input manually selected terms in the other iterations. In order to determine terms in this part of the process, we can also use other term extraction methods and terms/keywords from users.

3.2   Calculating Importance Indices of Text Mining in Each Period

As for importance indices of words and phrases in a corpus, there are some well-known indices. Term frequency divided by inversed document frequency (tf-idf) is one of the popular indices used for measuring the importance of the terms. tf-idf for each term $term_i$ can be defined as follows:

$$TFIDF(term_i, D_{t_j}) = TF(term, D_{t_j}) \times log\frac{|D_{t_j}|}{DF(term_i, D_{t_j})} \qquad (2)$$

where $TF(term, D_{t_j})$ is the frequency of each term $term_i$ in the set of t_jth documents D_{t_j}, and $|D_{t_j}|$ means the number of documents included in each period from t_{j-1} to t_j. $DF(term, D_{t_j})$ is the frequency of documents containing $term_i$.

As another importance index, we use Jaccard's matching coefficient [15][2]. Jaccard coefficient can be defined as follows:

$$Jaccard(term_i, D_{t_j}) = \frac{DF(w_1 \cap w_2 \cap ... \cap w_L, D_{t_j})}{DF(w_1 \cup w_2 \cup ... \cup w_L, D_{t_j})} \qquad (3)$$

where $DF(w_1 \cap w_2 \cap ... \cap w_L, D_{t_j})$ is equal to $DF(term_i, D_{t_j})$, because each $term_i$ consists of word or one more words $\{w_l | 1 \leq l \leq L\}$. $DF(w_1 \cup w_2 \cup ... \cup w_L, D_{t_j})$ means the frequency of documents that contains at least one word $w \in \{w_l | 1 \leq l \leq L\}$ that are included in the term $term_i$. Jaccard coefficient is originally defined as the ratio of the probability of an intersection divided by the probability of a union in a set of documents. In this framework, we applied this index by defining as the ratio of the frequency of an intersection divided by the frequency of a union in each set of documents D_{t_j}. Each value of Jaccard coefficient shows strength of co-occurrence of multiple words as an importance of the terms in the set of documents.

---

[2] Hereafter, we refer to this coefficient as "Jaccard coefficient".

	t_1	t_2	...	t_n
$term_1$	v_{11}	v_{12}	$... v_{1j} ...$	v_{1n}
$term_2$	v_{21}	v_{22}	$... v_{2j} ...$	v_{2n}
	v_{i1}	v_{i2}	$... v_{ij} ...$	v_{in}
$term_m$	v_{m1}	v_{m2}	$... v_{mj} ...$	v_{mn}

Fig. 2. Example of a dataset consisting of an importance index

In addition to the above two importance indices, we used simple appearance ratio of terms in a set of documents.

$$Odds(term_i, D_{t_j}) = \frac{DF(term_i, D_{t_j})}{|D_{t_j}| - DF(term_i, D_{t_j})} \tag{4}$$

where, $DF(term_i, D_{t_j})$ means the frequency of the appearance of each term $term_i$ in each set of documents D_{t_j}.

Fig.2 shows an example of the dataset consisting of an importance index for each year.

3.3 Extracting Temporal Patterns of Technical Terms Based on Importance Indices

Then, the framework provides the choice of some adequate trend extraction method to the dataset. In order to extract useful time-series patterns, there are so many conventional methods as surveyed in the literatures [16,17]. By applying an adequate time-series analysis method, users can find out valuable patterns by processing the values in rows in Fig.2.

Before calculating a similarity between the two sequences consisting of the values of importance indices of the two terms in D_{t_j}, we apply the following normalization to the values v_{ij} from $0 \leq y_{ij} \leq 1$.

$$y_{\cdot j} = \frac{v_{\cdot j} - min(v_{\cdot j})}{max(v_{\cdot j}) - min(v_{\cdot j})} \tag{5}$$

Then, we assign a similarity between two terms, $term_a$ and $term_b$, consisting of the two sequences, $\mathbf{a} = \{y_{aj} | 1 \leq j \leq n\}$ and $\mathbf{b} = \{y_{bj} | 1 \leq j \leq n\}$. For example, we can define the similarity between the two sequences based on Euclidian distance as the following:

$$Sim(\mathbf{a}, \mathbf{b}) = \sqrt{\sum_{j=1}^{n} (y_{aj} - y_{bj})^2} \tag{6}$$

The definition is used in the experiment in Section 4 with a clustering algorithm.

3.4 Assigning Meanings of the Terms and the Temporal Patterns Based on the Linear Trend to the Timeline

After identifying temporal patterns in the dataset, we apply the linear regression analysis technique in order to detect the degree of existing trends for each importance index.

The degree of each term $term_i$ is calculated as the following:

$$Deg(term_i) = \frac{\sum_{j=1}^{n} (y_i - \bar{y})(x_i - \bar{x})}{\sum_{j=1}^{n} (x_i - \bar{x})^2} \tag{7}$$

where the value in x-axis x_j means each timepoints as calculated $x_j = t_j - t_1$, and the value in y-axis y_j means the normalized value of v_{ij} of one importance index $Index_x$. In this definition, $\bar{x}$ is the average of overall period with the n time points, and $\bar{y}$ is the average of the values of each importance index for the period. Simultaneously, we calculate the intercept $Int(term)$ of each term $term_i$ as follows:

$$Int(term_i) = \bar{y} - Deg(term_i)\bar{x} \tag{8}$$

We observed the linear trends of each temporal pattern that are identified as temporal clusters. For each cluster c, the averaged degrees $Avg.Deg(c) = \sum_{term_i \in c} Deg(term_i)/num(term_i \in c)$ and intercepts $Avg.Int(c) = \sum_{term_i \in c} Int(term_i)/num(term_i \in c)$ of each term $term_i$ are used in the following experimentation.

4 Experiment: Extracting Temporal Patterns of Technical Terms by Using Temporal Clustering

In this experiment, we show the results temporal patterns by using the implementation of the method described in Section 3. As the input of temporal documents, we used the sets of the titles of the following two academic conferences[3]; AAAI and IJCAI.

We determine technical terms by using the term extraction method [19][4] for each entire set of documents.

---

[3] These titles are the part of the collection by DBLP [18].

[4] The implementation of this term extraction method is distributed in
   http://gensen.dl.itc.u-tokyo.ac.jp/termextract.html (in Japanese).

Table 1. Description of the titles of AAAI and IJCAI

Year	IJCAI # of titles	# of words	AAAI # of titles	# of words
1969			63	496
1971			66	551
1973			85	666
1975			148	1189
1977			235	1689
1980	95	665		
1981			219	1672
1982	104	754		
1983	92	693	271	2085
1984	69	431		
1985			263	1883
1986	199	1489		
1987	149	1093	244	1787
1988	150	1013		
1989			269	1967
1990	174	1234		
1991	160	1160	223	1635
1992	134	1005		
1993	135	1030	259	1986
1994	302	2261		
1995			346	2540
1996	305	2227		
1997	226	1697	254	1998
1998	206	1538		
1999	193	1535	233	1776
2000	228	1721		
2001			211	1611
2002	180	1366		
2003			348	2678
2004	194	1485		
2005	325	2497	350	2592
2006	385	3033		
2007	368	2958	491	3820
2008	356	2785		
TOTAL	4729	35670	4578	34621

Subsequently, the values of tf-idf, Jaccard coefficient, and Odds are calculated for each term in the annual documents. To the datasets consisting of temporal values of the importance indices, we extract temporal patterns by using k-means clustering. Then we apply the meanings of the clusters based on their linear trends calculated by the linear regression technique for the timeline.

4.1 Extracting Technical Terms

We use the titles of the two artificial intelligence (AI) related conferences as temporal sets of documents. The numbers of the titles as the documents is shown in Table 1.

As for the sets of documents, we assume each title of the articles to be one document. In order to consider the detailed differences of the spelling of words and phrases, we do not use any stemming technique to the documents.

By using the term extraction method with simple stop word detection for English, we extract technical terms from the two sets of the documents. From AAAI titles, the method extracted 5,629 terms, and 5,284 terms from IJCAI titles are extracted.

4.2 Extracting Temporal Patterns by Using k-Means Clustering

In order to extract temporal patterns of each importance index, we used k-means clustering implemented in Weka [20]. We set up the numbers of one percent of the terms as the number of clusters k as the 1% of the existing terms. The numbers of ks are set up 56 for AAAI and 53 for IJCAI respectively. By iterating less than 500 times, the system obtains clusters by using Euclidian distance between instances consisting of the values[5] of the same index. Since this implementation allows empty clusters, we got up to the k clusters for the datasets consisting of each index.

Table 2. The sum of squared errors of the clustering for the technical terms in the titles of the two conferences

Dataset	tf-idf		Jaccard Coefficient		Odds	
	# of Clusters	SSE	# of Clusters	SSE	# of Clusters	SSE
AAAI	32	50.81	52	921.12	32	4.21
IJCAI	31	29.69	48	568.00	31	2.61

Table 2 shows the result of the Sum of Seuared Error (SSE) of k-means clustering. As shown in this table, the SSE values of Jaccard coefficient are higher than the other two indices: tf-idf and Odds. The SSE values are calculated by using the following definition:

$$SSE(Index_x) = \sum_k \sum_n \sum_m (y_{ij} - c_k)^2, \tag{9}$$

where y_{ij} means the value of an importance index in a set of document D_{t_j}. The centroid of each cluster c_k is the averaged of the values that are included in the cluster. The value of each centroid is calculated as $c_k = Avg(y_{ij}|term_i \in c_k)$.

By using tf-idf and Odds, the system obtained the 32 clusters for AAAI and the 31 clusters for IJCAI respectively. The numbers of the obtained temporal patterns

---

[5] The system also normalized the values from 1 to 0 for each year.

as the clusters are less than the given k becase of the null clusters in the iterations. Since we did not selecte the terms with two or more words, the values of Jaccard coefficient of the terms with just one word, which are 0 or 1, are not suitable to make the clusters. Although the SSE evaluates numerical aspect of the obtained clusters, the meanings of the clusters should be considered for users.

4.3   Assigning Meanings of the Temporal Patterns Based on the Linear Trends

We obtained up to 1% of the existing terms for extracting temporal patterns as clusters of similar, the SSE is shown in Table 2. By using the averaged degree and the averaged intercept of the terms within each cluster c, we assign the following meanings to the temporal patterns:

- Popular
  - the averaged degree is positive, and the intercept is also positive.
- Emergent
  - the averaged degree is positive, and the intercept is negative.
- Subsiding
  - the averaged degree is negative, and the intercept is positive.

As shown in Table 3, the k-means clustering founds temporal clusters based on the temporal sequence of the three indices of the technical terms. The centroid terms mean the terms that have the highest FLR scores in each cluster.

As same as the result on AAAI titles, the k-means clustering founds 33 temporal clusters on IJCAI titles based on the temporal sequence of the three indices of the technical terms as shown in Table 4.

Based on the linear trends of the two indices, tf-idf detects the trends more sharply than Odds as shown in Table 3 and Table 4. Although the SSEs of the clusters of tf-idf are not better than Odds, the clusters of tf-idf can consider more clear meanings of the linear trends.

4.4   Details of a Temporal Pattern of the Key Technical Terms

By focusing on the result of the titles of IJCAI with tf-idf, in this section, we show the details of the temporal patterns, which are obtained as the temporal clusters. The averaged values of the tf-idf as temporal patterns are visualized in Fig. 3 and Fig. 4.

According to the meanings based on the linear trend, the clusters #1,#3, #10,#13, #16,#25, #26, and #27 have the patterns of tf-idf with the emergent trend based on the liner trend for the timeline as shown in Fig. 3. The terms included in these patterns are appeared in recent years in the conference. Fig. 4 also shows the patterns of tf-idf with the subsiding trend. Although these subsiding terms decrease their appearance in recent years, this represents that the achievements of the technical issues related to these terms have been obtaining new research topics. The appearances of these terms partly reflect the researchers interests, because these articles through the competitive peer-review process.

Table 3. The centroid information about the temporal patterns of the terms on the titles of AAAI from 1980 to 2008

Cluster ID	tf-idf					Jaccard Coefficient					Odds				
	num(c)	term	AvgDeg(c)	AvgInt(c)	Meaning	num(c)	term	AvgDeg(c)	AvgInt(c)	Meaning	num(c)	term	AvgDeg(c)	AvgInt(c)	Meaning
1	28	applications	-0.0009	0.0733	Subsiding	60	path	-0.0067	0.1906	Subsiding	13	constraint satisfaction	-0.0001	0.0333	Subsiding
2	295	agent modeling	0.0001	0.0018	Popular	58	semantic evaluation	-0.0006	0.0304	Subsiding	298	agent modeling	0.0000	0.0004	Popular
3	46	machine learning	0.0023	-0.0035	Emergent	57	causal explanation	0.0014	0.0037	Popular	41	local search	0.0008	0.0027	Popular
4	32	solving	0.0048	0.0704	Popular	46	plan evaluation	0.0024	-0.0128	Emergent	22	reinforcement learning	0.0011	0.0082	Popular
5	768	model of learning	0.0000	0.0029	Popular	49	analysis	0.0032	0.4305	Popular	771	model of learning	0.0000	0.0005	Popular
6	253	distributed learning	0.0003	-0.0005	Emergent	31	change	0.0089	-0.0412	Emergent	268	distributed learning	0.0001	-0.0001	Emergent
7	231	bayesian networks	0.0005	-0.0021	Emergent	134	stochastic optimization	0.0011	-0.0096	Emergent	211	bayesian networks	0.0001	-0.0003	Emergent
8	174	case-based reasoning	0.0000	0.0078	Popular	138	situation calculus	0.0037	-0.0179	Emergent	154	machine learning approach	0.0000	0.0013	Popular
9	193	information retrieval	-0.0004	0.0121	Subsiding	85	anytime coordination	0.0020	0.0013	Popular	177	default reasoning	-0.0001	0.0023	Subsiding
10	294	using models	0.0000	0.0028	Popular	35	local search	0.0008	0.0012	Popular	314	using models	0.0000	0.0007	Popular
11	361	natural language generation	0.0004	-0.0028	Emergent	40	automatic algorithm designer	-0.0037	0.0793	Subsiding	368	concept learning	0.0001	-0.0007	Emergent
12	178	learning 10	-0.0001	0.0088	Subsiding	104	knowledge acquisition	0.0000	0.0074	Popular	191	temporal reasoning	0.0000	0.0023	Popular
13	250	multi-agent system	0.0006	-0.0037	Emergent	95	expert advice	0.0058	-0.0461	Emergent	291	multi-agent system	0.0002	-0.0011	Emergent
14	99	qualitative reasoning	-0.0013	0.0322	Subsiding	154	multi-agent system	0.0006	-0.0051	Emergent	98	qualitative reasoning	-0.0002	0.0064	Subsiding
15	252	plan recognition	0.0003	0.0012	Popular	54	real-time performance	-0.0011	0.0271	Subsiding	273	expert systems	0.0001	0.0005	Popular
16	162	logic programming	-0.0004	0.0116	Subsiding	72	action descriptions	0.0020	-0.0182	Emergent	148	logic programming	-0.0001	0.0024	Subsiding
17	205	using qualitative reasoning	0.0004	0.0094	Subsiding	106	scheduling	0.0158	-0.0007	Emergent	181	using qualitative reasoning	-0.0001	0.0017	Subsiding
18	414	learning representation	0.0004	-0.0029	Emergent	169	sets	0.0081	-0.0429	Emergent	419	learning representation	0.0001	-0.0006	Emergent
19	263	active learning	0.0002	0.0017	Popular	55	conflict resolution	-0.0013	0.0477	Subsiding	291	case-based reasoning	0.0001	0.0005	Popular
20	44	semantic web	0.0022	-0.0105	Emergent	69	case-based reasoning	0.0015	-0.0123	Emergent	50	semantic web	0.0008	-0.0021	Emergent
21	110	knowledge based design system	-0.0011	0.0255	Subsiding	18	computing	0.0065	0.0719	Popular	104	knowledge based design system	-0.0002	0.0050	Subsiding
22	30	local search	0.0042	-0.0032	Emergent	112	simple robots	0.0038	-0.0323	Emergent	25	solving	0.0021	0.0065	Popular
23	189	using ai	-0.0001	0.0076	Subsiding	83	planner	-0.0009	0.0687	Subsiding	197	production systems	0.0000	0.0020	Popular
24	14	design	-0.0046	0.2281	Subsiding	91	image analysis	-0.0075	0.1950	Subsiding	15	planning	0.0011	0.0734	Popular
25	47	temporal reasoning	0.0003	0.0359	Popular	1710	learning 10	-0.0001	0.0041	Subsiding	33	applications	0.0002	0.0097	Popular
26	398	structure learning	0.0005	-0.0046	Emergent	23	constraint satisfaction	-0.0044	0.3338	Subsiding	432	structure learning	0.0002	-0.0013	Emergent
27	26	knowledge representation	-0.0027	0.1086	Subsiding	53	mobile robots	0.0001	0.0165	Popular	15	knowledge representation	-0.0011	0.0351	Subsiding
28	177	inductive learning	-0.0002	0.0079	Subsiding	93	machine learning	0.0011	-0.0076	Emergent	157	inductive learning	0.0000	0.0013	Popular
29	34	reinforcement learning	0.0026	0.0499	Popular	56	causal theories	0.0007	0.0001	Popular	20	design	0.0017	0.0212	Popular
30	45	mobile robot	0.0017	0.0138	Popular	105	reinforcement learning	0.0009	-0.0040	Emergent	49	mobile robot	0.0004	0.0052	Popular
31	14	planning	0.0013	0.2465	Popular	130	knowledge integration	0.0022	-0.0155	Emergent	2	learning	0.0123	0.0785	Popular
32	3	learning	0.0083	0.5570	Popular	85	fuzzy logic	0.0008	0.0190	Popular	1	a	0.0000	1.0000	Popular
33						93	spatial aggregation	0.0007	0.0181	Popular					
34						62	probabilistic planning	0.0003	0.0009	Popular					
35						65	classification	0.0124	0.1723	Popular					
36						140	object recognition	-0.0003	0.0630	Subsiding					
37						22	knowledge representation	-0.0026	0.0607	Subsiding					
38						41	production systems	-0.0006	0.0212	Subsiding					
39						36	anytime	0.0063	0.0930	Popular					
40						150	efficient algorithm	0.0006	-0.0042	Emergent					
41						93	mobile robot	-0.0021	0.0546	Subsiding					
42						71	automated design	0.0031	-0.0209	Emergent					
43						68	bayesian reasoning	0.0004	0.0024	Popular					
44						100	continuous domains	0.0020	-0.0019	Emergent					
45						124	information extraction	-0.0029	0.0854	Subsiding					
46						56	generating multiple	0.0001	0.0070	Popular					
47						29	default reasoning	-0.0013	0.0288	Subsiding					
48						84	qualitative simulation	0.0003	0.0040	Popular					
49						92	belief revision	-0.0029	0.0835	Subsiding					
50						21	ontology	0.0125	-0.0430	Emergent					
51						46	learning	-0.0053	0.6084	Subsiding					
52						36	temporal reasoning	-0.0028	0.0605	Subsiding					

Table 4. The centroid information about the temporal patterns of the terms on the titles of IJCAI from 1969 to 2007

Cluster ID	tf-idf					Jaccard Coefficient					Odds				
	num(c)	term	Avg.Deg(c)	Avg.Int(c)	Meaning	num(c)	term	Avg.Deg(c)	Avg.Int(c)	Meaning	num(c)	term	Avg.Deg(c)	Avg.Int(c)	Meaning
1	384	information integration	0.0003	-0.0031	Emergent	57	using	0.0045	0.4084	Popular	380	dynamic model	0.0001	-0.0006	Emergent
2	1	a	0.0000	1.0000	Popular	22	simulation	0.0034	0.0891	Popular	1	a	0.0000	1.0000	Popular
3	689	image interpretation	0.0002	-0.0011	Emergent	82	belief functions	-0.0003	0.0502	Subsiding	500	planning using	0.0001	-0.0008	Emergent
4	260	qualitative reasoning	0.0000	0.0030	Popular	70	data base	-0.0029	0.0811	Subsiding	266	qualitative reasoning	0.0000	0.0006	Popular
5	18	system	-0.0066	0.3159	Subsiding	78	robot control	0.0002	0.0060	Popular	11	learning	0.0041	0.0262	Popular
6	201	semantic networks	-0.0004	0.0161	Subsiding	105	polynomial time	0.0044	-0.0221	Emergent	191	semantic networks	-0.0001	0.0027	Subsiding
7	116	automatic programming	-0.0009	0.0286	Subsiding	65	ai	0.0099	0.0719	Popular	21	expert system	-0.0002	0.0134	Subsiding
8	264	concept learning	-0.0002	0.0078	Subsiding	120	object matching	0.0025	-0.0254	Emergent	245	concept learning	0.0000	0.0014	Popular
9	292	planning using	0.0001	0.0014	Popular	95	theorem proving	0.0020	-0.0010	Emergent	286	using classification	0.0000	0.0003	Popular
10	297	learning system	0.0003	-0.0006	Emergent	75	text understanding	0.0004	0.0359	Popular	319	learning system	0.0001	-0.0001	Emergent
11	97	robot control	-0.0013	0.0428	Subsiding	71	image interpretation	0.0007	-0.0080	Emergent	76	robot control	-0.0001	0.0039	Subsiding
12	38	heuristic search	-0.0011	0.0722	Subsiding	169	temporal reasoning	0.0003	-0.0027	Emergent	25	resolution	0.0005	0.0029	Popular
13	30	logic programming	0.0017	0.0030	Popular	48	knowledge acquisition	0.0005	-0.0014	Emergent	10	information extraction	0.0006	-0.0013	Emergent
14	266	expert systems	-0.0001	0.0052	Subsiding	69	artificial intelligence	-0.0057	0.2132	Subsiding	259	expert systems	0.0000	0.0009	Popular
15	260	causal reasoning	-0.0001	0.0051	Subsiding	21	constraint satisfaction	0.0088	0.0506	Popular	531	causal reasoning	0.0000	0.0010	Popular
16	37	scheduling	0.0019	-0.0115	Emergent	60	situation calculus	0.0016	-0.0196	Emergent	18	heuristic search	0.0008	-0.0040	Emergent
17	84	fuzzy logic	-0.0011	0.0337	Subsiding	1781	knowledge representatio	0.0000	0.0022	Popular	115	artificial intelligence	-0.0003	0.0100	Subsiding
18	34	action	0.0030	0.0100	Popular	60	using data	0.0002	-0.0001	Emergent	26	reinforcement learning	0.0005	0.0008	Popular
19	47	machine learning	0.0013	0.0041	Popular	68	expert system	-0.0005	0.0190	Subsiding	46	temporal reasoning	0.0003	-0.0017	Emergent
20	49	mobile robot	0.0006	0.0247	Popular	91	information extraction	0.0015	-0.0127	Emergent	27	uncertainty	0.0003	0.0058	Popular
21	267	inductive learning	0.0001	0.0026	Popular	47	prolog	0.0067	0.0477	Popular	255	semantic model	0.0000	0.0006	Popular
22	16	models	0.0048	0.0445	Popular	151	constructive induction	0.0053	-0.0169	Subsiding	35	ai	0.0010	0.0029	Popular
23	333	using knowledge	0.0001	0.0001	Popular	92	decision trees	-0.0016	0.0625	Subsiding	336	using knowledge	0.0000	0.0001	Popular
24	19	application	-0.0019	0.1414	Subsiding	117	model-based diagnosis	0.0014	0.0130	Popular	16	system	-0.0012	0.0730	Subsiding
25	295	semantic analysis	0.0003	-0.0019	Emergent	43	logic programming	0.0004	-0.0001	Emergent	345	semantic analysis	0.0001	-0.0004	Emergent
26	286	network learning	0.0005	-0.0031	Emergent	70	heuristic search	-0.0004	0.0193	Subsiding	285	network learning	0.0001	-0.0006	Emergent
27	371	using ai	0.0004	-0.0045	Emergent	17	inheritance	0.0052	0.0249	Popular	402	using ai	0.0001	-0.0010	Emergent
28	11	learning	0.0055	0.1987	Popular	91	learning	0.0150	0.0698	Popular	19	solving	0.0018	0.0029	Popular
29	22	solving	0.0010	0.0876	Popular	134	dynamic systems	0.0055	-0.0463	Emergent	54	machine learning	0.0003	0.0011	Popular
30	225	vision system	-0.0003	0.0088	Subsiding	42	domain constraints	-0.0003	0.0174	Subsiding	211	using process knowledge	0.0000	0.0011	Popular
31	18	expert system	-0.0005	0.0689	Subsiding	117	symbolic execution	-0.0011	0.0597	Subsiding	16	structure	-0.0001	0.0254	Subsiding
32						70	path planning	-0.0001	0.0099	Subsiding					
33						34	examples	0.0119	-0.0404	Emergent					
34						27	natural language	0.0011	0.2089	Popular					
35						73	semantic networks	-0.0009	0.0471	Subsiding					
36						90	stereo vision	0.0009	0.0082	Popular					
37						22	structure	0.0009	0.3700	Popular					
38						20	mobile robot	0.0010	-0.0035	Emergent					
39						104	reinforcement learning	0.0013	-0.0087	Emergent					
40						67	expert systems	-0.0001	0.0079	Subsiding					
41						94	concept decomposition	-0.0021	0.0658	Subsiding					
42						118	boosting	0.0037	-0.0424	Emergent					
43						55	language generation	0.0001	0.0108	Popular					
44						35	fuzzy logic	-0.0024	0.0652	Subsiding					
45						54	control structures	0.0006	0.0018	Popular					
46						73	pattern recognition	-0.0001	0.0341	Subsiding					
47						116	machine learning	0.0002	0.0006	Popular					
48						104	logic programs	0.0015	-0.0169	Emergent					

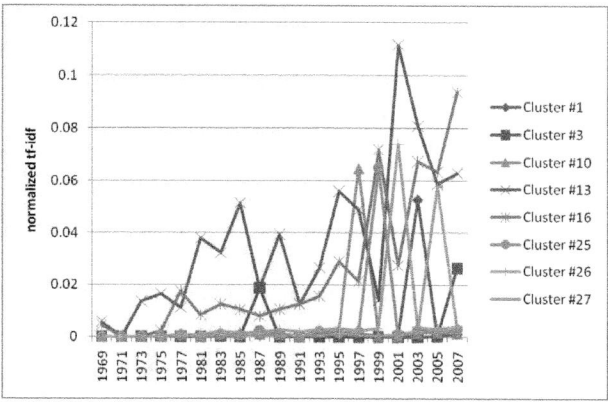

Fig. 3. The temporal patterns of tf-idf of terms with the emergent trend from the titles of IJCAI

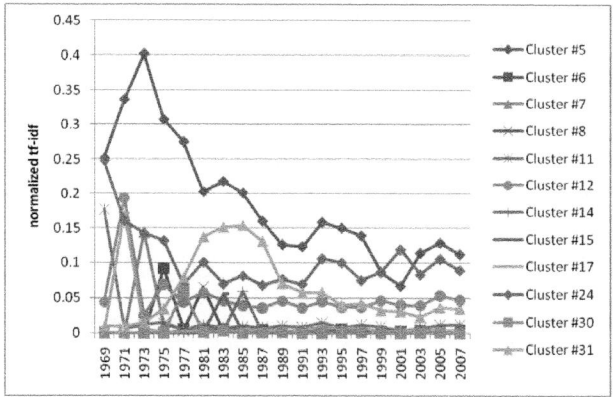

Fig. 4. The temporal patterns of tf-idf of terms with the subsiding trend from the titles of IJCAI

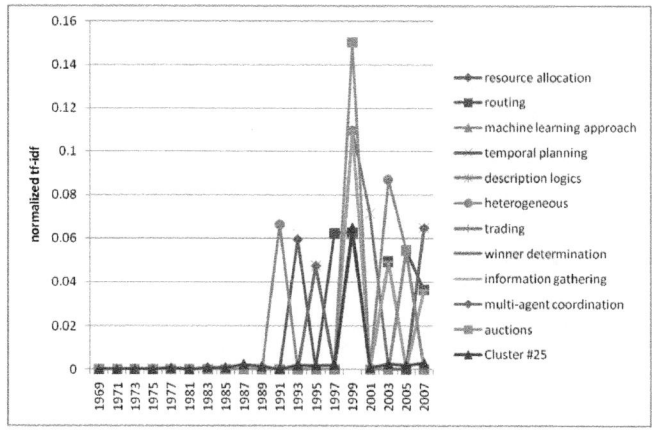

Fig. 5. Top ten emergent terms included in the pattern #25

Focusing on the detail of the emergent pattern, Fig.5 shows top ten terms included in the pattern #25, which includes the most recent burst words such as 'Auctions' detected by Kleinberg et al. in [10]. Although they were not used in the titles from '80s or more recently, researchers become to use these terms to express their key topics in these years.

5   Discussion

In this section, we discuss the availability of trend detection of our method by comparing the burst words that are detected their bursts after 1998 by [10]. The burst words are 'combinatorial', 'auctions', and 'reinforcement'.

In Section 4, the method detects the trends of the group of the terms that have similar temporal behaviors on each index through the periods. If the terms including the recent burst words show the increasing their importance, the importance indices detect their increases as the emergent patterns.

Table 5 shows the cluster IDs and the numbers of "Emergent" patterns based on the linear trends of the temporal patterns. Table 6 shows the same result on IJCAI titles and automatically extracted technical term candidates.

Table 5. The terms including the three recent burst words and their assigned patterns in the AAAI titles

Terms including the recent burst words	tf–idf	Jaccard	Odds
combinatorial auctions	Cluster #3	Cluster #31	Cluster #3
combinatorial optimization	Cluster #10	Cluster #36	Cluster #10
boosting combinatorial search	Cluster #15	Cluster #25	Cluster #15
iterative combinatorial auctions	Cluster #3	Cluster #25	Cluster #3
iterative auctions	Cluster #3	Cluster #25	Cluster #3
solve combinatorial optimization	Cluster #10	Cluster #27	Cluster #10
robust combinatorial auction protocol	Cluster #3	Cluster #25	Cluster #3
combinatorial auction	Cluster #3	Cluster #31	Cluster #3
power of sequential single–item auctions	Cluster #18	Cluster #25	Cluster #18
auctions	Cluster #3	Cluster #17	Cluster #20
solving concisely expressed combinatorial auction	Cluster #7	Cluster #30	Cluster #7
greedy–allocation combinatorial auction protocol	Cluster #11	Cluster #31	Cluster #11
expressive banner ad auctions	Cluster #26	Cluster #16	Cluster #26
real–world auctions	Cluster #26	Cluster #16	Cluster #26
reinforcement learning	Cluster #29	Cluster #30	Cluster #4
reinforcement learning perspective using	Cluster #10	Cluster #25	Cluster #10
reinforcement using supervised learning	Cluster #5	Cluster #25	Cluster #5
programming robots using reinforcement learning	Cluster #23	Cluster #25	Cluster #23
reinforcement learning algorithm	Cluster #2	Cluster #25	Cluster #2
efficient reinforcement learning	Cluster #5	Cluster #25	Cluster #5
reinforcement learning of coordination	Cluster #7	Cluster #25	Cluster #7
reinforcement learning algorithms	Cluster #10	Cluster #25	Cluster #10
reinforcement learning framework	Cluster #2	Cluster #25	Cluster #2
performance bounded reinforcement learning	Cluster #13	Cluster #25	Cluster #13
comparison of reinforcement learning methods	Cluster #10	Cluster #25	Cluster #10
cost–sensitive reinforcement learning	Cluster #23	Cluster #25	Cluster #23
dynamics of reinforcement learning	Cluster #15	Cluster #25	Cluster #15
average–reward reinforcement learning algorithm	Cluster #2	Cluster #25	Cluster #2
reputation–oriented reinforcement learning approach	Cluster #7	Cluster #25	Cluster #7
difficulty of modular reinforcement learning	Cluster #18	Cluster #25	Cluster #18
# of terms including the emergent pattern	16/30	8/30	15/30

As shown the tables, the terms including the two recent burst words, 'combinatorial' and 'auctions', are assigned to the emergent patterns by using this method. By looking the trends of the temporal patterns that are only less than 1% of the term candidates, we can get the emergent terms including the burst words.

Although the terms including 'reinforcement' are mostly not assigned to the popular patterns, the linear trends of each term based on the degree and the intercept of the indices shows the emergent trend that has the positive values for the degrees and the negative values for the intercepts. This indicates that the degree and the intercept of each term also support to find out required trends of terms by users.

By using our method, when a user only knows some keywords, the linear trends of the temporal patterns of the keywords help to find out similar terms with the given keywords. On the other hand, when the user knows terms with a remarkable trend, this method supports to confirm their trends with displaying the similar terms based on the temporal behavior of the index.

Table 6. The terms including the three recent burst words and their assigned patterns in the IJCAI titles

Terms including the recent burst words	tf-idf	Jaccard	Odds
combinatorial games approach	cluster #25	cluster #17	cluster #25
combinatorial auctions	cluster #25	cluster #29	cluster #25
combinatorial domains	cluster #3	cluster #11	cluster #3
computational complexity of combinatorial auctions	cluster #25	cluster #17	cluster #25
robust combinatorial auction protocol	cluster #26	cluster #17	cluster #26
characterization of strategy/false-name proof combinatorial auction protocols	cluster #1	cluster #17	cluster #1
specifying combinatorial	cluster #3	cluster #11	cluster #3
sequential auctions	cluster #25	cluster #39	cluster #25
managing combinatorial optimisation sub-	cluster #26	cluster #20	cluster #26
auctions	cluster #25	cluster #22	cluster #25
combinatorially	cluster #27	cluster #29	cluster #27
reinforcement learning	cluster #19	cluster #39	cluster #18
reinforcement learning approach	cluster #25	cluster #17	cluster #25
fuzzy reinforcement learning	cluster #9	cluster #17	cluster #9
neural reinforcement learning approach	cluster #25	cluster #17	cluster #25
efficient reinforcement learning	cluster #25	cluster #17	cluster #25
reinforcement learning of local shape	cluster #3	cluster #17	cluster #3
reinforcement algorithms using functional approximation	cluster #25	cluster #17	cluster #25
convergence of reinforcement learning	cluster #25	cluster #17	cluster #25
convergent reinforcement learning algorithm	cluster #10	cluster #17	cluster #10
delayed reinforcement learning	cluster #21	cluster #17	cluster #21
multiple-goal reinforcement learning	cluster #1	cluster #17	cluster #1
confidence based dual reinforcement q-routing	cluster #25	cluster #17	cluster #25
# of terms including the emergent pattern	20/23	7/23	20/23

6 Conclusion

In this paper, we evaluated the availability of a method to detect remarkable temporal patterns of technical terms appeared as the temporal behaviors of the importance indices in the sets of temporally published documents. We implemented the method with the automatic term extraction, the three importance indices, and the temporal pattern extraction by using k-means clustering. Then, we assigned the abstracted meanings of the extracted temporal patterns according to the linear trends of their centroids for the timeline based on the linear regression technique. The abstracted meanings that are assigned in this paper are "Emergent", "Subsiding", and "Popular".

The empirical results show that the temporal patterns of the importance indices can detect the trends of each term, according to their values for each temporal set of the titles of the two academic conferences that are famous on AI field. Regarding to the results, the temporal patterns indicate that we can find out not only the current key terms, but also historical keys of the research field by using the linear trends of the periodical values of the importance indices.

In addition, by comparing with the recent burst words that were detected by completely different technique, we evaluated the availability to distinguish the terms including these burst words as the emergent trend. In this comparison, our method detected more than half of the terms including the burst words as "Emergent". This shows the availability for detecting the emergent trend of the terms with their similar terms that was not achieved in the conventional burst word detection method.

In the future, we will apply other term extraction methods, importance indices, and trend detection method. As for importance indices, we are planning to apply evaluation metrics of information retrieval studies, probability of occurrence of the terms, and statistics values of the terms. Then, we will apply this framework to other documents from various domains.

References

1. Hurst, M.: Temporal text mining. In: AAAI Spring Symposium on Computational Approaches to Analyzing Weblogs, SS06–03–015 (2006)
2. Lent, B., Agrawal, R., Srikant, R.: Discovering trends in text databases, pp. 227–230. AAAI Press, Menlo Park (1997)
3. Kontostathis, A., Galitsky, L., Pottenger, W.M., Roy, S., Phelps, D.J.: A survey of emerging trend detection in textual data mining. A Comprehensive Survey of Text Mining, 185–222 (2003)
4. Abe, H., Tsumoto, S.: Detecting temporal trends of technical phrases by using importance indices and linear regression. In: Rauch, J., Raś, Z.W., Berka, P., Elomaa, T. (eds.) ISMIS 2009. LNCS, vol. 5722, pp. 251–259. Springer, Heidelberg (2009)
5. Lewis, D.D., Croft, W.B.: Term clustering of syntactic phrases. In: Proceedings of the 13th Annual International ACM SIGIR Conference on Research and Development in Information Retrieval SIGIR 1990, pp. 385–404. ACM, New York (1990)

6. Srinivasan, P., Ruiz, M.E., Kraft, D.H., Chen, J.: Vocabulary mining for information retrieval: rough sets and fuzzy sets. Inf. Process. Manage. 37, 15–38 (2001)
7. Swan, R., Allan, J.: Automatic generation of overview timelines. In: SIGIR 2000: Proceedings of the 23rd Annual International ACM SIGIR Conference on Research and Development in Information Retrieval, pp. 49–56. ACM, New York (2000)
8. Shannon, C.E.: A mathematical theory of communication. The Bell System Technical Journal 27, 379–423, 623–656 (1948)
9. Sparck Jones, K.: A statistical interpretation of term specificity and its application in retrieval. Document Retrieval Systems, 132–142 (1988)
10. Kleinberg, J.M.: Bursty and hierarchical structure in streams. Data Min. Knowl. Discov. 7(4), 373–397 (2003)
11. Mei, Q., Zhai, C.: Discovering evolutionary theme patterns from text: an exploration of temporal text mining. In: KDD 2005: Proceedings of the Eleventh ACM SIGKDD International Conference on Knowledge Discovery in Data Mining, pp. 198–207. ACM, New York (2005)
12. Wallach, H.M.: Topic modeling: beyond bag-of-words. In: ICML 2006: Proceedings of the 23rd International Conference on Machine Learning, pp. 977–984. ACM, New York (2006)
13. Abe, H., Tsumoto, S.: Detecting temporal patterns of importance indices about technical phrases. In: Velásquez, J.D., Ríos, S.A., Howlett, R.J., Jain, L.C. (eds.) KES 2009. LNCS, vol. 5712, pp. 252–258. Springer, Heidelberg (2009)
14. Abe, H., Tsumoto, S.: Comparing temporal behavior of phrases on multiple indexes with a burst word detection method. In: Sakai, H., Chakraborty, M.K., Hassanien, A.E., Slezak, D., Zhu, W. (eds.) RSFDGrC 2009. LNCS, vol. 5908, pp. 502–509. Springer, Heidelberg (2009)
15. Anderberg, M.R.: Cluster Analysis for Applications. Monographs and Textbooks on Probability and Mathematical Statistics. Academic Press, Inc., New York (1973)
16. Keogh, E., Chu, S., Hart, D., Pazzani, M.: Segmenting time series: A survey and novel approach. Data mining in Time Series Databases, pp. 1–22. World Scientific, Singapore (2003) (In an Edited Volume)
17. Liao, T.W.: Clustering of time series data: a survey. Pattern Recognition 38, 1857–1874 (2005)
18. The dblp computer science bibliography, http://www.informatik.uni-trier.de/~ley/db/
19. Nakagawa, H.: Automatic term recognition based on statistics of compound nouns. Terminology 6(2), 195–210 (2000)
20. Witten, I.H., Frank, E.: Data Mining: Practical Machine Learning Tools and Techniques with Java Implementations. Morgan Kaufmann, San Francisco (2000)

High Scent Web Page Recommendations Using Fuzzy Rough Set Attribute Reduction

Punam Bedi and Suruchi Chawla

Department of Computer Science, University of Delhi,
Delhi, India
pbedi@cs.du.ac.in, sur_chawla@rediffmail.com

Abstract. Information on the web is huge in size and to find the relevant information according to the information need of the user is a big challenge. Information scent of the clicked pages of the past query sessions has been used in the literature to generate web page recommendations for satisfying the information need of the current user. High scent information retrieval works on the bedrock of keyword vector of query sessions clustered using information scent. The dimensionality of the keyword vector is very high which affects the classification accuracy and computational efficiency associated with the processing of input queries and ultimately affects the precision of information retrieval. All the keywords in the keyword vector are not equally important for identifying the varied and differing information needs represented by clusters. Fuzzy Rough Set Attribute Reduction (FRSAR) has been applied in the presented work to reduce the high dimensionality of the keyword vector to obtain reduced relevant keywords resulting in improvement in space and time complexities. The effectiveness of fuzzy rough approach for high scent web page recommendations in information retrieval is verified with the experimental study conducted on the data extracted from the web history of Google search engine.

Keywords: Fuzzy rough set, Information retrieval, Information need, Information scent, Fuzzy similarity.

1 Introduction

Information on the web is growing at a rapid speed. Search engines search and retrieve the web pages relevant to the users information request. For retrieving information, the user expresses his information request in the form of input query on a search engine which in turn finds the web pages which are considered relevant to the information need of the user. Despite the advances in Information Retrieval (IR)/search technology, low precision (fraction of retrieved documents which are relevant) of search engine results is still a bottleneck. Low precision of search engine results is due to the fact that a large volume of documents are retrieved for the given query of which only a few are relevant. Effectiveness of information retrieval on the web can be improved if a higher proportion of the

J.F. Peters et al. (Eds.): Transactions on Rough Sets XIV, LNCS 6600, pp. 18–36, 2011.
© Springer-Verlag Berlin Heidelberg 2011

retrieved web pages are relevant to the information need of the user, i.e., if there is an increase in the precision.

In any information retrieval system on the web, a user enters the input query to retrieve the documents. Quite often, the input query entered by the user contains few and sometimes ambiguous keywords which are not sufficient to infer the information need of the user. Therefore out of thousands of documents retrieved by the search engine, only few are truly relevant to the user [2,17,20]. Thus there is a definite need to improve the precision of search results in order to satisfy the information need of the user effectively.

A query session is defined as set of clicked URLs associated with the user query. Information scent is the subjective sense of value and cost of accessing a page based on perceptual cues with respect to the information need of the user. Users tend to click URLs with high scent [10,18,31,32]. These high scent pages uniquely satisfy the information need of the user whereas low scent pages are less relevant to the information need of the user.

In the presented work, the information need of the query sessions of the past users is modeled using information scent and content of clicked pages. Each clicked URL in the query session is associated with the quantitative value called Information Scent which is the measure of its relevancy with respect to the information need of the user associated with the query session. Each query session is represented by weighted keyword vector with weights taken as information scent. Fuzzy Rough set attribute reduction has been used to reduce these keyword vectors representing the query sessions. The reduced keyword vectors contain only those keywords which are all imperative in distinguishing the different information needs associated with identified clusters.

These query sessions are then clustered using Leaders-Subleaders algorithm [37] to get the set of query sessions with similar information need. Each cluster is represented by mean weighted keyword vector. All this is done in offline processing mode. The input query vector is then used to find the cluster/subcluster which closely represent the information need [4] and recommend the high scent clicked web pages associated with the selected cluster during online processing phase [6,9] matching more closely to the information need of the user.

The rest of the paper is organized as follows. Section 2 introduces the concept of information scent. Section 3 explains the basic concepts of Fuzzy Rough approach for attribute reduction. The representation of query session keyword vectors along with their clustering using Leaders-Subleaders algorithm is discussed in section 4. This section also discusses the proposed approach for high scent web page recommendations using FRSAR. Experimental study is given in section 5 and section 6 concludes the paper.

2 Related Work

Extensive work has been done in the direction of personalization of the web search [2,34]. In literature, techniques such as query expansion [13] and relevance feedback [15,33,36] have been employed in the personalization domain. In these techniques, the user query is expanded with related words automatically in order to retrieve the relevant documents which may not tally with the original query exactly.

In [3,13], the query log has been used to suggest the terms for automatic query expansion using correlation between document terms and input query terms in query sessions of past users on the search engine. All the clicked documents of the query sessions which are related to any term of input query in current query session are selected and the terms in selected clicked documents are evaluated against the whole input query to determine its relevance as an expansion terms with respect to the input query.

Information scent in query session mining is used in [5,7,8,9] to improve the Information Retrieval (IR) precision. Information scent is the subjective sense of value and cost of accessing a page based on perceptual cues with respect to the information need of the user. Users tend to click URLs with high scent [10,18,31,32]. These high scent pages uniquely satisfy the information need of the user whereas low scent pages are less relevant to the information need of the user.

Attribute selection is important for reducing computational cost of classification. A lot of work has been done in literature on dimensionality reduction. However, existing work tends to destroy the underlying semantics of the features after reduction or require additional information about the given data set [21]. Rough set theory (RST) can be used as a tool to discover data dependencies and to reduce the number of attributes contained in a data set using the data alone and no additional information [21]. Rough Set Attribute Reduction (RSAR) has been used by Chouchoulas et al. [11] for attribute reduction without loss of information for the discrete information systems. However, data objects in the information system can have both real and crisp attributes. Hence RSAR is not capable of reducing the information system having data objects with real attributes. Duan et al. [14] have used Rough Fuzzy method to make personalized web search more effective. They used Fuzzy set discretization and Rough set attribute reduction for identifying the discerning keywords for focused web search. Jensen & Shen [21,22] used Fuzzy Rough set attribute reduction on the clustered query session keyword vector to identify those keywords which are all relevant to identify the differing information need associated with the identified clusters. The reduced set of keyword vector representing each cluster of query sessions minimizes the space complexity because of less memory requirement for storing the clusters mean weighted keyword vector. Time complexity is improved as it requires less time in computing the cluster which best represents the information need associated with the input query.

In the presented work, Fuzzy Rough set attribute reduction has been used to reduce the keyword vectors representing the query sessions. The reduced keyword vector contains only those keywords which are all imperative in distinguishing the different information needs associated with identified clusters. Each query session is represented by weighted keyword vector with weights taken as information scent. Time complexity will be improved in online processing in computing the cluster which best represent the information need associated with the input query. This improvement is brought about with reduced relevant attribute set obtained using Fuzzy Rough attribute reduction. Reduced set of attributes uniquely identifies the information need of the user associated with the input query after the removal of redundant and irrelevant attributes.

The k-means is the most popular centroid algorithm. In k-means clustering, the data set is partitioned into k subsets in such a way that all points in a given subset are closest to the same centre. Initially, it indiscriminately selects k of the instances to represent the clusters. On the basis of selected attributes, all the remaining instances are allotted to their nearest center. Once all the object instances are assigned to the cluster, the new centres of the clusters are computed by taking the average of all data points contained in that cluster. This process of computing the new mean continues till there is no change in the gravity centre of the cluster. In case k cannot be predicted, then iterations can be carried out to find a value of k which is most appropriate. The efficacy of this technique depends greatly on the objective function that measure the distance between instances. It is difficult to find the distance measure that can be applied on wide variety of data types. The k-means clustering algorithm was used in [6] to cluster the query session keyword vectors.

When large data sets are involved, the k-means has to perform lot of I/O operations and computations [16,19]. There are some clustering algorithms available in the literature which can be used for large datasets. These include Leader-Subleaders algorithm [37], Fast single-link clustering method [28], Distance based fast hierarchical clustering method for large datasets [27], TI-DBSCAN [23], and Neighborhood-based clustering by means of the triangle inequality [24]. Fast single-link clustering method uses summarization scheme known as data-sphere (DS) and uses the leaders clustering method to collect the statistics of each data sphere. Subsequently, these data spheres are used with the single-link clustering method to derive the clusters of data. Distance based fast hierarchical clustering method works for large datasets and uses leaders clustering method to derive a set of leaders of the dataset. Leaders are not further categorized into subLeaders further in this paper. TI-DBSCAN and Neighborhood-based clustering by means of the triangle inequality use the triangle inequality property to quickly reduce the neighborhood search in density-based clustering.

Our proposed work uses Leaders-Subleaders algorithm [37] to cluster the query sessions keyword vector because of its good performance and classification accuracy for large data sets and it requires only two database scans to find the clusters /subclusters of the given dataset. Leaders-Subleaders is an extension of Leaders algorithm [35] which is a hierarchical clustering algorithm. In Leaders algorithm entire data set is partitioned incrementally into clusters where each cluster is represented by the leader. In Leaders-Subleaders algorithm, data objects are clustered to create groups of data objects. Each cluster is further partitioned into subclusters where each subcluster is represented by subleader. Clusters/subclusters representation improves the classification accuracy and it has low computation cost. It performs better than Leaders algorithm.

Leaders-Subleaders is an incremental algorithm which creates L clusters with L Leaders and SL_i subleaders in the ith cluster. Subleaders are the representative of the subclusters and help in classifying the given test pattern more accurately. Leaders-Subleaders algorithm requires only two database scans to find the subclusters /clusters of the given dataset.

3   Information Scent

3.1   Introduction

The concept of information scent is derived from the Information Foraging theory which states that information scent refers to the cues used by information foragers to make judgments related to the selection of information sources to pursue and consume [26,30]. These cues include items such as web links or bibliographic citations that provide users with brief information about content that is not available immediately. The information scent cues direct the user to the information they are looking for and present to the user an overall sense of content of collections.

Chi et. al. [10] have developed Web User Flow by Information Scent (WUFIS) algorithm, which simulates how web users, with a given information need, will navigate through a web site. The user searches for information on the web by making choices on which links to follow driven by his need. The traversal decisions made by him during search approximate his information need. In order to compute the information need of the user using the model, the input should be the documents list which was found interesting by the user moving from one page to another and the output generated is the weighted list of keyword vector representing his information need. Inferring User Need by Information Scent (IUNIS) algorithm is used to infer the information need of the user from his traversed list of documents. This traversed history of user is stored implicitly in the web server usage logs. Inferring User Need by Information Scent algorithm [10] is the implicit user modeling method. The information foraging theory is the basis of the IUNIS algorithm. The algorithm is used to infer user information need by studying user browsing history.

On the web, users search for information by navigating from page to page along the web links. Their actions are guided by their information need. Information scent is the subjective sense of value and cost of accessing a page based on perceptual cues with respect to the information need of user. More the page is satisfying the information need of user, more will be the information scent associated to it. The interaction between user needs, user action and content of web can be used to infer information need from a pattern of surfing [10,31,32]. Information scent is used to quantitatively measure the sense of value of the clicked page with respect to the information need of the user. High information scent URLs are those clicked URLs in the query sessions that are close to the information need associated with the query sessions. High scent pages are uniquely clicked in the session for a given information need. For a given sequence of clicked documents in particular query session more unique is the page to the session relative to the entire set of query sessions of data set, more likely it is close to the information need of the current session and thus more is the information scent associated to it in determining the information need of the session. Another parameter that is taken in accessing the information scent of the clicked pages is the time spent on the clicked pages. The reason for considering the time factor is that the clicked page which consumes more user attention is more likely to satisfy his information need than the page which takes less time of user. Thus both the parameters decide the weightage of the pages in determining the information need associated to query sessions using Information Scent.

3.2   Proposed Information Scent Metric

The Inferring User Need by Information Scent (IUNIS) algorithm provides various combinations of parameters to quantify the information scent [10,11]. Information scent is used to model the information need of the query sessions of the user on the web and generate the query session keyword vectors. The query session keyword vectors are clustered to identify the similar information need query sessions. Each query session is defined as an input query along with the clicked URLs in its answer.

$$querysession = (query, (clicked\ URLs)+)$$

where *clicked URLs* are those URLs which user clicks before submitting another query.

The two factors, page access $PF \cdot IPF$ and Time are used to quantify the information scent. Here PF (Page Frequency) is the access frequency of the clicked page in the given query session and IPF (Inverse Page Frequency) is the ratio of total query sessions in the log to the number of query sessions in which this page is clicked. The Inverse page frequency is based on the assumption that the pages which are frequently accessed in many query sessions are not unique to the information need of the query session in which they are present. The second factor Time is the time spent on a page in a given query session. By including the Time, more weightage is given to those pages that consume more user attention.

Assuming that user has not left the page in between, the information scent S_{id} is calculated for each page P_{id} in a given session i as follows:

$$S_{id} = PF \cdot IPF(P_{id}) \cdot Time(P_{id}) \quad \forall d \in 1, \ldots n \tag{1}$$
$$PF \cdot IPF(P_{id}) = f_{P_{id}}/max(f_{P_{id}}) \cdot log(M/m_{P_{id}}) \quad \forall d \in 1, \ldots n \tag{2}$$

where n is the number of unique clicked web pages in query session i and $PF \cdot IPF(P_{id})$: PF corresponds to the page P_{id} normalized frequency $f_{P_{id}}$ in a given query session Q_i and IPF correspond to the ratio of total number of query sessions M in the whole log to the number of query sessions $m_{P_{id}}$ that contain the given page P_{id}.

$Time(P_{id})$: It is the ratio of time spent on the page P_{id} in a given session Q_i to the total duration of session Q_i.

3.3   Modeling the Information Need Using Information Scent

Information need of a user is modeled using information scent and content of clicked URLs in the query session. The weighted content vector of the page, P_{id} in the query session Q_i is given below:

$$P_{id} = Content_d \qquad\qquad \forall d \in 1, \ldots n \qquad\qquad (3)$$

where $Content_d$: The content vector of a page P_{id} is a keyword vector $(w_{1,d}, w_{2,d}, w_{3,d}, \ldots, w_{v,d})$, v is the is the number of terms in the vocabulary set V and Vocabulary V is a set of distinct terms found in all distinct clicked pages in the whole dataset relevant to a content feature.

A vector model is used for representing the content of each page P_{id} in all the query sessions. $TF \cdot IDF$ term weight is used to represent the content vector for a given page P_{id}. The $TF \cdot IDF$ weight is a real number indicating the relative importance of a term in a given document. Term Frequency (TF) is calculated as the number of times the term appears in the document and Inverse Document Frequency (IDF) is computed as the ratio of the number of all documents to the number of documents that contain the term.

The information scent associated with the given clicked page, P_{id} is calculated by using two factors namely $PF \cdot IPF$ page access and Time. Each query session Q_i is constructed as a linear combination of vector of each page, P_{id} scaled by the information scent weight S_{id} as given below:

$$Q_i = \sum_{d=1}^{n} S_{id} \cdot P_{id} \qquad\qquad (4)$$

In above equation, n is the number of distinct clicked pages in the session Q_i and S_{id} (information scent) is calculated for each page P_{id} using equations 1 and 2. Each query session Q_i is obtained as weighted vector using equation 4. This weighted query session vector models the information need associated with the query session Q_i.

4   Fuzzy Rough Approach for Attribute Reduction

4.1   Introduction to Rough Sets

Rough set theory (RST), introduced by Pawlak [29] in 1980, is a mathematical tool to deal with vagueness and uncertainty in data. RST is an extension of the classical set theory and is used for representing incomplete knowledge. Rough sets are sets that have fuzzy boundaries and cannot be precisely characterized using available set of attributes. The basic concept of the RST is the notion of approximation space, which is an ordered pair A= (U, R) where U is nonempty set of objects called universe and R is an equivalence relation on U, called indiscernibility relation. Rough set uses lower and upper approximations to approximate the vague concept. The lower approximation is a description of the domain objects which are known with certainty to belong to the vague concept, whereas the upper approximation is a description of the objects which possibly belong to the vague concept.

The data set is represented in the form of table where each row represents the data object and each column represents the attributes of the data object. This table is called an information system. More formally, it is a pair S= (U, A), where U is a non-empty finite set of objects called the universe and A is a non-empty finite set of attributes such that $a : U \rightarrow V_a$, $\forall a \in A$. The set V_a is called the value set of a.

Let S = (U, A) be an information system, then with any $B \subseteq A$ there is associated an equivalence relation $IND_S(B)$ given as follows:

$$IND_S(B) = \{(x, x^{'}) \in U \times U \mid \forall a \in B \quad for \ which \quad a(x) = a(x^{'})\} \quad (5)$$

Here $IND_S(B)$ is called the B-indiscernibility relation, i.e., if $(x, x^{'}) \in IND_S$ (B), then objects x and $x^{'}$ are indiscernible from each other by attributes from B. The equivalence classes of the B indiscernibility relation are denoted as $[x]_B$. An equivalence relation induces a partitioning of the universe. These partitions can be used to build new subsets of the universe. Let the concept $X \subseteq U$ can be approximated using only the information contained in B by constructing the B-lower and B-upper approximations of X, denoted $\underline{B}X$ and $\overline{B}X$ respectively, where

$$\underline{B}X = \{x \mid [x]_B \subseteq X\} \quad (6)$$

$$\overline{B}X = \{x \mid [x]_B \cap X \neq \phi\} \quad (7)$$

The objects in $\underline{B}X$ can be with certainty classified as members of X on the basis of knowledge in B, while the objects in $\overline{B}X$ can be only classified as possible members of X on the basis of knowledge in B. The set BNB(X)=$\overline{B}X - \underline{B}X$ is called the B-boundary region of X and thus consists of those objects that we cannot decisively classify into X on the basis of knowledge in B. The set $U - \overline{B}X$ is called the B-outside region of X and consists of those objects which can be with certainty classified as do not belonging to X (on the basis of knowledge in B). A set is said to be rough (respectively crisp) if the boundary region is non-empty (respectively empty).

4.2 Rough Set Attribute Reduction

One of the applications of the Rough Set Theory is the reduction of redundant discrete attributes of data set without loss of information. In order to apply the Rough set attribute reduction method, the dataset is represented in the form of an information system $S = (U, A)$ where U represents the universe of finite set of objects. Here $A = C \cup D$, where C is the set of condition attributes and D is the set of decision attributes and $C \cap D = \phi$ [25].

RSAR (Rough Set Attribute Reduction) uses the indiscernibility relation $IND(P)$ [38] which is equivalence relation defined on attribute set P, a subset of C.

$$IND(P) = \{(x, y) \mid x, y \in U \ \forall a \in P \ \ for \ which \ \ a(x) = a(y)\} \qquad (8)$$

$IND(P)$ divide the entire set U into set of equivalence classes $U/IND(P)$ where

$$U/IND(P) = \cup\{[x]_p \mid x \in U\},$$
$$[x]_p = \{y \mid y \in U \ \ and \ \ (x, y) \in IND(P)\} \qquad (9)$$

Let X be a subset of U, then the P lower approximation of the set X denoted as $\underline{P}X$ can be defined as

$$\underline{P}X = \cup\{Y \in U/IND(P) \mid Y \subseteq X\} \qquad (10)$$

The P positive region of D is the set of all objects from the universe U which can be classified with certainty to classes of U/D employing attributes from P.

$$POS_P(D) = \cup_{X \in U/D}\underline{P}X \qquad (11)$$

Attribute reduction using Rough set approach is achieved by finding the dependency of decision attributes D on condition attributes P. P is the subset of C and the dependency function [12] is defined as follows:

$$k = \gamma_P(D) \ = \mid POS_P(D) \mid / \mid U \mid \qquad (12)$$

where, $\mid U \mid$ represents the cardinality of the set U.

The value of k lies in [0, 1]. If k=1, then the set of attributes D depend totally on P, if $0 < k < 1$ then D depends partially on P and if k=0 then D does not depend on P.

The reduction algorithm using rough set finds Q (subset of C) which is the minimal set of attribute such that $\gamma_C(D)=\gamma_Q(D)$. This implies that Q describes the original data set without loss of information and is the reduced attribute set.

4.3   Fuzzy Rough Set Attribute Reduction (FRSAR)

Rough Set Attribute Reduction (RSAR) is well suited to a discrete information system S1=(U,A). Rough set theory in RSAR fails to find the similarity of data objects having real value attributes.

Fuzzy Rough set concept is used for the reduction of attributes in data set containing real attributes without loss of information. The crisp lower and upper approximations $\underline{P}X$ and $\overline{P}X$ become Fuzzy lower and upper approximation set with membership function $\mu_{\underline{P}X}(X)$ and $\mu_{\overline{P}X}(X)$ respectively. The P positive region of D, i.e., $POS_P(D)$ becomes fuzzy set whose membership function is defined by $\mu_{POS_P(D)}$. The dependency of attribute set D on P is given by $\gamma'_P(D)$.

Fuzzy Equivalence Classes. In order to apply FRSAR, fuzzy similarity relation R_P is used to determine the similarity of the two data objects which have real valued attributes. The concept of crisp equivalence classes can be extended by the inclusion of a fuzzy similarity relation R_P on the universe, which determines the extent to which two elements are similar in R_P. The usual properties of reflexivity $(\mu_{R_P}(x, x) = 1)$, symmetry $(\mu_{R_P}(x, y) = \mu_{R_P}(y, x))$, and transitivity $(\mu_{R_P}(x, z) \geq \mu_{R_P}(x, y) \wedge \mu_{R_P}(y, z))$ hold. Using the fuzzy similarity relation R_P, the fuzzy equivalence class $[x]_{R_P}$ for objects close to x can be defined as follows:

$$\mu_{[x]_{R_P}}(y) = \mu_{R_P}(x, y) \tag{13}$$

The following axioms should hold for a fuzzy equivalence class F [22] which represents $[x]_{R_P}$.

$$\exists x, \mu_F(x) = 1,$$

$$\mu_F(x) \wedge \mu_{R_P}(x, y) \leq \mu_F(y),$$

$$\mu_F(x) \wedge \mu_F(y) \leq \mu_{R_P}(x, y)$$

The first axiom corresponds to the requirement that an equivalence class is non-empty. The second axiom states that elements in $y's$ neighbourhood are in the equivalence class of y. The final axiom states that any two elements in F are related via R_P.

Fuzzy lower and upper approximations. In literature, the fuzzy P-lower and P-upper approximations are defined as

$$\mu_{\underline{P}X}(F_i) = inf_x \ max(1 - \mu_{F_i}(x), \mu_X(x)) \tag{14}$$

$$\mu_{\overline{P}X}(F_i) = sup_x \ min(\mu_{F_i}(x), \mu_X(x)) \tag{15}$$

where F_i denotes i^{th} fuzzy equivalence class belonging to U/P.

The fuzzy lower and upper approximation membership functions are defined as below:

$$\mu_{\underline{P}X}(x) = sup_{x \in U/P} \ min(\mu_F(x), inf_{y \in U} \ max\{1 - \mu_F(y), \mu_X(y)\}) \tag{16}$$

$$\mu_{\overline{P}X}(x) = sup_{x \in U/P} \ min(\mu_F(x), sup_{y \in U} \ min\{\mu_F(y), \mu_X(y)\}) \tag{17}$$

In implementation, not all $y \in U$ are required to be considered. Only those y where F(y) is non-zero, i.e. where object y is a fuzzy member of (fuzzy) equivalence class F are sufficient to be considered. The tuple $\langle \underline{P}X \mid \overline{P}X \rangle$ is called a Fuzzy - Rough set.

The membership of an object $x \in U$ to the fuzzy positive region is defined as follows:

$$\mu_{POS_P(D)}(x) = max_{x \in U/D}(\mu_{\underline{P}X}(x)) \tag{18}$$

Using the definition of fuzzy positive region, the dependency function is defined as

$$\gamma'(D) = \frac{\sum_{x \in U} \mu_{POS_p(D)}(x)}{\mid U \mid} \tag{19}$$

5   Use of FRSAR in Information Retrieval

Fuzzy Rough approach [22] is used in this paper for the keyword reduction of query session vectors representing the information need of the query sessions on the web.

5.1   Clustering a Query Session

Information need associated with the query session is modeled using information scent and content of clicked URLs. Each query session is represented by keyword vector weighted by information scent as given by the equation 4.

The Leaders-Subleaders clustering algorithm [37] has been applied to query sessions which are real valued vectors. Leaders-Subleaders is an incremental algorithm which creates L clusters with L leaders and SL_i subleaders in the ith cluster. Subleaders are the representative of the subclusters and help in classifying the given test pattern more accurately. Leaders-Subleaders algorithm requires only two database scans to find the clusters/subclusters of the given dataset. The similarity function chosen is cosine similarity which calculates the extent to which two query sessions are similar.

In algorithm 1, we have defined U as a finite set of query session keyword vectors called Universe and A as the set of attributes describing a query session, where $A = (C \cup D)$ and (U, A) is an information system S1. Decision column D represents the label of cluster to which query session belongs and each cluster uniquely represents the specific information need. The query session keyword vectors are stored in (U, A). Each query session in clusters is stored as row in (U, A) and C columns represent the keywords of weighted keyword vector of a query session. A particular cell S1(row, col) represent the weight of keyword represented by column col of query session vector which is labeled by row. The membership function of fuzzy similarity relation R_p to find the extent of similarity of query session keyword vector is defined below.

$$\mu_{R_P}(x, y) = \{\cos(x, y) : x, y \in U\} \tag{20}$$

Here x and y are keyword vector of query session with attributes in set P and with weights as information scent. Here $\cos(x, y)$ calculates the cosine similarity between vectors x and y.

Fuzzy Rough set attribute reduction algorithm operates on (U, A) information system to reduce the size of keyword vectors without loss of information represented by query sessions in (U, A).

Algorithm 1. Query session vector clustering using Leaders-Subleaders algorithm

---

Leaders Computation

Input   : Query session vector dataset, Threshold .

Output: Leaders list associated with clusters of query sessions where each cluster is represented by a leader.

Select any query session vector as the initial leader and add it to the leaders list and set leader counter LC=1

for *all query session vectors not yet processed* **do**

> select the query session vector Q
>
> calculate the similarity of Q with all the leaders
>
> select the leader which has maximum similarity represented by max
>
> **if** *max $\geq$ Threshold* **then**
>
> > assign it to the selected leader
> >
> > mark the cluster number associated with the selected leader for Q
> >
> > add it to the members list of this cluster
> >
> > increment the member count of this cluster
>
> **else**
>
> > add Q to leaders list
> >
> > increment leader counter LC=LC+1

Subleaders Computation

Input   : leaders list associated with clusters of query sessions where each cluster is represented by the leader , SubThreshold: SubThreshold > Threshold value for similarity of query session vector in subleaders

Output: Leaders-subleaders list associated with clusters of query sessions.

for *i=1 to LC* **do**

> Initialize subleaders list SL_i with any query session vector Q_i in the ith cluster.
>
> Set counter SL_iCount=1;
>
> **for** *j= 2 to | $Cluster_i$ |* **do**
>
> > Calculate the similarity of Q_{ij} with all subleaders in SL_i
> >
> > Select the subleader with maximum similarity represented by max1
> >
> > **if** *max1 >= SubThreshold* **then**
> >
> > > assign it to the selected subleader.
> > >
> > > mark the subcluster number associated with the selected subleader for Q_{ij}.
> > >
> > > add it to the members list of this subcluster.
> > >
> > > increment the member count of this subcluster.
> >
> > **else**
> >
> > > add Q_{ij} to the subleaders list SL_i .
> > >
> > > set SL_i Count= SL_i Count +1

5.2   Fuzzy Rough Set Attribute Reduction of Keyword Vector

Algorithm 2 is used to generate the reduct R, a subset of C using dependency function $\gamma'_R(D)$ which is measure of dependency of decision attribute set D on R [22].

In this algorithm, |Keywords| represent the count of all distinct keywords of clicked URLs present in the data set after all stopword removal and stemming

---

Algorithm 2. Fuzzy rough reduction of the keyword vector

R=C where $\mathbf{C} = \{k_1, k_2, \ldots k_{|Keywords|}\}$, $D = \{1, 2, \ldots, | Cluster |\}$,
$\gamma'_{best} = \gamma'_R(D)$, $Y = \{\}$
repeat
    $\gamma'_{prev} = \gamma'_{best}$
    $\forall x \in (C - Y)$
    $T = R - \{x\}$
    if $\gamma'_{R-\{x\}}(D) = \gamma'_R(D)$ **then**
        $R = T$,
        $\gamma'_{best} = \gamma'_T(D)$,
        $Y = Y \cup \{x\}$
until $\gamma'_{prev} \neq \gamma'_{best}$;
return R

Algorithm 3. High scent web page recommendations

Offline Processing
Clustered query sessions are represented in the form of information system
$S1 = (U, A)$ where $A = (C \cup D)$ and C are set of keywords of keyword vector
representing all query sessions and D is the class label of the cluster to which
query session belongs.
Apply the Fuzzy Rough Set Attribute Reduction (FRSAR) to reduce the
dimensionality of information system using fuzzy similarity relation for query
sessions keyword vector in FRSAR approach given in section 4.3.
Use reduced set of attributes R to define each cluster mean keyword vector.
Online processing
The input query is represented in the keyword vector scaled to the dimension of
reduced set R.
foreach *input query session vector q* **do**
    Calculate the similarity of q with all leader vectors associated with their
    clusters.
    Select the leader with the maximum similarity represented as max.
    Calculate the similarity of q with all subleaders of the selected leader.
    Select the subleader with the maximum similarity represented as max1.
    if $max > max1$ **then**
        └ use the cluster associated with selected leader.
    else
        └ use the subcluster associated with selected subleader.
The high scent web pages associated with the selected cluster will be
recommended for a given input query.

---

using porter stemming algorithm. |Clusters| represent the count of clusters obtained in query session mining.

Algorithm 2 generates R by incrementally removing the least informative attribute from the full set of attributes C till there is no change in the value of dependency function. This process continues until no more attributes can be removed without reducing the total number of discernible objects in the data. The proposed method of high scent web page recommendations using query sessions keyword vector reduction with FRSAR in query session mining is given in algorithm 3.

6   Experimental Study

Experiment was performed on the data collected from web history of Google search engine. The data set was generated by users who had expertise in specific domains mainly academics, entertainment and sports. The web history of Google search engine contains the following fields for each clicked URLs.

1. Time of the Day 2. Query terms 3. Clicked URLs

On submission of input query, Google search engine returns a result page consists of URLs retrieved for a given query along with the content information about URLs. In the experiment only those query sessions in the data set were selected which had at least one click in their answer. Query sessions considered consist of query terms along with clicked URLs. The numbers of distinct URLs in the collected data set were found to be 2245. The data set was pre-processed to get 250 query sessions. The data set generated from web history was loaded into database format for further processing.

The experiment was performed on Pentium IV PC with 1 GB RAM under Windows XP using JADE (Java Agent Development Environment) and Oracle database. Web Sphinx crawler was used to fetch the clicked documents of query sessions in the data set. Each query session as transformed into the vector representation using information scent and content of clicked URLs and this vector was stored in the database. The Leaders-Subleaders algorithm was executed where the Threshold value for leaders computation was set to 0.5 and the SubThreshold value for subleaders computation was set to 0.75. The similarity of vectors was measured using cosine formula for weighted term vector. Clusters of 250 query sessions were stored in the form of information system $S1 = (U, A)$, where $A = (C \cup D)$. The initial dimensionality of C was 1123, i.e., 1123 keyword attributes were representing each clustered query session vector in information system S1. D was the class label of the clusters to which query sessions vector belongs. The dimensionality of reduced set of attributes obtained using Fuzzy Rough attribute reduction algorithm was 351 which is 31% of original set of attributes. The number of reduced set of attributes has been increased when applied on query sessions clustered using Leaders-Subleaders algorithm as compared to clustering using k-means algorithm in [6]. As the level of clustering increases due to sub clustering within the clustering using two database scans through the dataset, the number of class labels increases. The number of distinguishing attributes needed to identify different class labels increase which help in effectively identifying the information need associated with the clusters and sub clusters after the elimination of redundant and irrelevant attributes.

In order to analyze the effectiveness of keyword reduction in query session mining in satisfying the information need of the users in information retrieval, the performance of both the approaches, i.e., with and without using FRSAR was evaluated using randomly selected test input queries which were categorized

Table 1. Sample list of untrained queries

Category	Queries
Untrained Set	cgi perl tutorial, sql tutorial, tutorial oracle, Movie song, Arena football, South Dakota wrestling Major league baseball tryouts, Free download mp3, vcd files, mpeg movies.

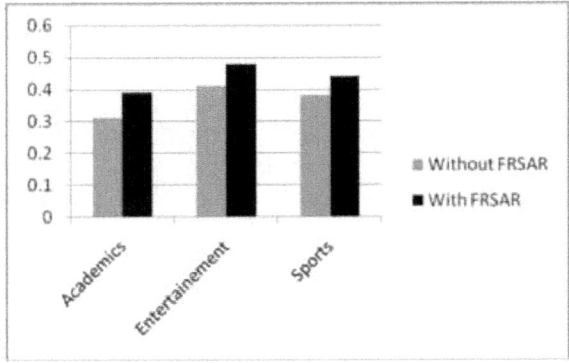

Fig. 1. Average precision for set of untrained input queries without FRSAR and with FRSAR

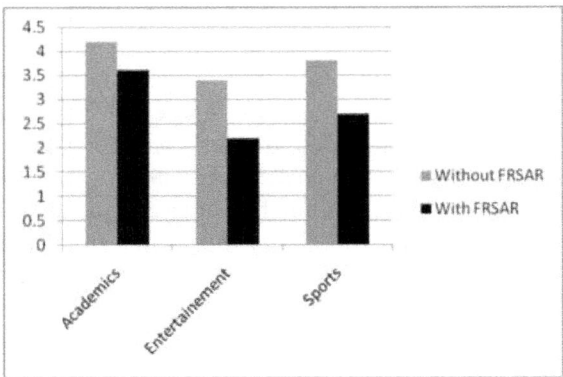

Fig. 2. Average response time for set of untrained input queries without FRSAR and with FRSAR

as untrained queries set. The untrained queries were those queries which did not have session associated with them in data set and are categorized as unseen queries. Some of the sample untrained queries are shown in Table 1.

The experiment was performed using 39 untrained queries distributed in each of the domain. The precision was evaluated on untrained set of queries belonging to each of the domain considered for both without FRSAR and with FRSAR. The average precision was calculated for first two result pages. Users marked the relevant documents within the list of URLs retrieved using Google search engine along with web page recommendation for a given query. Fig 1 shows the average precision calculated for untrained queries. The average precision is improved for untrained queries using keyword reduction with FRSAR. The reason for this is that dimensionality reduction identifies those attributes which uniquely identify the different information needs represented by clusters of query sessions.

Fig 2 shows that the dimensionality reduction has the significant impact on the time complexity of online processing phase. The time to classify the input query is reduced significantly. This effect is reflected in the average response time of web page recommendations for input queries with FRSAR on Google search engine. The online processing time decreases when compared without using FRSAR which can be useful for a system like search engine which require quick response time.

The storage requirement for clustered data set has been reduced due to reduced keyword vector obtained using FRSAR. Fuzzy rough set attribute reduction is used for dimensionality reduction before online processing phase uses the attributes belonging to the resultant reduct. Hence, the computational cost of Fuzzy Rough set attribute reduction has no impact on the run time efficiency of online processing.

The obtained results were also analyzed using paired t-test for average precision and average response time before and after attribute reduction with 38 degrees of freedom (d.f.) for a combined sample as well as for all three categories (Academics, Entertainment and Sports) each with 12 d.f. The observed value of t for average precision was -4.870 for a combined sample. Value of t for paired difference of average precision was -2.407 for academics, -3.162 for entertainment and -3.094 for sports categories. It was observed that the computed t value for paired difference of average precision lie outside the 95% confidence interval in each case. Hence Null hypothesis was rejected and alternate hypothesis was accepted in each case and it was concluded that average precision improve significantly using FRSAR.

Similarly t values were computed for paired difference of average response time before and after attribute reduction. These values obtained for the experiment were 8.285, 5.552, 4.969, 4.587 respectively for a combined sample with 38 d.f., for academics with 12 d.f., entertainment with 12 d.f. and sports with 12 d.f. It was observed that the computed t value for paired difference of average response time is greater than the tabled t value for 95% confidence level in each case. Null hypothesis was rejected and alternate hypothesis was accepted in each case and it was concluded that average response time reduces significantly using FRSAR.

7   Conclusion

Fuzzy Rough Set Attribute Reduction (FRSAR) approach to improve the information retrieval precision has been presented in this paper. Fuzzy Rough sets are used to reduce the high dimensionality of query session keyword vectors without loss of information. Fuzzy rough attribute reduction uniquely identifies the information need of the user associated with the input query after removing redundant and irrelevant attributes from associated cluster/subcluster. Information scent of the clicked pages of the past query sessions has been used to generate the web page recommendations for satisfying the information need of the current user. Query session vectors are clustered using Leaders-Subleaders algorithm. Fuzzy similarity relation is used to measure the similarity between keywords to distinguish between the clusters. Two level clustering of query session keyword vectors contribute in the improvement of identification of information need of the user query. Experiments were conducted on the data extracted from the web history of Google search engine. Fuzzy rough set attribute reduction is applied in offline processing phase. Reducts produced by offline processing phase were used in online processing phase. Experimental study conducted confirms the improvement in information retrieval precision and response time which was also verified statistically.

References

1. Allan, J.: Incremental relevance feedback for information filtering. In: Proceedings of the 19th Annual International ACM SIGIR Conference on Research and Development in Information Retrieval, pp. 270–278 (1996)
2. Baeza-Yates, R., Ribeiro-Neto, B.: Modern Information Retrieval. Addison-Wesley, Reading (1999)
3. Baeza-Yates, R., Hurtado, C.A., Mendoza, M.: Query recommendation using query logs in search engines. In: Favela, J., Menasalvas, E., Chávez, E. (eds.) AWIC 2004. LNCS (LNAI), vol. 3034, pp. 164–175. Springer, Heidelberg (2004)
4. Bazan, J., Peters, J.F., Skowron, A., Nguyen, H.S., Szczuka, M.: Rough set approach for pattern extraction from classifiers. Electronics Notes in Theoretical Computer Science 82(4), 20–29 (2003)
5. Bedi, P., Chawla, S.: Improving information retrieval precision using query log mining and information scent. Information Technology Journal 6(4), 584–588 (2007)
6. Bedi, P., Chawla, S.: Use of fuzzy rough set attribute reduction in high scent web page recommendations. In: Sakai, H., Chakraborty, M.K., Hassanien, A.E., Ślęzak, D., Zhu, W. (eds.) RSFDGrC 2009. LNCS, vol. 5908, pp. 192–200. Springer, Heidelberg (2009)
7. Chawla, S., Bedi, P.: Improving information retrieval precision by finding related queries with similar information need using information scent. In: Proceedings ICETET 2008. The 1st International Conference on Emerging Trends in Engineering and Technology, pp. 486–491 (2008)
8. Chawla, S., Bedi, P.: Personalized web search using information scent. In: Proceedings CISSE 2007 - International Joint Conferences on Computer, Information and Systems Science and Engineering, Springer, Heidelberg (2007)

9. Chawla, S., Bedi, P.: Finding hubs and authorities using information scent to improve the information retrieval precision. In: Proceedings ICAI 2008 -The 2008 International Conference on Artificial Intelligence, WORLDCOMP 2008, July 14-17 (2008)

10. Chi, E.H., Pirolli, P., Chen, K., Pitkow, J.: Using information scent to model user information needs and actions on the Web. In: Proceedings ACM CHI 2001 Conference on Human Factors in Computing Systems, pp. 490–497 (2001)

11. Chouchoulas, A., Shen, Q.: Rough set aided keyword reduction for text categorization. Applied Artificial Intelligence 15(9), 843–873 (2001)

12. Cornelis, C., Jensen, R., Hurtado, G., Ślęzak, D.: Attribute selection with fuzzy decision reducts. Information Sciences 180, 209–224 (2010)

13. Cui, H., Wen, J.R., Nie, J.Y., Ma, W.Y.: Query expansion by mining user logs. IEEE Transactions on Knowledge and Data Engineering 15(4), 829–839 (2003)

14. Duan, Q., Miao, D., Zhang, H., Zheng, J.: Personalized web retrieval based on rough fuzzy method. Journal of Computational Information Systems 3(2), 203–208 (2007)

15. Efron, M.: Query expansion and dimensionality reduction: Notions of optimality in Rocchio relevance feedback and latent semantic indexing. Information Processing and Management 44, 163–180 (2008)

16. Gordon, A.D.: How many clusters? An investigation of five procedures for detecting nested cluster structure. In: Hayashi, C., Ohsumi, N., Yajima, K., Tanaka, Y., Bock, H., Baba, Y. (eds.) Data Science, Classification, and Related Methods, Springer, Tokyo (1998)

17. Gudivada, V.N., Raghavan, V.V., Grosky, W., Kasanagottu, R.: Information retrieval on World Wide Web. IEEE Expert., 58–68 (1997)

18. Heer, J., Chi, E.H.: Identification of web user traffic composition using multi-modal clustering and information scent. In: Proceedings of Workshop on Web Mining, SIAM Conference on Data Mining, pp. 51–58 (2001)

19. Jain, A.K., Murty, M.N., Flyn, P.J.: Data clustering: A review. ACM Computing Surveys 31(3), 264–323 (1999)

20. Jansen, J., Spink, A., Bateman, J., Saracevic, T.: Real life information retrieval: a study of user queries on the web. ACM SIGIR Forum 32(1), 5–17 (1998)

21. Jensen, R., Shen, Q.: Semantic-preserving dimensionality reduction: rough and fuzzy-rough based approach. IEEE Transactions on Knowledge and Data Engineering 16(12) (2004)

22. Jensen, R., Shen, Q.: Fuzzy-rough attribute reduction with application to web categorization. Fuzzy Sets and Systems 141(3), 469–485 (2004)

23. Kryszkiewicz, M., Lasek, P.: TI-DBSCAN: Clustering with DBSCAN by Means of the Triangle Inequality. In: Szczuka, M., Kryszkiewicz, M., Ramanna, S., Jensen, R., Hu, Q. (eds.) RSCTC 2010. LNCS, vol. 6086, pp. 60–69. Springer, Heidelberg (2010)

24. Kryszkiewicz, M., Lasek, P.: A neighborhood-based clustering by means of the triangle inequality. In: Fyfe, C., Tino, P., Charles, D., Garcia-Osorio, C., Yin, H. (eds.) IDEAL 2010. LNCS, vol. 6283, pp. 284–291. Springer, Heidelberg (2010)

25. Nguyen, H.S.: On the decision table with maximal number of reducts. Electronic Notes in Theoretical Computer Science 82(4), 198–205 (2003)

26. Olston, C., Chi, E.H.: ScentTrails: Integrating browsing and searching on the World Wide Web. ACM Transactions on Computer-Human Interaction 10, 177–197 (2003)

27. Patra, B.K., Hubballi, N., Biswas, S., Nandi, S.: Distance based fast hierarchical clustering method for large datasets. In: Szczuka, M., Kryszkiewicz, M., Ramanna, S., Jensen, R., Hu, Q. (eds.) RSCTC 2010. LNCS, vol. 6086, pp. 50–59. Springer, Heidelberg (2010)
28. Patra, B.K., Nandi, S.: Fast single-link clustering method based on tolerance rough set model. In: Sakai, H., Chakraborty, M.K., Hassanien, A.E., Ślęzak, D., Zhu, W. (eds.) RSFDGrC 2009. LNCS (LNAI), vol. 5908, pp. 414–422. Springer, Heidelberg (2009)
29. Pawlak, Z., Grzymala-Busse, J., Slowinski, R., Ziarko, W.: Rough sets. Communications of the ACM 38(11), 88–95 (1995)
30. Pirolli, P., Card, S.K.: Information foragings. Psychological Review 106, 643–675 (1999)
31. Pirolli, P.: Computational models of information scent-following in a very large browsable text collection. In: Proceedings ACM CHI 1997- Conference on Human Factors in Computing Systems, pp. 3–10 (1997)
32. Pirolli, P.: The use of proximal information scent to forage for distal content on the World Wide Web. In: Working with Technology in Mind: Brunswikian, Resources for Cognitive Science and Engineering, Oxford University Press, Oxford (2004)
33. Rocchio, J.J.: Relevance feedback in information retrieval, pp. 313–343. Prentice Hall, Englewood Cliffs (1971)
34. Salton, G., McGill, M.: An Introduction to Modern Information Retrieval. McGraw-Hill, New York (1983)
35. Spath, H.: Cluster Analysis Algorithms for Data Reduction and Classification. Ellis Horwood, Chichester (1980)
36. Vechtomova, O., Karamuftuoglu, M.: Elicitation and use of relevance feedback information. Information Processing and Management 42, 191–206 (2006)
37. Vijaya, P.A., Murty, M.N., Subramanian, D.K.: Leaders-Subleaders: an efficient hierarchical clustering algorithm for large data sets. Pattern Recognition Letters 25(4), 505–513 (2004)
38. Zhong, N., Dong, J., Ohsuga, S.: Using rough set with heuristics for feature selection. Journal of Intelligent Information System 16, 199–214 (2001)

A Formal Concept Analysis Approach
to Rough Data Tables

Bernhard Ganter and Christian Meschke*

Institut für Algebra, TU Dresden, Germany
{Bernhard.Ganter,Christian.Meschke}@tu-dresden.de

Abstract. In order to handle very large data bases efficiently, the data warehousing system ICE [1] builds so-called *rough tables* containing information that is abstracted from certain blocks of the original table. In this article we propose a formal description of such rough tables. We also investigate possibilities of mining them for implicational knowledge. The underyling article is an extended version of [2].

1 Introduction

Consider a large data table. It has rows, describing certain *objects*, and columns for *attributes* which these objects may have. The entry in row g and column m gives the *attribute value* that attribute m has for object g. By "large" we mean that the table has many rows, perhaps 10^9, or more. Even for a moderate number of attributes the size of such a table may be in the terabytes.

Data analysis on such a table faces complexity problems and requires a good choice of strategy. In the present paper we investigate an approach by Infobright using rough objects and granular data, and combine it with methods from Formal Concept Analysis.

Infobright Community Edition (ICE) [1,3] is an open source data warehousing system which is optimized to obtain high compression rates and to process ana-lytic queries very quickly. ICE chops the stream of rows into so-called *rough rows*, each subsuming 65536 rows. The rough rows divide the columns into so-called *data packs*. Each data pack gets stored in a compressed form. For processing a query one does not want to decompress all data packs. Therefore ICE cre-ates a so-called *data pack node* to every data pack. A data pack node contains meta–information about the corresponding data pack. If for instance the column contains numeric values, the data pack nodes could consist, e.g., of minimum, maximum and the sum of the data pack values. The *rough table* is the data table that has the rough rows as rows, the same attributes as the original large data table, and the data pack nodes as values.

In order to sound the possibilities of getting interesting information about the original data table from the rough table, Infobright offered a contest [4] for which they provided a rough table with 15259 rough rows (the original table has

---

* Supported by the DFG research grant no. GA 216/10-1.

J.F. Peters et al. (Eds.): Transactions on Rough Sets XIV, LNCS 6600, pp. 37–61, 2011.
© Springer-Verlag Berlin Heidelberg 2011

one billion rows) and 32 attributes. Furthermore, Infobright invited to propose ways to do data mining in such rough tables.

Our approach is a systematic one. Our focus is on "what can be done" rather than on "how to get quick results". Although it is likely that a large data table will contain erroneous and imprecise data, we first concentrate on the case of precise data. Approximative and fault-tolerant methods shall later be build on this basis.

2   Formal Concept Analysis

Formal Concept Analysis was introduced by Rudolf Wille in 1982 [5]. Since that time it became an established field of mathematics. It is usually assigned to be an applied theory which is not false because of numerous applications in a broad spectrum of areas. But its well developed theoretical foundations are also closely related to lattice theory, a field of pure mathematics. In this section we give a brief introduction to Formal Concept Analysis with a special focus on so-called *attribute implications*. For a more detailled introduction we refer the reader to [6].

Unsurprisingly the most fundamental notion of Formal Concept Analysis is that of a *formal concept*. It is motivated by classical philosophy where a concept is determined by two things: its *extent* and its *intent*. Thereby, the extent is the set of all objects belonging to the concept and the intent is the set of attributes that somehow constitute the concept. The extent and the intent are closely related in the following way. The intent is the set of attributes the objects from the extent have in common and the extent is the set of all objects that share all attributes from the intent.

Let us take for instance the concept *bicycle*. The extent is then the set of all bicycles and the intent contains all attributes bicycles have in common, like for instance having two wheels, a frame, pedals, a chain or being a means of transportation. If it is not clear what the underlying universal sets of objects and attributes are, it is not possible to give a precise definition of the notion of a concept. It depends on the context.

Definition 1. *A **formal context** is a triple (G, M, I) where G and M are sets and $I \subseteq G \times M$ is a binary relation from G to M. The elements of G are normally called **objects** and the elements of M are usually referred to as **attributes**. The relation I is called the **incidence** relation and an incidence $(g, m) \in I$, commonly written down as gIm, is usually interpreted in the way that the object g **has** the attribute m.*

Formal contexts are usually visualised by cross tables (like for instance in Fig. 1) where the rows correspond to objects, the columns correspond to attributes and crosses mark the incidence relation. For a set $A \subseteq G$ of objects and a set $B \subseteq M$ of attributes we define

$$A^I := \{m \in M \mid aIm \text{ for all } a \in A\}, \text{ and}$$
$$B^I := \{g \in G \mid gIb \text{ for all } b \in B\}.$$

Hence, we have in both directions mappings between the powerset of G and the powerset of M, which are both denoted by $(\cdot)^I$. If there is the danger of mixing up the two so-called **derivation operators**, one should use two different notions. The latter is typically the case when G and M are not disjoint.

Definition 2. *A pair* (A, B) *where* $A \subseteq G$ *and* $B \subseteq M$ *is called a* **formal concept** *if* $A^I = B$ *and* $A = B^I$. A *is then called the* **extent** *and* B *is called the* **intent** *of the formal concept* (A, B).

The set of all extents of a given formal context is denoted by $\mathrm{Ext}(G, M, I)$. It forms a closure system on G. This closure system is generated by the **attribute extents**

$$m^I := \{m\}^I \qquad (m \in M).$$

This means that every column of the corresponding cross table is an extent and one receives all extents by building all possible intersections of these columns. Note that G itself corresponds to the empty intersection and hence always is an extent. Dually, the set $\mathrm{Int}(G, M, I)$ of all intents forms a closure system on M which is generated by the **object intents**

$$g^I := \{g\}^I \qquad (g \in G).$$

The closure operator corresponding to the closure system of all extents is the mapping $A \mapsto A^{II}$. Furthermore, the two derivation operators form a *Galois connection* as statement (4) of the following Proposition 1 shows.

Proposition 1. *For* $A, A_1, A_2 \subseteq G$ *and* $B, B_1, B_2 \subseteq M$ *it holds that*

(1) $A_1 \subseteq A_2 \implies A_1^I \supseteq A_2^I$, *(1')* $B_1 \subseteq B_2 \implies B_1^I \supseteq B_2^I$,

(2) $A \subseteq A^{II}$, *(2')* $B \subseteq B^{II}$,

(3) $A^I = A^{III}$, *(3')* $B^I = B^{III}$,

$$(4)\ A \subseteq B^I \iff A \times B \subseteq I \iff B \subseteq A^I.$$

Proof. See [6]. □

A concept is always determined by either of its two components: its extent or its intent. Since furthermore the set of all extents forms a closure system – and hence a complete lattice – the set of all concepts of a formal context forms a complete lattice if one orders them as follows:

$$(A_1, B_1) \le (A_2, B_2) :\iff A_1 \subseteq A_2.$$

The set of all concepts ordered by this subconcept-superconcept-order relation is then called the **concept lattice**. Furthermore, the Basic Theorem of Formal Concept Analysis (see [6] Theorem 3) says that every complete lattice is isomorphic to a concept lattice.

Fig. 1 shows a small example of a formal context (G, M, I), where G is the set of integer numbers from 1 to 10 and M is the 5-element set containing the

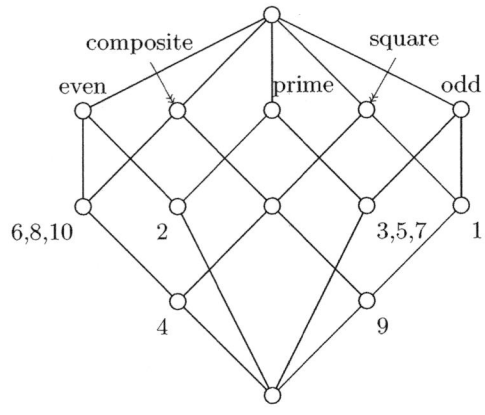

Fig. 1. A formal context and its concept lattice

properties even, odd, composite, prime and square. The corresponding concept lattice is displayed on the right. One reads the diagram as follows. Every node represents a concept (A, B). One reads the extent A from the diagram by collecting all objects that label nodes below (A, B). Thereby, *below* means less or equal in the displayed order. Dually, one reads the intent B from the diagram by collecting all attributes that label nodes above (A, B). For instance the unlabelled concept in the middle of the diagram is

$$(\{4, 9\}, \{\text{composite, square}\}).$$

For further details about concept lattices, like for instance how to determine all concepts of a given context, or how to draw a diagram of a concept lattice, we refer the reader to [6]. The reason for this very informal discussion of concept lattices at this point simply is that we will not need them explicitly in the following. But the reader should note that the diagram of a concept lattice contains exactly the same information as the underlying cross table but has the advantage that certain dependencies between objects and attributes are much easier to find out. Such dependencies are for instance the *attribute implications*.

Definition 3. *Let (G, M, I) be a formal context. A pair of subsets P and C of M is called an **attribute implication** and is usually written down as $P \to C$ (instead of (P, C)). Thereby, P is called the **premise** and C is called the **conclusion** of the implication. One says a subset $X \subseteq M$ **respects** $P \to C$ if*

$$X \supseteq P \implies X \supseteq C$$

*holds. One says the implication $P \to C$ **holds** in the formal context (G, M, I) if all object intents g^I $(g \in G)$ respect $P \to C$. An object $g \in G$ is called a **counter-example** for $P \to C$ if its object intent g^I does not respect $P \to C$.*

Proposition 2. *An implication $P \to C$ holds in (G, M, I) if and only if $C \subseteq P^{II}$ which again is equivalent to $P^I \subseteq C^I$. It then automatically holds in the set of all concept intents as well.*

Proof. See [6] Proposition 19. □

Even for very small contexts the number of attribute implications that hold tends to be huge. This is because a lot of implications hold for trivial reasons. If for instance the premise P is a superset of the conclusion C, the implication $P \to C$ holds for trivial reasons. Another reason is that the set of implications that hold in a given context contains a lot of redundancies one is usually not interested in, because there are syntactic rules of inference (the *Armstrong rules*, see for instance [6] Proposition 21) on how to derive implications from others. So one is usually interested in a minimal base set of implications from which one can derive all valid ones.

The most popular such set being used in Formal Concept Analysis is the so-called *Duquenne-Guigues-base* (also called *stem base*) and some of its variations. For details about the stem base (e.g. definition, how to determine it) we refer the reader again to the literature. For our example from Fig. 1 the stem base looks as follows:

$$
\begin{array}{c}
\text{even, square} \to \text{composite} \\
\text{prime, square} \to \text{even, odd, composite} \\
\text{even, odd} \to \text{composite, prime, square} \\
\text{odd, composite} \to \text{square} \\
\text{composite, prime} \to \text{even, odd, square}
\end{array}
$$

When working with attribute implications one laxly omits the set brackets. Furthermore, the reader should note that the second, third and the last one of the implications of the stem base are of the form $P \to M \setminus P$. In this example these implications hold because there are no objects having all attributes from P. In other words the corresponding extent P^I is empty. Especially if a formal context is large and therefore has an immense number of implications holding in it, one is preferably interested in implications whose premise is satisfied by a large number of objects.

Definition 4. *Let (G, M, I) be a formal context. An **association rule** $P \to C$ consists of two attribute sets: the **premise** P and the **conclusion** C. We call*

$$
\begin{aligned}
supp(B) &:= & |B^I| & \quad \text{the **support** of } B \subseteq M, \\
supp(P \to C) &:= supp(P \cup C) & & \quad \text{the **support** of the rule } P \to C, \text{ and} \\
conf(P \to C) &:= \frac{supp(P \cup C)}{supp(P)} & & \quad \text{the **confidence** of the rule } P \to C.
\end{aligned}
$$

If in the latter the support of the premise is 0, one defines the confidence of the association rule to be 1. Furthermore, for given thresholds

$$
\sigma \in \{0, 1, \ldots, |G|\} \quad and \quad \alpha \in [0, 1]
$$

*one says an association rule **holds** in (G, M, I) if its support and its confidence both exceed the respective thresholds, i.e, if*

$$supp(P \to C) \geq \sigma \quad and \quad conf(P \to C) \geq \alpha.$$

*One says an attribute set (or an association rule) is **frequent** if its support is greater or equal σ.*

Note that syntactically attribute implications and association rules are the same, namely pairs of attribute sets. The reason one has to use two different notions are the two different meanings of *holding in* a formal context. The attribute implications that hold in a formal context are precisely those association rules that hold with $\sigma = 0$ and $\alpha = 1$.

The confidence of a rule $P \to C$ clearly contains the information of how many objects sharing the attributes of the premise P also share the attributes of the conclusion C. The standard example for association rules comes from market basket analysis: If a customer buys *diapers*, he then in 60% of the cases also buys *beer*. Diapers and beer are bought together in 0.5% of all cases. So the association rule

$$\text{diapers} \to \text{beer}$$

has a support of $0.005 \cdot |G|$ and a confidence of 0.6. Note that the support of an attribute set $B \subseteq M$ is often defined in a *proportional way* as

$$\frac{|B^I|}{|G|}.$$

For reasons that might become clear in Section 6 we choose to define it the *absolute way.*

In the following we shortly explain how one can compute the association rules of a given formal context with given thresholds σ and α for the minimal support and the minimal confidence. We do this with the help of an example. For a detailled overview to this topic we refer the reader to [7,8].

The most expensive part of calculating the association rules is the computation of the *frequent concept intents*. Thereby, a concept (A, B) is said to be **frequent** if its intent B is frequent, i.e., if the cardinality of $A \, (= B^I)$ is greater or equal than the threshold σ.

The frequent concepts can be understood as the *most important* ones. They form an order filter in the concept lattice. The set of all frequent concepts is often called an *iceberg concept lattice*. The explanation of this figurative notion is the following. The concept lattice is thought to be an iceberg swimming in the ocean. In this picture the minimum support threshold σ symbolizes the water surface. Thus, the frequent concepts form the visible peak of the iceberg and the non-frequent ones lie below the water surface and, hence, are not visible to the viewer.

	Group of 77	Non-aligned	LLDC	MSAC	OPEC	ACP
Afghanistan	×	×	×	×		
Algeria	×	×			×	
Angola	×	×				×
Antigua and Barbuda	×					×
Argentina	×					
Bahamas	×					×
Bahrain	×	×				
Bangladesh	×	×	×	×		
Barbados	×	×				×
Belize	×	×				×
Benin	×	×	×	×		×
Bhutan	×	×	×			
Bolivia	×	×				
Botswana	×	×	×			×
Brazil	×					
Brunei						
Burkina Faso	×	×	×	×		×
Burundi	×	×	×	×		×
Cambodia	×	×		×		
Cameroon	×	×		×		×
Cape Verde	×	×	×	×		×
Central African Rep.	×	×	×	×		×
Chad	×	×	×	×		×
Chile	×					
China						
Colombia	×	×				
Comoros	×	×	×			×
Congo	×	×				×
Costa Rica	×					
Cuba	×	×				
Djibouti	×	×	×			×
Dominica	×	×				×
Dominican Rep.	×					×

	Group of 77	Non-aligned	LLDC	MSAC	OPEC	ACP
Ecuador	×	×			×	
Egypt	×	×		×		
El Salvador	×			×		
Equatorial Guinea	×	×	×			×
Ethiopia	×	×	×	×		×
Fiji	×					×
Gabon	×	×			×	×
Gambia	×	×	×	×		×
Ghana	×	×	×	×		×
Grenada	×	×				×
Guatemala	×			×		
Guinea	×	×	×	×		×
Guinea-Bissau	×	×	×	×		×
Guyana	×	×		×		×
Haiti	×		×	×		×
Honduras	×			×		
Hong Kong						
India	×	×		×		
Indonesia	×	×			×	
Iran	×	×			×	
Iraq	×	×			×	
Ivory Coast	×	×		×		×
Jamaica	×	×				×
Jordan	×	×				
Kenya	×	×		×		×
Kiribati				×		×
Korea-North	×	×	×			
Korea-South	×					
Kuwait	×	×			×	
Laos	×	×	×	×		
Lebanon	×	×				
Lesotho	×	×	×	×		×
Liberia	×	×				×

The abbreviations stand for: LLDC := Least Developed Countries, MSAC := Most Seriously Affected Countries, OPEC := Organization of Petrol Exporting Countries, ACP := African, Caribbean and Pacific Countries.

Fig. 2. Membership of developing countries in supranational groups. (Part 1).

	Group of 77	Non-aligned	LLDC	MSAC	OPEC	ACP		Group of 77	Non-aligned	LLDC	MSAC	OPEC	ACP
Libya	×	×			×		Senegal	×	×		×		×
Madagascar	×	×	×	×		×	Seychelles	×	×				×
Malawi	×	×	×			×	Sierra Leone	×	×	×	×		×
Malaysia	×	×					Singapore	×	×				
Maledives	×	×	×				Solomon Islands	×					×
Mali	×	×	×	×		×	Somalia	×	×	×	×		×
Mauretania	×	×	×	×		×	Sri Lanka	×	×		×		
Mauritius	×	×				×	St Kitts						
Mexico	×						St Lucia	×	×				×
Mongolia		×					St Vincent& Grenad.	×					×
Morocco	×	×					Sudan	×	×	×	×		×
Mozambique	×	×		×		×	Surinam	×	×				×
Myanmar	×		×	×			Swaziland	×	×				×
Namibia	×					×	Syria	×	×				
Nauru							Taiwan						
Nepal	×	×	×	×			Tanzania	×	×	×	×		×
Nicaragua	×	×					Thailand	×					
Niger	×	×	×	×		×	Togo	×	×	×			×
Nigeria	×	×			×	×	Tonga	×					×
Oman	×	×					Trinidad and Tobago	×	×				×
Pakistan	×	×		×			Tunisia	×	×				
Panama	×	×					Tuvalu			×			×
Papua New Guinea	×					×	Uganda	×	×	×	×		×
Paraguay	×						United Arab Emirates	×	×			×	
Peru	×	×					Uruguay	×					
Philippines	×						Vanuatu	×	×	×			×
Qatar	×	×			×		Venezuela	×	×			×	
Réunion							Vietnam	×	×	×			
Rwanda	×	×	×	×		×	Yemen	×	×	×	×		
Samoa	×		×	×		×	Zaire	×	×	×			×
São Tomé e Principe	×	×	×			×	Zambia	×	×	×			×
Saudi Arabia	×	×			×		Zimbabwe	×	×				×

Fig. 2. Membership of developing countries in supranational groups. (Part 2). Source: *Lexikon Dritte Welt*, Rowohlt-Verlag, Reinbek 1993.

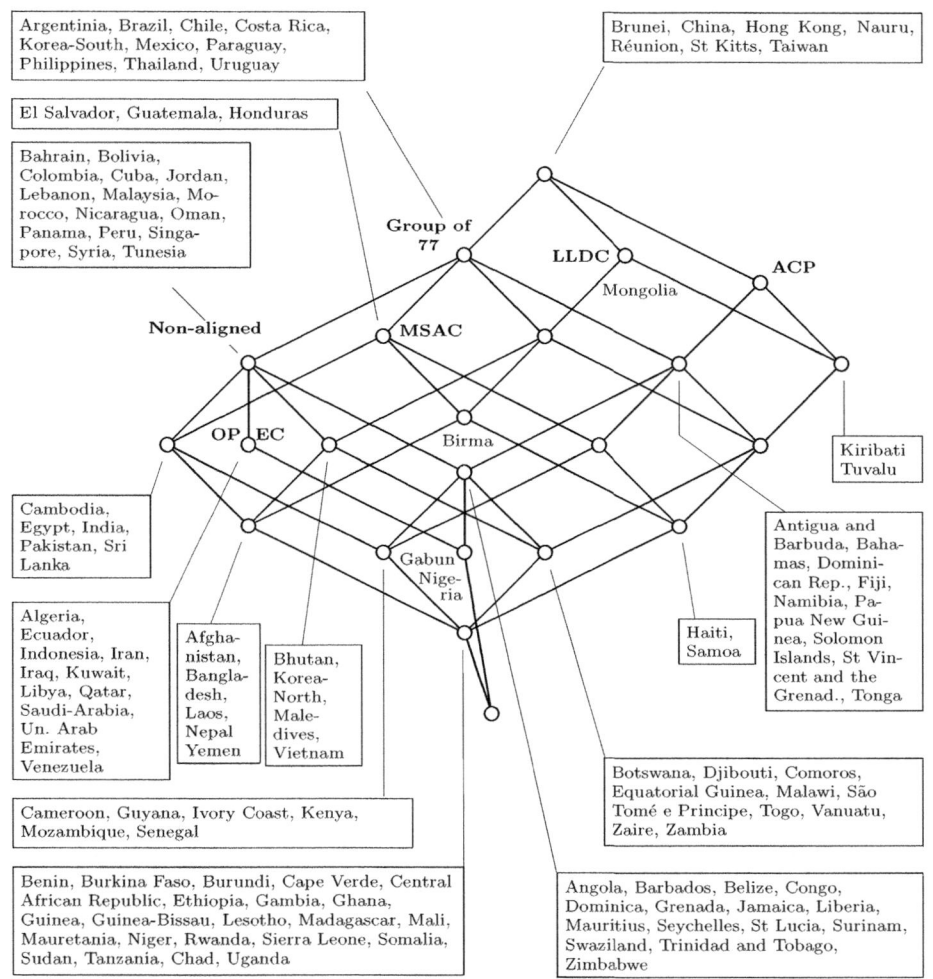

Fig. 3. Concept lattice of the context of developing countries from Fig. 2

There are efficient ways to compute the intents of the frequent concepts (of course without computing the whole concept lattice). One way is a variation of the *Next-closure* algorithm (see [6, Theorem 6]). Another way is the *Titanic* algorithm to compute iceberg concept lattices (see [8]). In the frame of this brief introduction we do not describe how the algorithms work in detail.

Fig. 4 shows the iceberg lattice of the concept lattice displayed in Fig. 3 for the minimal support $\sigma := 40$. Hence, the figure shows all concepts whose extent conatins 40 objects or more. In the diagram the support of the ten frequent concepts is written nearby the corresponding nodes. For instance, we can read from the diagram that the concept with the intent {Group of 77, Non-aligned} has the support 95. In other words there are 95 objects sharing the attributes

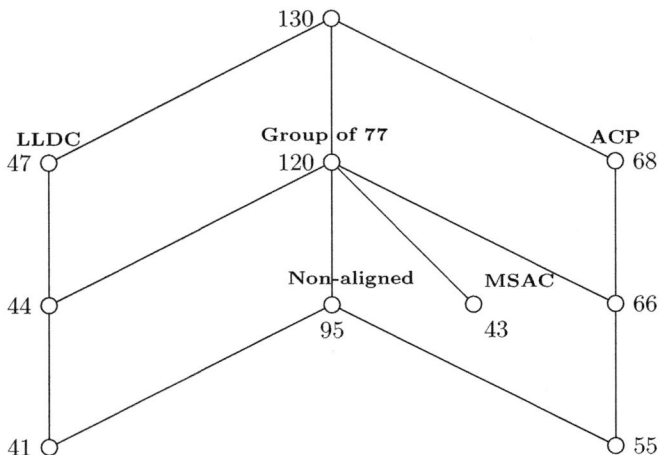

Fig. 4. The iceberg concept lattice of the formal context displayed in Fig. 2 with the minimal support $\sigma = 40$

Group of 77 and *Non-aligned*. Obviously, the iceberg concept lattice is not a lattice. But one can fix this by enriching it with a bottom elment.

One obtains the valid association rules in two steps. At first one determines the association rules that have the confidence 1, and afterwards one calculates the ones with a confidence smaller than 1. The association rules with confidence 1 (which are the valid attribute implications) can for instance be determined by calculating the frequent *pseudo intents*. This yields to a subset of the *Duquenne-Guigues base* (see for instance [6]) from which one can derive all frequent implications.

The valid association rules with a confidence smaller than 1 can be read from the iceberg lattice as follows. At first one has to choose a minimal confidence threshold. For our example we chose $\alpha := 0.8$. For attribute sets P and C the two rules $P \rightarrow C$ and $P \rightarrow P \cup C$ share the same support and confidence. Furthermore, one can obviously infer one from the other. Thus, one can w.l.o.g assume that one is just interested in rules $P \rightarrow C$ with $P \subseteq C$. The so-called *Luxenburger base* (see [7]) of association rules with a support < 1 now consists of those rules $P \rightarrow C$ where $P \subsetneq C$ are neighboured in the iceberg lattice and where the confidence

$$\text{conf}(P \rightarrow C) = \frac{\text{supp}(P \cup C)}{\text{supp}(P)} = \frac{\text{supp}(C)}{\text{supp}(P)}$$

exceeds the minimum conficence threshold α. This yields to an obvious algorithm to determine this Luxenburger base. For every edge in the diagram of the iceberg lattice one calculates the corresponding confidence. For our example this was done in Fig. 5. Whenever the confidence exceeds the threshold $\alpha = 0.8$

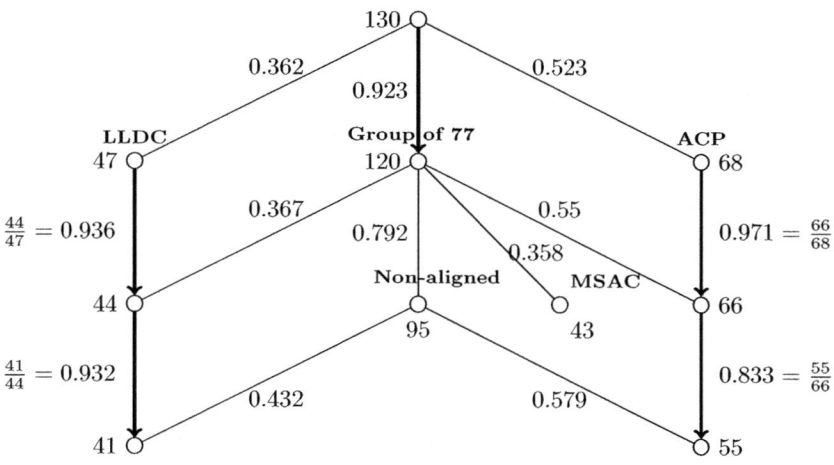

Fig. 5. A visualization of the Luxenburger base

one highlights the line as a thick arrow. Hence, the thick arrows represent the association rules of the Luxenburger base.

We see that in our example the Luxenburger base consists of five association rules. The rule {LLDC} → {LLDC,Group of 77} for instance has a support of 44 and a confidence of 0.936. Hence, it is a valid rule for our thresholds σ and α. For obvious reasons one writes these rules down in the form $P \to C \backslash P$. Furthermore, one omits set brackets. One receives the following base of association rules, which is by the way the same result the free FCA software tool Concept Explorer[1] delivers.

[95]	Non-aligned	=[100%]⇒ Group of 77 [95]
[43]	MSAC	=[100%]⇒ Group of 77 [43]
[68]	ACP	=[97%]⇒ Group of 77 [66]
[47]	LLDC	=[94%]⇒ Group of 77 [44]
[44] Group of 77, LLDC		=[93%]⇒ Non-aligned [41]
[130]		=[92%]⇒ Group of 77 [120]
[66] Group of 77, ACP		=[83%]⇒ Non-aligned [55]

3   Partial Formal Contexts

Now that the reader is familiar with the basic notions of Formal Concept Analysis we already expand them to the more sophisticated situation of *incomplete knowledge*. To encode the granulation process mentioned in the introduction we use the notion of a partial formal context. The information we are mining is in

---

[1] http://sourceforge.net/projects/conexp/

the form of implications or, more loosely, of association rules. Our aim is to infer such rules in the full data set from rules in the granulated data.

Definition 5. *A **partial formal context** (G, M, i) consists of two sets G and M together with a mapping $i : G \times M \to \{\times, \bullet, ?\}$.*

As in the case of formal contexts we call the elements of G the *objects* of the partial formal context, those of M the *attributes*. We read $i(g, m)$ as follows:

$$i(g, m) = \begin{cases} \times & \text{the object } g \text{ has the attribute } m, \\ \bullet & \text{the object } g \text{ does not have the attribute } m, \\ ? & \text{it is unknown if object } g \text{ has attribute } m. \end{cases}$$

Partial formal contexts have been considered under different aspects by several authors [9,10,11]. A partial formal context (G, M, j) is said to **extend** (G, M, i) if one can build it from (G, M, i) by replacing question marks "?" by crosses "$\times$" or dots "$\bullet$", i.e., if it holds that

$$i^{-1}(\{\times\}) \subseteq j^{-1}(\{\times\}) \quad \text{and} \quad i^{-1}(\{\bullet\}) \subseteq j^{-1}(\{\bullet\}).$$

Partial formal contexts which are maximal w.r.t. to this *extension order* are called **complete**. A formal context (G, M, I) in the usual sense, where $I \subseteq G \times M$ is a relation, is called a **completion** of a partial formal context (G, M, i) iff

$$i^{-1}(\{\times\}) \subseteq I \subseteq i^{-1}(\{\times, ?\}).$$

We say that an attribute implication $A \to B$, where $A, B \subseteq M$, **holds** in a partial formal context (G, M, i) iff it holds in every completion. An equivalent condition is that the following holds for every object $g \in G$:

if $i(g, m) \in \{\times, ?\}$ for all $m \in A$ then $i(g, n) = \times$ for all $n \in B$.

An implication $A \to B$ **is refuted** by the partial formal context (G, M, i) if it holds in no completion. This is equivalent to the existence of an object g with

$$i(g, m) = \times \text{ for all } m \in A \text{ and } i(g, n) = \bullet \text{ for some } n \in B.$$

In order to better handle canonical formal contexts related to the partial context (G, M, i) we define for $S \subseteq \{\times, \bullet, ?\}$

$$i_S := \{(g, m) \in G \times M \mid i(g, m) \in S\} = i^{-1}(S).$$

We allow to leave away the set brackets of S. For instance we write $i_{\times,?}$ instead of $i_{\{\times,?\}}$.

Proposition 3. *Let (G, M, i) and (G, M, j) be partial formal contexts such that (G, M, j) extends (G, M, i). Then*

– *every implication that holds in* (G, M, i) *also holds in* (G, M, j), *and*

– *every implication that is refuted by* (G, M, i) *is also refuted by* (G, M, j).

Proof. For every implication $A \rightarrow B$ that holds in (G, M, i) it follows for $g \in G$ that

$$A \subseteq g^{j\times,?} \implies A^{\complement} \supseteq g^{j\bullet} \supseteq g^{i\bullet} \implies A \subseteq g^{i\times,?} \implies B \subseteq g^{i\times} \subseteq g^{j\times}.$$

The second item follows immediately from the observation that every object that refutes an implication in (G, M, i) also refutes this implication in (G, M, j). □

4   Partial Contexts Obtained from Streams

There is a natural way how partial formal contexts arise from complete ones. Let (G, M, I) be a formal context and let $\mathcal{F}$ be a family of nonempty subsets of the object set G, i.e. $\mathcal{F} \subseteq \mathfrak{P}_{>0}(G) := \mathfrak{P}(G) \setminus \{\emptyset\}$. We obtain a partial formal context $(\mathcal{F}, M, i)$ by defining for every **block** $F \in \mathcal{F}$

$$i(F, m) := \begin{cases} \times & \text{if } F \subseteq m^I, \\ \bullet & \text{if } F \cap m^I = \emptyset, \\ ? & \text{else.} \end{cases}$$

We refer to $(\mathcal{F}, M, i)$ as the $\mathcal{F}$-**granulated** partial context to (G, M, I). Note that this reflects the situation of Infobright's rough tables from the contest [4] and is only formulated in a different language. For further details we refer the reader to the following Section 5.

Proposition 4. *Let* $(\mathfrak{P}_{>0}(G), M, i)$ *be constructed from* (G, M, I) *as defined above, for the special case that* $\mathcal{F} := \mathfrak{P}_{>0}(G)$. *Then*

– *an implication that holds in* $(\mathfrak{P}_{>0}(G), M, i)$ *also holds in* (G, M, I) *and*

– *an implication is refuted by* $(\mathfrak{P}_{>0}(G), M, i)$ *iff it does not hold in* (G, M, I).

Proof. The rows of (G, M, I) correspond to the rows belonging to singleton blocks $\{g\}$ in $(\mathfrak{P}_{>0}(G), M, i)$ in the following way: $\{g\}^{i\times} = g^I$, $\{g\}^{i?} = \emptyset$ and $\{g\}^{i\bullet} = g^{\overline{I}}$. Hence, the first item and "$\Leftarrow$" of the second item are trivial. The remaining direction of the second item is a special case of the following Proposition 5. □

Proposition 5. *For every* $\mathcal{F} \subseteq \mathfrak{P}_{>0}(G)$ *it is true that*

– *no implication refuted by* $(\mathcal{F}, M, i)$ *holds in* (G, M, I).

If $\mathcal{F}$ *is a covering of* G *then it is true that*

– *every implication that holds in* $(\mathcal{F}, M, i)$ *also holds in* (G, M, I).

Proof. Let $F \in \mathcal{F}$ be a block that refutes $A \to B$ in $(\mathcal{F}, M, i)$. Then it holds that $A \subseteq F^{i\times} = F^I$ and $B \not\subseteq F^{i\times,?}$. Hence, it follows $B \not\subseteq F^{i\times} = F^I$ which implies that $A \to B$ cannot hold in (G, M, I), since F^I is an intent containing the premise A, but not containing B.

Let $A \to B$ be an implication that holds in $(\mathcal{F}, M, i)$ and let $g \in G$. Since $\mathcal{F}$ is a covering there is a block $F \in \mathcal{F}$ containing g. Hence, it holds that $g^I \subseteq F^{i\times,?}$ which implies

$$A \subseteq g^I \implies A \subseteq F^{i\times,?} \implies B \subseteq F^{i\times} \implies B \subseteq g^I. \qquad \square$$

Now suppose that (G, M, I) is given as a stream of rows, and is chopped into data packs as described in the introduction. For each pack we take notes only if each object in the pack does have the attribute, in which case we note an "×" for the pack, or if no object in the pack has that attribute. We then note down "•". If some have and some do not, we note a question mark. This is a very strict rule, and we refer to it as **hard granulation**. Its disadvantage is that its outcome can drastically be changed by a single value in the pack. It shares this property with logical analysis: If a given logical formula does or does not hold in the original data, may be decided by a single counterexample. Proposition 5 above shows our possibilities to argue about implicational information of (G, M, I) based only on the granulated context $(\mathcal{F}, M, i)$. It is therefore necessary to investigate the circumstances under which an implication holds in or is refuted by $(\mathcal{F}, M, i)$. For both concerns it is sufficient to just take a look at implications of the form $A \to b$, where $A \subseteq M$ and $b \in M$.

Proposition 6. *For $F \in \mathcal{F}$ the following three statements are equivalent:*

(a) F refutes $A \to b$ in $(\mathcal{F}, M, i)$,

(b) $F \subseteq A^I \setminus b^I$,

(c) every single object $g \in F$ refutes $A \to b$ in (G, M, I).

Proof. F refutes $A \to b$ iff it holds that $A \subseteq F^{i\times} = F^I$ and $b \in F^{i\bullet} = F^{\neq}$, which again is equivalent to $F \subseteq A^I$ and $F \subseteq b^{I\complement}$. The rest follows immediately. $\qquad \square$

The preceding propositions clarify under which conditions an implication $A \to b$ is refuted by the granulated context $(\mathcal{F}, M, i)$. If we insist on a definite answer, an answer that proves a refutation in the full data set on basis of the granulated data, these seem to be the natural conditions. But how likely is it that these conditions are satisfied? We attempt to give a first estimation. Obviously, the number $r := |A^I \setminus b^I|$ of all objects from the original data table (G, M, I) that share all attributes from A but do not have attribute b has to be large enough. Let k be a fixed number and let $n := |G|$ be the number of objects. For the probability that a block F of cardinality k refutes $A \to b$ the following holds:

$$P(F \text{ refutes } A \to b) = \frac{\binom{r}{k}}{\binom{n}{k}} = \frac{r \cdot (r-1) \cdot \ldots \cdot (r-k+1)}{n \cdot (n-1) \cdot \ldots \cdot (n-k+1)} \leq \left(\frac{r}{n}\right)^k. \qquad (1)$$

We now assume that all $F \in \mathcal{F}$ have the same cardinality k. With the inequality from above we can conclude the following upper approximation of the probability that a partial context $(\mathcal{F}, M, i)$ refutes $A \to b$:

$$P((\mathcal{F}, M, i) \text{ refutes } A \to b) \leq \sum_{F \in \mathcal{F}} P(F \text{ refutes } A \to b) \leq |\mathcal{F}| \cdot \left(\frac{r}{n}\right)^k.$$

If we for instance assume that 95% of all objects in (G, M, I) refute $A \to b$ and that $\mathcal{F}$ contains one million blocks, i.e., $\frac{r}{n} = 95\%$ and $|\mathcal{F}| = 1.000.000$, we get that already for relatively small block sizes of $k \geq 539$ the probability that $(\mathcal{F}, M, i)$ refutes $A \to b$ is smaller than one part of a million.

Proposition 7. *For $A, B \subseteq M$ the following four statements are equivalent:*

(a) $A \to B$ holds in $(\mathcal{F}, M, i)$,

(b) for all $F \in \mathcal{F}$ the implication $A \subseteq F^{i\times, \varrho} \implies B \subseteq F^{i\times}$ holds,

(c) for all $F \in \mathcal{F}$ the implication $A \subseteq \bigcup_{g \in F} g^I \implies B \subseteq \bigcap_{g \in F} g^I$ holds,

(d) for all $F \in \mathcal{F}$ the implication $(\forall a \in A : F \nsubseteq a^I) \implies F \subseteq B^I$ holds.

Proof. Omitted.   □

If one takes a look at the third condition it becomes obvious that the bigger the block sizes $|F|$ are, the more likely it becomes that the premisses are valid, and the less likely it becomes that the conclusions hold. Hence, if the number of the blocks and the sizes of the blocks are relatively large, we do not expect a lot of attribute implications to hold in $(\mathcal{F}, M, i)$: The probability that a *single* block F of cardinality k fulfills the implication from (d) is

$$P((\exists a \in A : F \subseteq a^I) \text{ or } F \subseteq B^I) \leq \sum_{a \in A} P(F \subseteq a^I) + P(F \subseteq B^I)$$

$$\leq |A| \cdot \left(\frac{n - |A^I|}{n}\right)^k + \left(\frac{|B^I|}{n}\right)^k.$$

Thereby the second inequation follows analogously to inequation (1). Let us assume that $A \to B$ is a *nontrivial* implication that holds in (G, M, I), i.e. $\emptyset \neq A^I \subseteq B^I \neq G$. Then for a large block size k this probability tends to be very small. Hence, for a large number of blocks it is far more improbable that $A \to B$ holds in $(\mathcal{F}, M, i)$.

5   The Contest Data Set

The Infobright data set does not come as a formal context right away, but needs some (uncritical) transformation. The formalisation of a data table which we use is that of a *many-valued context* (G, M, W, J), where G is a set of objects, M

a set of many-valued attributes, W a set of attribute values and J is a ternary incidence relation satisfying

$$(g, m, v) \in J \text{ and } (g, m, w) \in J \text{ implies } v = w.$$

The standard interpretation of $(g, m, v) \in J$ is that the value of attribute m for object g is v. The value the object g has with respect to attribute m is commonly denoted with $m(g)$. To better distinguish such many-valued contexts from the formal contexts introduced first we shall refer to these sometimes as *one-valued*.

One of the standard techniques in Formal Concept Analysis expresses many-valued contexts as one-valued ones by means of *conceptual scales*. With conceptual scaling, every many-valued attribute is represented by several one-valued attributes, and the incidence to these depends on the respective attribute value. Details can be found in [6], but for the moment it suffices to know that with this technique a data table can be transformed to a (one-valued) formal context, and this transformation can be done object-wise, one after another. As a consequence, we may transform a stream of objects with many-valued attributes into a stream of objects in a formal context. To keep things simple, we summarize: Conceptual scaling associates to each column m of the data table a set of attributes (the "scale attributes for the many-valued attribute m").

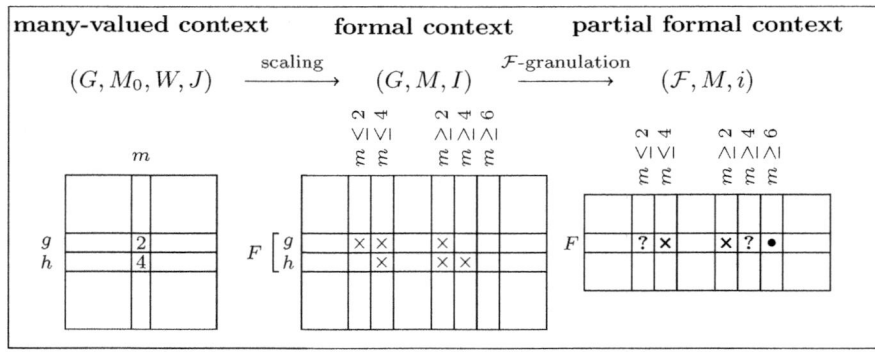

Fig. 6. A schematic illustration of interordinal scaling and $\mathcal{F}$-granulation

In the case of Infobright's contest data set we may think of the underlying, very large data table as a many-valued context (G, M_0, W, J) in which for every attribute $m \in M_0$ the set

$$W_m := m[G] = \{w \in W \mid (g, m, w) \in J \text{ for some } g \in G\}$$

of all values occurring in the column of m are ordered linearly in a canonical way. Depending on the data type of the attribute m this canonical order $\leq_m$ can for instance be the natural order of numbers or the alphabetical order of character strings. If one transforms this data table (G, M_0, W, J) into the formal context

(G, M, I) via scaling every attribute from M_0 *interordinally*, this formal context (G, M, I) directly yields to the granulated partial formal context $(\mathcal{F}, M, i)$ which contains exactly the same information as the contests rough table from [4].

We leave out the details about the interordinal scaling of the original data table (G, M_0, W, J). We refer the reader to Fig. 6 to get an idea on how it works. The problem with the contest data set is that for almost every attribute $m \in M_0$ it holds that for almost every rough row $F \in \mathcal{F}$ the minimal and maximal m-values in F are exactly the overall minimal and maximal m-values, i.e.,

$$\min_{f \in F} m(f) \;=\; \min_{g \in G} m(g) \quad \text{and} \quad \max_{f \in F} m(f) \;=\; \max_{g \in G} m(g).$$

This yields to the effect that $(\mathcal{F}, M, i)$ is almost full of question marks, which minimizes the chances to verify or to refute some interesting attribute implications.

6   Soft Granulation

There is a reason why the approach of the previous section led to rather disappointing results: Our definition of the granulation process was too rigid. We defined that a block has a certain object if *all* members of a pack have the attribute, etc. As an example from the Infobright data, we mention the *minimum* parameter: It expresses that all members of the pack have a value greater or equal this one.

For a rough estimation, such parameters that can drastically be changed by a single member of the block seem inappropriate. It seems more promising to work with parameters which reflect the "tendency" of the data packs. The simplest suggestion is counting: Let us record for each data pack (F, m) the number of objects having the attribute. This yields us to the follwing

Definition 6. *Let G and M be sets and let $\mathcal{F}$ be a partition of G. For every mapping $j : \mathcal{F} \times M \to \mathbb{N}_0$ we call the triple $(\mathcal{F}, M, j)$ a **soft-granulated (formal) context**. Furthermore, we call a formal context (G, M, J) a **realizer** of $(\mathcal{F}, M, j)$ if for every block $F \in \mathcal{F}$ and every attribute $m \in M$ it holds that*

$$j(F, m) \;=\; |F \cap m^J|.$$

*Given a context (G, M, I) there is obviously a unique soft-granulated context $(\mathcal{F}, M, i)$ having (G, M, I) as a realizer. It is called the **soft granulated context to** (G, M, I). Furthermore, we call*

$$supp(F, m) \;:=\; |F \cap m^I|$$

*the **support** and*

$$nsupp(F, m) \;:=\; |F \setminus m^I| \;=\; |F| - supp(F, m)$$

*the **negative support** of the data pack (F, m). Let $(\mathcal{F}, M, i)$ be a soft-granulated context to a formal context (G, M, I), let $B, P, C \subseteq M$ be sets of attributes and let $F \in \mathcal{F}$ be a block. Then we say*

- B is **possible** in F if for every $b \in B$ it holds that $i(F, b) > 0$,
- B **holds** in F if for every $a \in A$ it holds that $i(F, m) = |F|$,
- B is **refuted** by F if it is not possible in F,
- the implication $P \to C$ **holds** in $(\mathcal{F}, M, i)$ if it holds in every realizer,
- the implication $P \to C$ is **refuted** by $(\mathcal{F}, M, i)$ if there is no realizer in which $P \to C$ holds.

Let us reconsider our setting. We have a very large but finite formal context (G, M, I). This context is not totally known to us. What we know about (G, M, I) are the following two things: for a fixed partition $\mathcal{F}$ of the object set G we have the $\mathcal{F}$-soft-granulated context $(\mathcal{F}, M, i)$ to (G, M, I). This means that for every data pack $(F, m) \in \mathcal{F} \times M$ we know the number of crosses $i(F, m) = \mathrm{supp}(F, m)$ and the number of empty cells $|F| - i(F, m) = \mathrm{nsupp}(F, m)$. The second thing we assume to have is some background knowledge. We assume to have a list $\mathcal{L}$ of implications that hold in (G, M, I). These *background implications* typically result from a scaling process.

What we are now looking for is a way to explore as many knowledge as possible about the original data table (G, M, I) just using the granulated context $(\mathcal{F}, M, i)$ and the background implications from $\mathcal{L}$. Thereby we are mining in $(\mathcal{F}, M, i)$ for association rules that hold in (G, M, I). Let us reconsider that an association rule holds in a formal context if its support and it confidence exceed the thresholds σ and α. We now generalise the aforementioned definitions about implications in soft-granulated contexts to the setting of background implications and of association rules.

Definition 7. *Let $\mathcal{L}$ be a set of implications over M and let $(\mathcal{F}, M, i)$ be a soft-granulated context. A realizer (G, M, J) is now called a $\mathcal{L}$-**realizer** of $(\mathcal{F}, M, i)$ if all implications from $\mathcal{L}$ hold in it. We now say an association rule $\mathcal{L}$-**holds** in $(\mathcal{F}, M, i)$ if it holds in every $\mathcal{L}$-realizer. An association rule is said to be $\mathcal{L}$-**refuted** by $(\mathcal{F}, M, i)$ if there is no $\mathcal{L}$-realizer in which it holds. If the background knowledge is well-understood one may omit the preposition $\mathcal{L}$ and just say "holds in" and "refuted by".*

In order to verify or to refute a given association rule we must be able to approximate its support and its confidence just using the available information from the soft-granulated context and the background knowledge. Let us first take a look at the case without background knowledge, i.e., where $\mathcal{L} = \emptyset$. If we ignore the background implications from $\mathcal{L}$, we receive the following *worst case* approximations of the support of an attribute set $A \subseteq M$ restricted to a block $F \in \mathcal{F}$

$$\underline{\mathrm{supp}}(F, A) := \left\lceil |F| - \sum_{a \in A} \mathrm{nsupp}(F, a) \right\rceil$$

$$\overline{\mathrm{supp}}(F, A) := \begin{cases} \min_{a \in A} \mathrm{supp}(F, a) & \text{if } A \neq \emptyset, \\ |F| & \text{else.} \end{cases}$$

where for $x \in \mathbb{R}$ we define $\lceil x \rceil := \max\{0, x\}$. If we now define

$$\underline{\mathrm{supp}}(A) := \sum_{F \in \mathcal{F}} \underline{\mathrm{supp}}(F, A) \quad \text{and} \quad \overline{\mathrm{supp}}(A) := \sum_{F \in \mathcal{F}} \overline{\mathrm{supp}}(F, A)$$

we obviously receive an lower and an upper approximation of the support of A. Furthermore, we define for $A, B \subseteq M$

$$\underline{\mathrm{conf}}(A \to B) := \frac{\underline{\mathrm{supp}}(A \cup B)}{\underline{\mathrm{supp}}(A)} \quad \text{and} \quad \overline{\mathrm{conf}}(A \to B) := \frac{\overline{\mathrm{supp}}(A \cup B)}{\underline{\mathrm{supp}}(A)}.$$

It is not hard to see that the following results holds:

Proposition 8. *For $A, B \subseteq M$ it holds that:*

$$\underline{supp}(A) \quad \leq \quad supp(A) \quad \leq \quad \overline{supp}(A),$$

$$\underline{conf}(A \to B) \leq conf(A \to B) \leq \overline{conf}(A \to B).$$

Proof. Omitted.   □

Even though these approximations are very coarse in most cases, they are tight in the sense that there are cases where equality holds.

Corollary 1. *If for an association rule $A \to B$ it holds that*

$$\sigma \leq \underline{supp}(A \cup B) \quad \text{and} \quad \alpha \leq \underline{conf}(A \to B),$$

the rule also holds in (G, M, I).

In the end of Section 2 we shortly described how one can determine the valid association rules of a given formal context with the help of the iceberg concept lattice. The question we want to answer in the following is, if there is a similar approach for the case of the soft-granulated contexts. In the classical case described in Section 2 we used the frequent intents to find the valid association rules. If there is a similar approach, one might for instance consider $\underline{\mathrm{supp}}$-frequent attribute sets that are also *closed in some sense*. Thereby, an attribute set $A \subseteq M$ is said to be $\underline{\mathrm{supp}}$-**frequent** if the inequality $\underline{\mathrm{supp}}(A) \geq \sigma$ holds. But what could be a useful definition of *being closed*?

The Titanic algorithm [7] makes use of the fact that one can totally describe the intent closure operator by just using the support function $\mathrm{supp}(\cdot)$. This works in the way that one can saturate an attribute set $A \subseteq M$ with all elemente $a \in M \setminus A$ that do not yield to a decrease of the support value, i.e., elements m with

$$\mathrm{supp}(A) = \mathrm{supp}(A \cup \{m\}).$$

The result of this saturation delivers A^{II} which is the smallest concept intent containing A. The following Proposition 9 shows that the analogous $\underline{\mathrm{supp}}$-saturation also yields to a closure operator and that this closure operator behaves in the same way with respect to $\underline{\mathrm{supp}}(\cdot)$ as the intent closure operator behaves with $\mathrm{supp}(\cdot)$.

Proposition 9. *The mapping* $\lceil \cdot \rceil_{\mathcal{F}} : \mathfrak{P}(M) \to \mathfrak{P}(M)$ *defined by*

$$\lceil A \rceil_{\mathcal{F}} := A \cup \{m \in M \setminus A \mid \underline{supp}(A) = \underline{supp}(A \cup \{m\})\}$$

is a closure operator on M. *It is compatible with the the approximated support function* $\underline{supp}(\cdot)$ *in the sense that for* $A, B \subseteq M$ *the following two implications hold:*

(i) $A \subseteq B$ *and* $\underline{supp}(A) = \underline{supp}(B) \implies \lceil A \rceil_{\mathcal{F}} = \lceil B \rceil_{\mathcal{F}}$,

(ii) $\lceil A \rceil_{\mathcal{F}} = \lceil B \rceil_{\mathcal{F}} \implies \underline{supp}(A) = \underline{supp}(B)$.

Proof. The proof has been omitted since we do not need the result in the following. □

An attribute implication $A \to B$ holds in the original formal context (G, M, I) if and only if $B \subseteq A^{II}$ holds (see Prop. 2). Hence, one can determine the validity of an attribute implication by calculating the intent closure of the premise. In the first place one might hope that the closure operator $\lceil \cdot \rceil_{\mathcal{F}}$ is related in the same way to the implications that hold for sure in the soft-granulated context $(\mathcal{F}, M, i)$. More formally, this hope means: does an attribute implication $A \to B$ hold in the soft-granulated context if and only if $B \subseteq \lceil A \rceil_{\mathcal{F}}$ holds?

Unfortunately, the answer to this question is no. Hence, in comparism to the classical case one has to calculate *all* $\underline{supp}$-frequent attribute sets. But there are two reasons why in this case this not that bad. The first one is that for the desired applications where (G, M, I) is very large, and where the block sizes are very large, too, it is extremely unlikely that for a set A with $\underline{supp}(A) > 0$ there is some $m \in M \setminus A$ such that A and $A \cup \{m\}$ share the same $\underline{supp}$-value. And this also holds when $A \to m$ holds in (G, M, I) (even with a high support). In other words, in practise one expects the closure operator $\lceil \cdot \rceil_{\mathcal{F}}$ to be of the form that very small subsets of M that have a nonempty $\underline{supp}$-value are mapped on itself again. Of course, those sets that have 0 as its $\underline{supp}$-value are mapped onto M. Hence, it makes nearly no difference if one computes all $\underline{supp}$-frequent attribute sets, or *just* the closed ones. The second reason is that due to the coarseness of the approximation $\underline{supp}(\cdot)$ of supp$(\cdot)$ it is very likely that the cardinality of $\underline{supp}$-frequent subsets is quite limited, which again yields to a much smaller number of those (compared to the number of supp-frequent subsets).

We will now explain how one can derive valid association rules from the soft-granulated context. As in Section 2 we do this with the help of an example. Fig. 7 shows a possible soft granulation of the formal context from Fig. 2. Thereby, the original cross-table was chopped into 13 blocks which all have cardinality 10. In comparism to the original case described in Section 2 we now choose smaller minimal thresholds for the support and confidence. We choose $\sigma := 30$ and $\alpha := 0.7$. Fig. 8 shows the 11 $\underline{supp}$-frequent subsets of M. Furthermore, the figure contains the approximation intervals

$$[\underline{supp}(A), \overline{supp}(A)]$$

		Group of 77	Non-aligned	LLDC	MSAC	OPEC	ACP
F_1	Afghanistan – Belize	10	7	2	2	1	5
F_2	Benin – Cameroon	9	8	5	5	0	5
F_3	Cape Verde – Cuba	9	7	4	3	0	5
F_4	Djibouti – Gabon	10	7	3	3	2	7
F_5	Gambia – Hong Kong	9	6	5	8	0	7
F_6	India – Korea-North	9	9	2	3	3	4
F_7	Korea-South – Malaysia	10	9	4	3	2	4
F_8	Maledives – Namibia	9	6	5	4	0	5
F_9	Nauru – Paraguay	9	7	2	3	1	2
F_{10}	Peru – Seychelles	9	7	3	3	2	5
F_{11}	Sierra Leone – Surinam	9	7	3	4	0	7
F_{12}	Swaziland – Tuvalu	8	6	3	1	0	6
F_{13}	Uganda – Zimbabwe	10	9	6	2	2	5

Fig. 7. A soft-granulated context of the formal context from Fig. 2. Thereby, the object set of cardinality 130 was chopped into 13 blocks $F_1, \ldots, F_{13}$ of cardinality 10.

of the frequent sets A. Since for the empty set and for the singleton sets the approximations obviously deliver the precise support, the intervals are of the form $[s, s]$. In order to keep the notations easy, we just write $[s]$ instead of $[s, s]$.

In the same way we derived the Luxenburger base of association rules in Fig. 5 from the iceberg concept lattice in Fig. 4 we can now derive a generating set of the association rules we can find in the soft-granulated context $(\mathcal{F}, M, i)$ by evaluating the possible confidences of the edges in the diagram. Thereby,

$$\underline{\text{conf}}(A \to B) = \frac{\underline{\text{supp}}(A \cup B)}{\overline{\text{supp}}(A)} \quad \text{and} \quad \overline{\text{conf}}(A \to B) = \frac{\overline{\text{supp}}(A \cup B)}{\underline{\text{supp}}(A)}$$

are the obvious approximations of the confidence of $A \to B$. And analogously to the procedure done in Fig. 5 we highlight the rules whose $\underline{\text{conf}}$-value exceeds the threshold $\alpha = 0.7$ as a thick arrow. This procedure is displayed in Fig. 10. The rules we found are put together in Fig. 9.

Let us take for instance the final rule from 9. It must be read in the following way. At least 70.8%, but at most 79.2% of all objects that have the attribute *Group of 77* also have the attribute *Non-aligned*. The total number of objects that share both attributes lies between 85 and 95.

In summary we note that the described procedure consists of two steps. At first on calculates all $\underline{\text{supp}}$-frequent subsets of M. This can efficiently be done by a Titanic- or a Next-Closure-like algorithm. Afterwards one derives association rules from it in a similar fashion as one derives the Luxenburger base from the iceberg lattice. The only difference is that one now computes the confidence

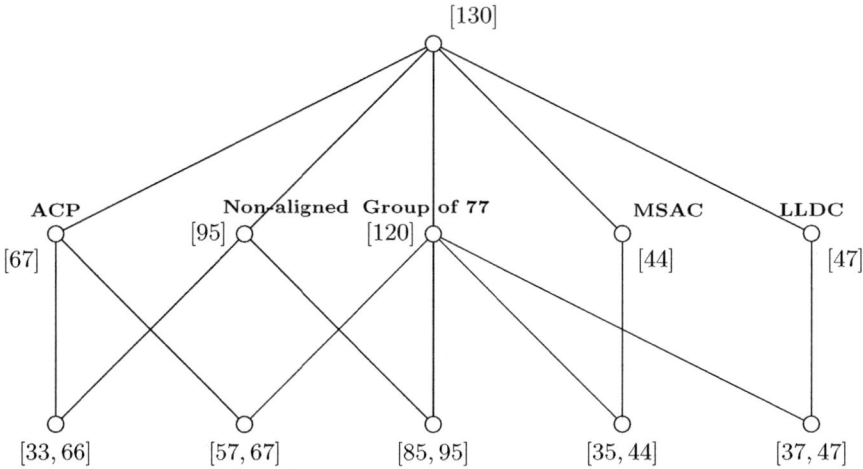

Fig. 8. The <u>supp</u>-frequent attribute sets of the soft-granulated context from Fig. 7 for the minimal support threshold $\sigma = 30$

[130]		$=[0.731] \Rightarrow$	Non-Aligned	[95]
[130]		$=[0.923] \Rightarrow$	Group of 77	[120]
[67]	ACP	$=[0.851, 1] \Rightarrow$	Group of 77	[57, 67]
[44]	MSAC	$=[0.795, 1] \Rightarrow$	Group of 77	[35, 44]
[47]	LLDC	$=[0.787, 1] \Rightarrow$	Group of 77	[37, 47]
[95]	Non-Aligned	$=[0.895, 1] \Rightarrow$	Group of 77	[85, 95]
[120]	Group of 77	$=[0.708, 0.792] \Rightarrow$	Non-Aligned	[85, 95]

Fig. 9. The association rules derived from the soft-granulated context in Fig 7

values in an interval arithmetic. It is in the nature of things that one will just find association rules where the premise and the conclusion are both very small. But this is not a drawback since the simplicity of a rule does not imply that it is less interesting. Furthermore, the procedure is doable even for very large underlying formal contexts. And of course, the finer the chosen granulation is the more precise the resulting association rules will be.

Now we are going to discuss the question on how to involve the background implications. What are the best possible approximations of the support of an attribute set just having our aforementioned setting? Obviously, one can (at least in theory) calculate all $\mathcal{L}$-realizers (G, M, J) and receives then the best approximations by the minimal and maximal supports of A occuring in these realizers. But since one usually knows the context (G, M, I) but just does not want to use it (too often), this approach is ridiculous. One could instead just use the original very large formal context (G, M, I).

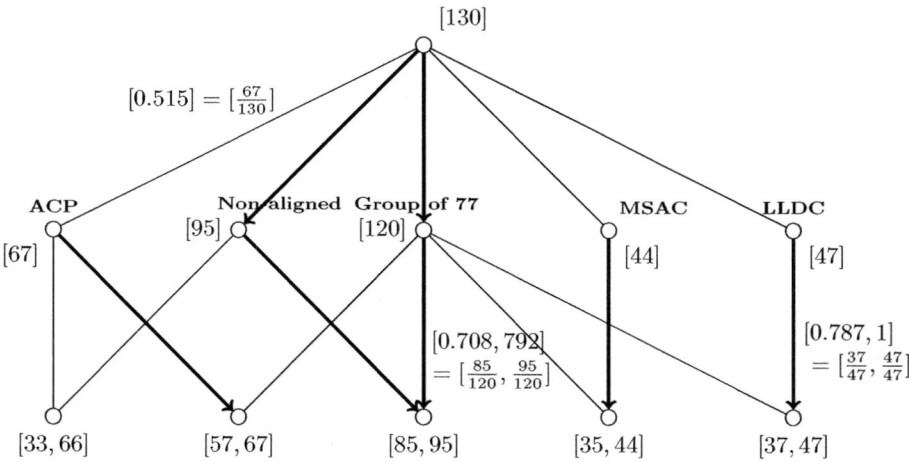

Fig. 10. A visualization of a generating set of the association rules one can derive from the soft-granulated context in Fig. 7 for $\sigma = 30$ and $\alpha = 0.7$. See also Fig. 5. In order to keep the diagram readable we did not label all edges with the corresponding confidence. Instead we just evaluated three example edges in the picture.

The question is, whether or not there is a doable way to calculate these *best* approximations. We cannot answer this question so far, and can just present a pair of approximations of the support which is doable with respect to the calculation but which is in general not as good as the best approximations.

Definition 8. *Let $\mathfrak{C}$ be a closure system on M with the corresponding closure operator $\overline{\cdot}$. A subset B of $C \subseteq M$ is called a **base** of C if it is a minimal preimage of $\overline{C}$ under $\overline{\cdot}$, i.e., if B is minimal w.r.t. $\overline{B} = \overline{C}$.*

Definition 9. *The set of background implications $\mathcal{L}$ yields to a closure operator $\mathcal{L}^*$ on M given by*

$$\mathcal{L}^*(X) := \bigcup_{n \in \mathbb{N}} \mathcal{L}^n(X),$$

where

$$\mathcal{L}(X) := X \cup \bigcup \{B \mid A \to B \in \mathcal{L} \text{ and } A \subseteq X\}.$$

*The closure system belonging to $\mathcal{L}^*(\cdot)$ is denoted by $\mathfrak{C}_{\mathcal{L}}$. Bases regarding to $\mathfrak{C}_{\mathcal{L}}$ are called $\mathcal{L}$-**bases**.*

Instead of taking the approximation $\underline{supp}(F, A)$ for a block $F \in \mathcal{F}$ and an attribute set $A \subseteq M$ one can also count $\underline{supp}(B, F)$ for every $\mathcal{L}$-base B of A. But it appears that there are cases where a superset C of A has bases which deliver higher $\underline{supp}(F, \cdot)$ values then bases of A itself. Since it is not doable to check all bases of supersets of A, one can for instance use $\mathcal{T}^*(A)$ as an upper bound for A, where

$$\mathcal{T} := \{\{b\} \to A \mid A \to B \in \mathcal{L} \text{ and } b \in B\}$$

is the transposed list of the list of implications

$$\mathcal{L}' := \{A \to \{b\} \mid A \to B \in \mathcal{L} \text{ and } b \in B\},$$

which is obviously equivalent to the list $\mathcal{L}$ of background implications. Hence, we receive the following approximations that consider the background implications:

$$\underline{\mathrm{supp}}_{\mathcal{L}}(F, A) := \max\{\underline{\mathrm{supp}}(F, B) \mid B \text{ is a } \mathcal{L}\text{-base of a } C \text{ with } A \subseteq C \subseteq \mathcal{T}^*(A)\},$$

$$\overline{\mathrm{supp}}_{\mathcal{L}}(F, A) := \overline{\mathrm{supp}}(F, \mathcal{L}^*(A)).$$

We now suggest to use the improved approximations of the support functions one receives by

$$\underline{\mathrm{supp}}_{\mathcal{L}}(A) := \sum_{F \in \mathcal{F}} \underline{\mathrm{supp}}_{\mathcal{L}}(F, A) \quad \text{and} \quad \overline{\mathrm{supp}}_{\mathcal{L}}(A) := \sum_{F \in \mathcal{F}} \overline{\mathrm{supp}}_{\mathcal{L}}(F, A)$$

to calculate the a Luxenburger-type again as it was described above for the case without background knowledge. Since we have not elaborated this approach so far, we cannot deliver any practical results. In order to make this approach applicable, one for sure needs a clever implementation. One for instance could start calculating the $\underline{\mathrm{supp}}_{\mathcal{L}}$-frequent $\mathcal{L}$-bases since they will be needed to compute the lower approximation of the support quickly.

7   Conclusion

We proposed a way to describe the rough tables occurring at the data warehousing system ICE. We did that from the standpoint of Formal Concept Analysis and tried to mine these rough tables for implicational knowledge. We argued that it is very unlikely that the very rigid *minimum* and *maximum* parameters as for instance used in the contest data set [4] will yield to satisfying results. We constituted that – having in mind the data mining in rough tables – in the process of building the data pack nodes it is worth to create more sophisticated parameters that allow to give a better estimation of the distribution of the values in the data packs (like counting the number of incidences in the data packs of the scaled data table).

Ongoing work has to include the following issues: Are there (in the case with background knowlege) better approximations of the support then the ones proposed? Furthermore, one has to develop some efficient implementations of the proposed approach. And most important, experimental results are needed to find out whether or not the soft granulation described in Section 6 will lead to satisfying results in practice.

References

1. Infobright: Home page at, http://www.infobright.org
2. Ganter, B., Meschke, C.: A formal concept analysis approach to rough data tables. In: Sakai, H., Chakraborty, M.K., Hassanien, A.E., Slezak, D., Zhu, W. (eds.) RSFDGrC 2009. LNCS, vol. 5908, pp. 117–126. Springer, Heidelberg (2009)
3. Infobright: Community Edition: Technology White Paper, http://www.infobright.org/wiki/662270f87c77e37e879ba8f7ac2ea258/
4. Infobright: Infobright challenge at PReMI 2009 (2009), http://web.iitd.ac.in/premi09/infobright.pdf
5. Wille, R.: Restructuring lattice theory: an approach based on hierarchies of concepts. In: Rival, I. (ed.) Ordered Sets, Boston, pp. 445–470 (1982); seminal publication on formal concept analysis
6. Ganter, B., Wille, R.: Formal concept analysis mathematic foundations. Springer, Heidelberg (1999)
7. Lakhal, L., Stumme, G.: Efficient Mining of Association Rules Based on Formal Concept Analysis. In: Ganter, B., Stumme, G., Wille, R. (eds.) Formal Concept Analysis. LNCS (LNAI), vol. 3626, pp. 180–195. Springer, Heidelberg (2005)
8. Stumme, G., Taouil, R., Bastide, Y., Pasquier, N., Lakhal, L.: Computing iceberg concept lattices with TITANIC. Data & Knowledge Engineering 42(2), 189–222 (2002)
9. Baader, F., Ganter, B., Sattler, U., Sertkaya, B.: Completing description logic knowledge bases using formal concept analysis. In: Golbreich, C., Kalyanpur, A., Parsia, B. (eds.) OWLED. CEUR Workshop Proceedings, vol. 258, CEUR-WS.org (2007)
10. Bělohlávek, R., Vychodil, V.: What is a fuzzy concept lattice? In: Vaclav Snasel, R.B. (ed.) OWLED. CEUR Workshop Proceedings, vol. 162, CEUR-WS.org (2005)
11. Burmeister, P., Holzer, R.: On the treatment of incomplete knowledge in formal concept analysis. In: Ganter, B., Mineau, G.W. (eds.) ICCS 2000. LNCS (LNAI), vol. 1867, Springer, Heidelberg (2000)
12. Pawlak, Z.: Rough Sets: Theoretical Aspects of Reasoning about Data. Kluwer Academic Publishers, Norwell (1992)

Rough Multiset and Its Multiset Topology

K.P. Girish and Sunil Jacob John

Department of Mathematics, National Institute of Technology
Calicut 673 601, Kerala, India
girikalam@yahoo.com, sunil@nitc.ac.in

Abstract. This article introduces the notion of multiset topology (*M*-topology) and points out the concept of open multisets (mset, for short). Multiset topologies are obtained by using multiset relation. Rough multiset is introduced in terms of lower and upper approximations. We use a multiset topological concept to investigate Pawlaks rough set theory by replaing its universe by multiset. The multiset topology induced by a multiset relation is used to generalize the rough multiset concept. Properties of rough multisets are obtained. Comparison between our approach and previous approaches are given and a generalized approximation mset space is a multiset topological space for any reflexive mset relation.

Keywords: Multiset, Rough multiset, Multiset Relation, Multiset Topology, Approximation operators.

1 Introduction

The advances in science and technology have given rise to a wide range of problems where the objects under analysis are characterized by many diverse features (attributes), which may be quantitative and qualitative. Furthermore, the same objects may exist in several copies with different values of attributes and their convolution is either impossible or mathematically correct. Examples of such problems are the classification of multicriteria alternatives estimated by several experts, the recognition of graphic symbols, text document processing and so on. A convenient mathematical model for representing multiattribute objects is a multiset or a set with repeating elements or bag (alternate name for multiset) [1,2,3,4,5,6,25]. The most essential property of multisets is the multiplicity of the elements that allows us to distinguish a multiset from a set and consider a multiset as a qualitatively new mathematical concept.

In 1982, Pawlak introduced the concept of a rough set. This concept is fundamental to the examination of granularity in knowledge. It is a concept which has many applications in data analysis [12,15,17,18,19,21,22,27,28,29,30,33]. The idea is to approximate a subset of a universal set by a lower approximation and upper approximation. These two approximations are exactly the interior and the closure of the set with respect to a certain topology τ (see, e.g.,[26]) on a collection U of imprecise data acquired from any real-life field. The base of the topology τ is formed by equivalence classes of an equivalence relation defined on

J.F. Peters et al. (Eds.): Transactions on Rough Sets XIV, LNCS 6600, pp. 62–80, 2011.
© Springer-Verlag Berlin Heidelberg 2011

U using available information about data. The aim of this paper is to develop the above concepts in the context of multisets.

The authors have given a new dimension to Pawlaks rough set theory by replacing its universe by Multisets. This is called a rough Multiset and is an useful structure in modeling an information multisystem. The process involved in the intermediate stages of reactions in chemical systems is a typical example of a situation which gives rise to Multiset relations. In an information multisystem the objects are repeated more than once. An information multisystem is represented using rough Multisets and is more convenient than ordinary rough sets.

Rough Multisets are defined in terms of lower mset approximation and upper mset approximation with the help of equivalence mset relations introduced by the authors in [10]. Grzymala-Busse introduced the concept of rough Multisets using multi-relations in [11]. A Multi-relation is a relation connecting two objects in which a pair is repeated more than once. But a Multiset relation according to the authors is in entirely different concept when compared with the concept multi-relation introduced by Grzymala-Busse [11]. A multiset relation is a relation connecting two objects in which each object is repeated more than once and the pair is also repeated more than once. That is, in a multi-relation 'x is related to y' with the pair (x, y) repeated more than once but in the case of a Multiset relation 'x repeated m times' is related to 'y repeated n times' (i.e. m/x is related to n/y) with the pair $(m/x, n/y)$ repeated more than once. Thus a multi-relation is a relation connecting the elements in the sets but a Multiset relation is a relation connecting the elements in the Multisets.

The concept of topological structures and their generalizations are one of the most powerful notions in branches of sciences such as chemistry, physics and information systems [7,15,16,28,34,36]. In most applications the topology is employed out of a need to handle the qualitative information. In any information system, some situations may occur, where the respective counts of objects in the universe of discourse are not single. In such situations we have to deal with collections of information in which duplicates are significant. In such type of situations multiset relations is very helpful for connecting the informations in an information system. In fact, topological structures on multisets are generalized methods for measuring the similarity and dissimilarity between the objects in multiset as universes.

This paper begins with the introduction of multisets and brief survey of Yagers theory of bags [31,35]. After the introduction of multisets we are defined different types of collection of multisets and operations under such collections. The notion of multiset relation and multiset functions are introduced and the basic concepts of general topology on general set [8,14,32] to topology on multiset are established. Introduce post-class and pre-class from the mset relations and multiset topologies are generated by using multiset relations. Rough multiset is introduced in terms of lower and upper approximations. These two approximations are exactly the interior and the closure the mset with respect to a certain M-topology τ on an mset M of imprecise data. The M-base of the M-topology τ is formed by m-equivalence classes of an equivalence mset relation R defined on M using the available information about data. Following the connection between rough multiset concepts and M-topological notions, we investigate new definitions for the rough multisets. We obtained many properties from the new definition of rough multiset. The m-equivalence class may be replaced by an element of the M-base $\{\langle m/x \rangle : x \in {}^m M\}$ of the M-topology τ. Comparison between our approach and previous approaches are given. Finally we show that the generalized approximation mset space is an M-topological space for any reflexive mset relation.

2   Preliminaries and Basic Definitions

In this section, a brief survey of the notion of msets introduced by Yager [35], the different types of collections of msets and the basic definitions and notions of Relations and Functions in Multiset Context introduced by K P Girish and Sunil Jacob John [9,10] are presented.

Definition 1. *An mset M drawn from the set X is represented by a function Count M or C_M defined as $C_M : X \to N$ where N represents the set of non negative integers.*

Here $C_M(x)$ is the number of occurrences of the element x in the mset M. We present the mset M drawn from the set $X = \{x_1, x_2, \ldots, x_n\}$ as $M = \{m_1/x_1, m_2/x_2, \ldots, m_n/x_n\}$ where m_i is the number of occurrences of the element $x_i, i = 1, 2, \ldots, n$ in the mset M. However those elements which are not included in the mset M have zero count.

Example 2. *Let $X = \{a, b, c, d, e\}$ be any set. Then $M = \{2/a, 4/b, 5/d, 1/e\}$ is an mset drawn from X.*

Clearly, a set is a special case of an mset.

Let M and N be two msets drawn from a set X. Then, the following are defined [2,5,10,34].

(i) $M = N$ if $C_M(x) = C_N(x) \; \forall \; x \in X$.
(ii) $M \subseteq N$ if $C_M(x) \leq C_N(x) \; \forall \; x \in X$.
(iii) $P = M \cup N$ if $C_P(x) = \max\{C_M(x), C_N(x)\} \; \forall \; x \in X$.
(iv) $P = M \cap N$ if $C_P(x) = \min\{C_M(x), C_N(x)\} \; \forall \; x \in X$.
(v) $P = M \oplus N$ if $C_P(x) = C_M(x) + C_N(x) \; \forall \; x \in X$.
(vi) $P = M \ominus N$ if $C_P(x) = \max\{C_M(x) - C_N(x), 0\} \; \forall \; x \in X$.

Where $\oplus$ and $\ominus$ represents mset addition and mset subtraction respectively.

Let M be an mset drawn from a set X. The support set of M denoted by $M*$ is a subset of X and $M* = \{x \in X : C_M(x) > 0\}$. i.e., $M*$ is an ordinary set. $M*$ is also called root set.

An mset M is said to be an empty mset if for all $x \in X$, $C_M(x) = 0$.

The cardinality of an mset M drawn from a set X is denoted by $\text{Card}(M)$ or $|M|$ and is given by $\text{Card } M = \sum_{x \in X} C_M(x)$.

Definition 3. *[13] A domain X, is defined as a set of elements from which msets are constructed. The mset space $[X]^m$ is the set of all msets whose elements are in X such that no element in the mset occurs more than m times.*

The set $[X]^\infty$ is the set of all msets over a domain X such that there is no limit on the number of occurrences of an element in an mset. If $X = \{x_1, x_2, \ldots, x_k\}$ then $[X]^m = \{\{m_1/x_1, m_2/x_2, \ldots, m_k/x_k\}: \text{for } i = 1, 2, \ldots, k; m_i \in \{0, 1, 2, \ldots, m\}\}$.

Definition 4. *[13] Let X be a support set and $[X]^m$ be the mset space defined over X. Then for any mset $M \in [X]^m$, the complement M^c of M in $[X]^m$ is an element of $[X]^m$ such that $C_M^c(x) = m - C_M(x) \ \forall \ x \in X$.*

Remark 5. *Using Definition 4, the mset sum can be modified as follows:*

$$C_{M1 \oplus M2}(x) = \min\{m, C_{M1}(x) + C_{M2}(x)\} \ \forall \ x \in X.$$

Notation 6. *Let M be an mset from X with x appearing n times in M. It is denoted by $x \in {}^n M$. $M = \{k_1/x_1, k_2/x_2, \ldots, k_n/x_n\}$ where M is an mset with x_1 appearing k_1 times, x_2 appearing k_2 times and so on. $[M]_x$ denotes that the element x belongs to the mset M and $|[M]_x|$ denotes the cardinality of an element x in M.*

A new notation can be introduced for the purpose of defining Cartesian product, Relation and its domain and co-domain. The entry of the form $(m/x, n/y)/k$ denotes that x is repeated m-times, y is repeated n-times and the pair (x, y) is repeated k times. The counts of the members of the domain and co-domain vary in relation to the counts of the x co-ordinate and y co-ordinate in $(m/x, n/y)/k$. For this purpose we introduce the notation $C_1(x, y)$ and $C_2(x, y)$. $C_1(x, y)$ denotes the count of the first co-ordinate in the ordered pair (x, y) and $C_2(x, y)$ denotes the count of the second co-ordinate in the ordered pair (x, y).

Throughout this paper M stands for a multiset drawn from the multiset space $[X]^m$. We can define the following types of submets of M and collection of submsets from the mset space $[X]^m$.

Definition 7 (Whole submsets). *A submset N of M is a whole submset of M with each element in N having full multiplicity as in M. i.e., $C_N(x) = C_M(x)$ for every x in N.*

Definition 8 (Partial Whole submsets). *A submset N of M is a partial whole submset of M with at least one element in N having full multiplicity as in M. i.e., $C_N(x) = C_M(x)$ for some x in N.*

Definition 9 (Full submsets). *A submset N of M is a full submset of M if M and N having same support set with $C_N(x) \leq C_M(x)$ for every x in N. i.e., $M* = N*$ with $C_N(x) \leq C_M(x)$ for every x in N.*

Note 1. Empty set $\varnothing$ is a whole submset of every mset but it is neither a full submset nor a partial whole submset of any nonempty mset M.

Example 10. *Let $M = \{2/x, 3/y, 5/z\}$ be an mset. Following are the some of the submsets of M which are whole submsets, partial whole submsets and full submsets.*

(a) A submset $\{2/x, 3/y\}$ is a whole submset and partial whole submset of M but it is not a full submset of M.

(b) A submset $\{1/x, 3/y, 2/z\}$ is a partial whole submset and full submset of M but it is not a whole submset of M.

(c) A submset $\{1/x, 3/y\}$ is a partial whole submset of M which is neither whole submset nor full submset of M.

Definition 11 (Power Whole Msets). *Let $M \in [X]^m$ be an mset. The power whole mset of M denoted by $PW(M)$ is defined as the set of all whole submsets of M. i.e., for constructing power whole submsets of M, every element of M with its full multiplicity behaves like an element in a classical set. The cardinality of $PW(M)$ is 2^n where n is the cardinality of the support set (root set) of M.*

Definition 12 (Power Full Msets). *Let $M \in [X]^m$ be an mset. Then the power full msets of M, $PF(M)$, is defined as the set of all full submsets of M. The cardinality of $PF(M)$ is the product of the counts of the elements in M.*

Note 2. PW(M) and PF(M) are ordinary sets whose elements are msets.

If M is an ordinary set with n distinct elements, then the power set $P(M)$ of M contains exactly 2^n elements. If M is a multiset with n elements (repetitions counted), then the power set $P(M)$ contains strictly less than 2^n elements because singleton submsets do not repeat in $P(M)$. In the classical set theory, Cantors power set theorem fails for msets. It is possible to formulate the following reasonable definition of a power mset of M for finite mset M that preserves Cantors power set theorem.

Definition 13 (Power Mset). *Let $M \in [X]^m$ be an mset. The power mset $P(M)$ of M is the set of all sub msets of M. We have $N \in P(M)$ if and only if $N \subseteq M$. If $N = \Phi$, then $N \in {}^1P(M)$; and if $N \neq \Phi$, then $N \in {}^kP(M)$ where $k = \prod_z \binom{|[M]_z|}{|[N]_z|}$, the product $\prod_z$ is taken over distinct elements of z of the mset N and $|[M]_z| = m$ iff $z \in {}^mM$, $|[N]_z| = n$ iff $z \in {}^nN$, then*

$$\binom{|[M]_z|}{|[N]_z|} = \binom{m}{n} = \frac{m!}{n!(m-n)!}.$$

The power set of an mset is the support set of the power mset and is denoted by $P^*(M)$. The following theorem shows the cardinality of the power set of an mset.

Theorem 14. *[23] Let $P(M)$ be a power mset drawn from the mset $M = \{m_1/x_1, m_2/x_2, \ldots, m_n/x_n\}$ and $P^*(M)$ be the power set of an mset M. Then $\mathrm{Card}(P^*(M)) = \prod_{i=1}^n (1 + m_i)$.*

Example 15. *Let $M = \{2/x, 3/y\}$ be an mset. The collection PW(M)= $\{\{2/x\},$ $\{3/y\}, M, \varnothing\}\}$ is a power whole submset of M. The collection PF(M) = $\{\{2/x, 1/y\}, \{2/x, 2/y\}, \{2/x, 3/y\}, \{1/x, 1/y\}, \{1/x, 2/y\}, \{1/x, 3/y\}\}$ is a power full submset of M. The collection P(M) = $\{3/\{2/x, 1/y\}, 3/\{2/x, 2/y\},$ $6/\{1/x, 1/y\}, 6/\{1/x, 2/y\}, 2/\{1/x, 3/y\}, 1/\{2/x\}, 1/\{3/y\}, 2/\{1/x\}, 3/\{1/y\},$ $3/\{2/y\}, M, \varnothing\}\}$ is the power mset of M. The collection $P^*(M) = \{\{2/x, 1/y\},$ $\{2/x, 2/y\}, \{1/x, 1/y\}, \{1/x, 2/y\}, \{1/x, 3/y\}, \{2/x\}, \{3/y\}, \{1/x\}, \{1/y\},$ $\{2/y\}, M, \varnothing\}\}$ is the support set of P(M).*

Note 3. Power mset is an mset but its support set is an ordinary set whose elements are msets.

Definition 16. *The maximum mset is defined as Z where $C_Z(x) = \max\{C_M(x) : x \in {}^k M, M \in [X]^m$ and $k \leq m\}$.*

Operations under collection of msets

Let $[X]^m$ be an mset space and $\{M_1, M_2, \ldots\}$ be a collection of msets drawn from $[X]^m$. Then the following operations are possible under an arbitrary collection of msets.

(i) The union $\bigcup_{i \in I} M_i = \{C_{\cup M_i}(x)/x : C_{\cup M_i}(x) = \max\{C_{M_i}(x) : x \in X\}\}$.

(ii) The intersection $\bigcap_{i \in I} M_i = \{C_{\cap M_i}(x)/x : C_{\cap M_i}(x) = \min\{C_{M_i}(x) : x \in X\}\}$.

(iii) The mset addition $\bigoplus_{i \in I} M_i = \{C_{\oplus M_i}(x)/x : C_{\oplus M_i}(x) = \sum_{i \in I} C_{M_i}(x), x \in X\}$.

(iv) The mset complement
$M^c = Z \ominus M = \{C_{M^c}(x)/x : C_{M^c}(x) = C_Z(x) - C_M(x), x \in X\}$.

Definition 17. *Let M_1 and M_2 be two msets drawn from a set X, then the Cartesian product of M_1 and M_2 is defined as*

$$M_1 \times M_2 = \{(m/x, n/y)/mn : x \in {}^m M_1, y \in {}^n M_2\}.$$

We can define the Cartesian product of three or more nonempty msets by generalizing the definition of the Cartesian product of two msets.

Definition 18. *A sub mset R of $M \times M$ is said to be an mset relation on M if every member $(m/x, n/y)$ of R has a count, product of $C_1(x, y)$ and $C_2(x, y)$. We denote m/x related to n/y by m/x R n/y.*

The Domain and Range of the mset relation R on M is defined as follows:

$$\text{Dom } R = \{x \in {}^r M : \exists y \in {}^s M \text{ such that } r/x \text{ R } s/y\}$$
$$\text{where } C_{\text{Dom}R}(x) = \sup\{C_1(x, y) : x \in {}^r M\}.$$
$$\text{Ran } R = \{y \in {}^s M : \exists x \in {}^r M \text{ such that } r/x \text{ R } s/y\}$$
$$\text{where } C_{\text{Ran}R}(y) = \sup\{C_2(x, y) : y \in {}^s M\}.$$

Example 19. *Let* $M = \{8/x, 11/y, 15/z\}$ *be an mset. Then* $R = \{(2/x, 4/y)/8,$
$(5/x, 3/x)/15,\ (7/x, 11/z)/77,\ (8/y, 6/x)/48,\ (11/y, 13/z)/143,\ (7/z, 7/z)/49,$
$(12/z, 10/y)/120,\ (14/z, 5/x)/70\}$ *is an mset relation defined on* M.
Here Dom $R = \{7/x, 11/y, 14/z\}$ *and Ran* $R = \{6/x, 10/y, 13/z\}$.
Also $S = \{(2/x, 4/y)/5,\ (5/x, 3/x)/10,\ (7/x, 11/z)/77,\ (8/y, 6/x)/48,$
$(11/y, 13/z)/143,\ (7/z, 7/z)/49,\ (12/z, 10/y)/120,\ (14/z, 5/x)/70\}$ *is a submset*
of $M \times M$ *but* S *is not an mset relation on* M *because* $C_S((x, y)) = 5 \neq 2 \times 4$
and $C_S((x, x)) = 10 \neq 5 \times 3$. *i.e., count of some elements in* S *is not a product*
of $C_1(x, y)$ *and* $C_2(x, y)$.

Definition 20

> *(i) An mset relation* R *on an mset* M *is reflexive if* $m/x \ R \ m/x$ *for all* m/x
> *in* M *and irreflexive iff* $m/x \ R \ m/x$ *never holds.*
>
> *(ii) An mset relation* R *on an mset* M *is symmetric if* $m/x \ R \ n/y$ *implies* n/y
> $R \ m/x$.
>
> *(iii) An mset relation* R *on an mset* M *is transitive if* $m/x \ R \ n/y$, $n/y \ R \ k/z$
> *then* $m/x \ R \ k/z$.

An mset relation R *on an mset* M *is called an equivalence mset relation if it is*
reflexive, symmetric and transitive and R *on* M *is a tolerance mset relation if*
it is reflexive and symmetric.

Example 21. *Let* $M = \{3/x, 5/y, 3/z, 7/r\}$ *be an mset. Then the mset relation*
given by $R = \{(3/x, 3/x)/9, (3/z, 3/z)/9, (3/x, 7/r)/21, (7/r, 3/x)/21, (5/y, 5/y)/25,$
$(3/z, 3/z)/9, (7/r, 7/r)/49, (3/z, 3/x)/9, (3/z, 7/r)/21, (7/r, 3/z)/21\}$ *is an equiv-*
alence mset relation.

Definition 22. *An mset relation* f *is called an mset function if for every ele-*
ment m/x *in* $Dom f$, *there is exactly one* n/y *in* $Ran f$ *such that* $(m/x, n/y)$ *is*
in f *with the pair occurring the product of* $C_1(x, y)$ *and* $C_2(x, y)$.

For functions between arbitrary msets it is essential that images of indistinguishable elements of the domain must be indistinguishable elements of the range, but
the images of the distinct elements of the domain need not be distinct elements
of the range.

Example 23. *Let* $M_1 = \{8/x, 6/y\}$ *and* $M_2 = \{3/a, 7/b\}$ *be two msets. Then an*
mset function from M_1 *to* M_2 *may be defined as* $f = \{(8/x, 3/a)/24, (6/y, 7/b)/42\}$.

3 Multiset Topologies

Definition 24. *Let* $M \in [X]^m$ *and* $\tau \subseteq P^*(M)$. *Then* τ *is called a multiset*
topology if τ *satisfies the following properties.*

(1) $\varnothing$ *and* M *are in* τ.
(2) The union of the elements of any sub collection of τ *is in* τ.
(3) The intersection of the elements of any finite sub collection of τ *is in* τ.

Mathematically a multiset topological space is an ordered pair (M, τ) consisting of an mset $M \in [X]^m$ and a multiset topology $\tau \subseteq P^*(M)$ on M. Note that τ is an ordinary set whose elements are msets. Multiset Topology is abbreviated as an M-topology.

If M is an M-topological space with M-topology τ, we say that a submset U of M is an open mset of M if U belongs to the collection τ. Using this terminology, one can say that an M-topological space is an mset M together with a collection of submsets of M, called open msets, such that $\varnothing$ and M are both open and the arbitrary unions and finite intersections of open msets are open.

Example 25. *Let M is any mset in $[X]^m$. The collection $P^*(M)$, the support set of the power mset of M is an M-topology on M and is called the discrete M-topology.*

Example 26. *The collection consisting of M and Φ only, is an M-topology called indiscrete M-topology, or trivial M-topology.*

Example 27. *If M is any mset, then the collection PW(M) is an M-topology on M.*

Example 28. *The collection PF(M) is not an M-topology on M, because Φ does not belong to PF(M), but $PF(M) \cup \{\Phi\}$ is an M-topology on M.*

Example 29. *The collection τ of partial whole submsets of M is not an M-topology. Let $M = \{2/x, 3/y\}$. Then $A = \{2/x, 1/y\}$ and $B = \{1/x, 3/y\}$ are partial submsets. Now $A \cap B = \{1/x, 1/y\}$, but it is not a partial whole submsets. Thus τ is not closed under finite intersection.*

4   *M*-Basis and Sub *M*-Basis

Definition 30 (M-basis). *If M is an mset, an M-basis for an M-topology on M is collection $\mathbf{B}$ of partial whole submsets of M (called M-Basis element) such that*

  (i) *For each $x \in {}^m M$, for some $m > 0$, there is atleast one M-Basis element $B \in \mathbf{B}$ containing m/x. i.e., for each mset in $\mathbf{B}$ there is at least one element with full multiplicity as in M.*
 (ii) *If m/x belongs to the intersection of two M-Basis elements M_1 and M_2, then there is an M-Basis element M_3 containing m/x such that $M_3 \subseteq M_1 \cap M_2$. i.e., there is an M-basis element M_3 containing an element with full multiplicity as in M and that element must be in M_1 and M_2 also.*

Remark 31. *If $\mathbf{B}$ satisfies the conditions of definition 30, then an M-topology τ generated by $\mathbf{B}$ can be defined as follows:- A submset U of M is said to be an open mset in M (i.e., to be an element of τ) if for each $x \in {}^k U$, there is an M-Basis element $B \in \mathbf{B}$ such that $x \in {}^k B$ and $B \subseteq U$. Note that each M-basis element is itself an element of τ.*

Definition 32 (Sub M-basis). *A sub collection* **S** *of* τ *is called a sub M-basis for* τ *the collection of all finite intersections of members of* **S** *is an M-basis for* τ.

Definition 33. *The M-topology generated by the sub M-basis* **S** *is defined to be the collection* τ *of all unions of finite intersections of elements of* **S**.

Note that the empty intersection of the members of sub M-base is the universal mset.

5 M-Topology of mset Relation

Definition 34. *Let* R *be an mset relation on* M. *The post-mset of* $x \in {}^mM$ *is defined as* $m/x\ R = \{n/y : \exists$ *some* k *with* $k/x\ R\ n/y\}$ *and the pre-mset of* $x \in {}^rM$ *is defined as* $R\ r/x = \{p/y : \exists$ *some* q *with* $p/y\ R\ q/x\}$.

Definition 35. *If* R *is an mset relation defined on* M, *then the post-class and pre-class are defined by* $P_+ = \{m/x\ R : x \in {}^mM\}$ *and* $P_- = \{R\ m/x : x \in {}^mM\}$.

Note 4. M-Topology generated by the post-class and pre-class are denoted by τ_1 and τ_2 respectively.

Example 36. *Let* $M = \{3/a, 4/b, 2/c, 5/d\}$ *and* $R = \{(3/a, 3/a)/9, (2/a, 3/b)/6,$ $(2/b, 4/d)/8, (2/c, 3/d)/6, (2/d, 2/a)/4\}$ *be an mset relation on* M, *then the post-msets are* $3/a\ R = 2/a\ R = \{3/a, 3/b\}; 2/b\ R = \{4/d\}; 2/c\ R = \{3/d\}; 2/d\ R = \{2/a\}$. *From the post-msets we get the post-class* $P_+ = \{\{3/a, 3/b\}, \{4/d\}, \{2/a\}\}$ *and an M-base* $\beta_+ = \{\{3/a, 3/b\}, \{4/d\}, \{2/a\}, \varnothing, M\}$. *M-Topology generated by the post-class* $\tau_1 = \{\varnothing, M, \{2/a\}, \{4/d\}, \{3/a, 3/b\}, \{2/a, 4/d\}, \{3/a, 3/b, 4/d\}\}$.
Similarly the pre-msets are $R\ 3/a = R\ 2/a = \{3/a, 2/d\}, R\ 3/b = \{2/a\},$ $R\ 4/d = R\ 3/d = \{2/b, 2/c\}$. *From the pre-msets we get the pre-class* $P_- = \{\{3/a, 2/d\}, \{2/a\}, \{2/b, 2/c\}\}$ *and an M-base* $\beta_- = \{\{2/a\}, \{3/a, 2/d\}, \{2/b, 2/c\}, \varnothing, M\}$. *M-Topology generated by the pre-class* $\tau_2 = \{\varnothing, M, \{2/a\}, \{3/a, 2/d\}, \{2/b, 2/c\}, \{2/a, 2/b, 2/c\}\}$.

Definition 37. *Let* R *be any binary mset relation on* M, *an mset* $< m/x > R$ *is the intersection of all post-msets containing* m/x. *i.e.*,

$$< m/x > R = \begin{cases} \displaystyle\bigcap_{y \in {}^n m/x\ R} m/x\ R, & \text{if } \exists\ m/x \text{ in } M \text{ such that } y \in {}^n m/x\ R \\ \phi, & \text{otherwise.} \end{cases}$$

Also, $R < m/x >$ *is the intersection of all pre-msets containing* m/x. *i.e.*,

$$R < m/x > = \begin{cases} \displaystyle\bigcap_{y \in {}^n R\ m/x} R\ m/x, & \text{if } \exists\ m/x \text{ in } M \text{ such that } y \in {}^n R\ m/x \\ \phi, & \text{otherwise.} \end{cases}$$

Remark 38. *Let* R *be a reflexive mset relation on* M, *then the class* $\{\langle m/x \rangle R : x \in {}^mM\}$ *is a Sub M-base for the M-topology on* M *and the M-topology generated by this sub M-base is denoted by* τ_3.

Example 39. *Let* $M = \{3/a, 4/b, 2/c, 5/d\}$ *and* R *be a reflexive mset relation,* $R = \{(3/a, 3/a)/9, (4/b, 4/b)/16, (2/c, 2/c)/4, (5/d, 5/d)/25, (2/a, 3/b)/6,$ $(3/b, 3/d)/9, (2/c, 3/d)/6\}$, *then* $\langle 3/a \rangle R = \{3/a, 3/b\}$, $\langle 4/b \rangle R\, R = \{3/b\}$, $\langle 2/c \rangle R$ $R = \{2/c, 3/d\}$ *and* $\langle 5/d \rangle R\ R = \{3/d\}$. *From this msets we get the class* $\{\{3/a, 3/b\}, \{3/b\}, \{2/c, 3/d\}, \{3/d\}\}$ *and the M-topology generated by this class is* $\tau_3 = \{\varnothing, M, \{3/b\}, \{3/d\},\ \{3/a, 3/b\}, \{2/c, 3/d\}, \{3/b, 3/d\}, \{3/a, 3/b, 3/d\},$ $\{3/b, 2/c, 3/d\}\}$.

6   Rough Multisets and Its M-Topology

Let M be an mset and R be an equivalence mset relation on M. R generates a partition $M/R = \{M_1, M_2, \ldots, M_m\}$ on M where $M_1, M_2, \ldots, M_m$ are the m-equivalence classes generated by the equivalence mset relation R. An m-equivalence class in R containing an element $x \in {}^m M$ is denoted by $[m/x]$. The pair (M, R) is called an approximation mset space.

Definition 40. *For any* $N \subseteq M$, *we can define the lower mset approximation and upper mset approximation of* N *defined by*

$$R_L(N) = \{m/x : [m/x] \subseteq N\} \ and$$
$$R_U(N) = \{m/x : [m/x] \cap N \neq \varPhi\}$$

respectively. The pair $(R_L(N), R_U(N))$ *is referred to as the rough mset of* N.
  Boundary region of an mset N *is denoted by* $Bd_R(N)$, *and it is defined as*

$$Bd_R(N) = R_U(N) \ominus R_L(N).$$

  The rough mset $(R_L(N), R_U(N))$ *gives rise to a description of* N *under the present knowledge, that is the classification of* M.

Generalized approximation mset space by using post-mset concepts as follows:

Definition 41. *For any* $N \subseteq M$, *a pair of lower and upper mset approximations* $R_L(N)$ *and* $R_U(N)$, *are defined by;*

$$R_L(N) = \{m/x : m/x\ R \subseteq M\} \ and$$
$$R_U(N) = \{m/x : m/x\ R \cap M \neq \varPhi\}.$$

The pair $(R_L(N), R_U(N))$ *is referred to as rough mset of* N.

Obviously, if R is an equivalence mset relation, then $m/x\ R = [m/x]$. In addition, this definition is equivalent to definition 40.
  An M-topological space can be described by using a pair of interior and closure operators. There may exist some relationships between an M-topological space and rough multiset. In fact, the lower and upper mset approximation operators in an approximation mset space can be interpreted as a pair of interior and closure operators in an M-topological space (M, τ).

Definition 42 (Closure). *Given a submset A of an M-topological space M, the closure of an mset A is defined as the intersection of all closed msets containing A and is denoted by $Cl(A)$. i.e.,*

$$cl(A) = \cap\{K \subseteq M : K \text{ is a closed mset and } A \subseteq K\} \text{ and}$$
$$C_{cl(A)}(x) = \min\{C_K(x) : A \subseteq K\}.$$

Definition 43 (Interior). *Given a submset A of an M-topological space M, the interior of an mset is defined as the union of all open msets contained in A and is denoted by $Int(A)$. i.e.,*

$$Int(A) = \cup\{K \subseteq M : G \text{ is an open mset and } G \subseteq A\} \text{ and}$$
$$C_{Int(A)}(x) = \max\{C_G(x) : G \subseteq A\}.$$

Definition 44 (Closure (resp. interior) Operator on Multiset). *Let (M, τ) be an M-topological space, a closure (resp. interior) operator $cl : P^*(M) \to P^*(M)$ (resp. $Int : P^*(M) \to P^*(M)$) satisfy Kuratowski axioms if and only if for every $A, B, \in P^*(M)$ the following hold:*

(i) $cl(\varnothing) = \varnothing$ (resp. $Int(M) = M$)
(ii) $A \subseteq cl(A)$ (resp. $Int(A) \subseteq M$)
(iii) $cl(A \cup B) = cl(A) \cup cl(B)$ (resp. $Int(A \cap B) = Int(A) \cap Int(B)$)
(iv) $cl(cl(A)) = cl(A)$ (resp. $Int(Int(A)) = Int(A)$).

The following theorem states that a reflexive and transitive mset relation is sufficient for the mset approximation operators to be a pair of interior and closure operators on multisets.

Theorem 45. *If R is a reflexive and transitive mset relation on M, then the pair of lower and upper mset approximations is a pair of interior and closure operators satisfying Kuratowski axioms of M-topological spaces.*

Remark 46. *If R is a reflexive and transitive mset relation on M, then the M-topology generated by the lower and upper mset approximation operators is the same as the M-topology τ_1 generated by the post-msets as a sub M-base.*

Another notion of generalized approximation mset space by using the post-msets concepts are as follows:

Definition 47. *Let R be any mset relation on M, the lower and upper mset approximations on $N \subseteq M$ according to R are then defined as*

$$R_U(N) = \{m/x : \langle m/x \rangle R \cap N \neq \Phi\} \text{ and}$$
$$R_L(N) = \{m/x : \langle m/x \rangle R \subseteq M\}.$$

Obviously, if R is an equivalence mset relation on M, then $\langle m/x \rangle R = [m/x]$. In addition, these definitions are equivalent to the original definitions.

Proposition 48. *For any binary mset relation R on a nonempty mset M the following conditions hold:*

(i) *For every $N \subseteq M$, $R_L(N) = [R_U(N^c)]^c$*

(ii) *$R_L(M) = M$*

(iii) *For any sub msets M_1 and M_2 of M, $R_L(M_1 \cap M_2) = R_L(M_1) \cap R_L(M_2)$.*

(iv) *If $M_1 \subseteq M_2$ then $R_L(M_1) \subseteq R_L(M_2)$.*

(v) *For any sub msets M_1 and M_2 of M, $R_L(M_1 \cup M_2) \supseteq R_L(M_1) \cup R_L(M_2)$.*

(vi) *For any sub msets M_1 and M_2 of M, $R_L(M_1 \oplus M_2) \supseteq R_L(M_1) \oplus R_L(M_2)$.*

(vii) *For any sub msets M_1 and M_2 of M, $R_L(M_1 \ominus M_2) = R_L(M_1) \ominus R_L(M_2)$.*

(viii) *For every $N \subseteq M$, $R_L(N) \subseteq R_L(R_L(N))$.*

Proof

(i)

$$[R_U(N^c)]^c = \{x \in {}^m M : \langle m/x \rangle \cap N^c \neq \varnothing\}^c$$
$$= \{x \in {}^m M : \langle m/x \rangle \cap N^c = \varnothing\}$$
$$= \{x \in {}^m M : \langle m/x \rangle \subseteq N\}$$
$$= R_L(N).$$

(ii) Since for every $x \in {}^m M . \langle m/x \rangle \subseteq M$, $x \in {}^m R_L(M)$. Thus $M \subseteq R_L(M)$. Also since $R_L(M) \subseteq M$. Hence $R_L(M) = M$.

(iii)

$$R_L(M_1 \cap M_2) = \{x \in {}^m M : \langle m/x \rangle \subseteq M_1 \cap M_2\}$$
$$= \{x \in {}^m M : \langle m/x \rangle \subseteq M_1\} \cap \{x \in {}^m M : \langle m/x \rangle \subseteq M_2\}$$
$$= R_L(M_1) \cap R_L(M_2).$$

(iv) If $M_1 \subseteq M_2$ and $x \in {}^m R_L(M_1)$ then $\langle m/x \rangle \subseteq M_1$ and so $\langle m/x \rangle \subseteq M_2$, hence $x \in {}^m R_L(M_2)$. Thus $R_L(M_1) \subseteq R_L(M_2)$.

(v) Since $M_1 \subseteq M_1 \cup M_2$ and $M_2 \subseteq M_1 \cup M_2$, so $R_L(M_1) \subseteq R_L(M_1 \cup M_2)$ and $R_L(M_2) \subseteq R_L(M_1 \cup M_2)$, hence $R_L(M_1 \cup M_2) \supseteq R_L(M_1) \cup R_L(M_2)$.

(vi)

$$R_L(M_1 \oplus M_2) = \{x \in {}^m M : \langle m/x \rangle \subseteq M_1 \oplus M_2\}$$
$$\supseteq \{x \in {}^m M : \langle m/x \rangle \subseteq M_1\} \oplus \{x \in {}^m M : \langle m/x \rangle \subseteq M_2\}$$
$$= R_L(M_1) \oplus R_L(M_2).$$

(vii)

$$R_L(M_1 \ominus M_2) = \{x \in {}^m M : \langle m/x \rangle \subseteq M_1 \ominus M_2\}$$
$$= \{x \in {}^m M : \langle m/x \rangle \subseteq M_1\} \ominus \{x \in {}^m M : \langle m/x \rangle \subseteq M_2\}$$
$$= R_L(M_1) \ominus R_L(M_2).$$

(viii) Since $R_L(N) = \{x \in {}^m M : \langle m/x \rangle \subseteq N\}$ and $R_L(R_L(N)) = \{x \in {}^m M : \langle m/x \rangle \subseteq R_L(N)\}$, to prove that $R_L(N) \subseteq R_L(R_L(N))$ it is enough to prove that $\langle m/x \rangle \subseteq R_L(N)$.
Let

$$y \in {}^n \langle m/x \rangle \Rightarrow \langle n/y \rangle = \langle m/x \rangle \subseteq N$$
$$\Rightarrow \langle n/y \rangle \subseteq N$$
$$\Rightarrow y \in {}^n R_L(N)$$
$$\Rightarrow \langle m/x \rangle \subseteq R_L(N)$$
$$\Rightarrow R_L(N) \subseteq R_L(R_L(N)). \qquad \square$$

The following example gives the results (v) and (vi) in proposition 48 is not true in general.

Example 49. Let $M = \{3/x, 5/y, 3/z, 7/r, 4/s\}$ be an mset. Then the mset relation given by $R = \{(3/x, 3/x)/9, (3/x, 5/y/15, (5/y, 3/z)/15, (3/z, 3/x)/9, (5/y, 7/r)/35, (7/r, 4/s)/28, (4/s, 4/s)/16, (4/s, 5/y)/20\}$. Then, $\langle 3/x \rangle R = \{3/x\}$, $\langle 5/y \rangle R = \{5/y\}, \langle 3/z \rangle R = \langle 7/r \rangle R = \{3/z, 7/r\}$ and $\langle 4/s \rangle R = \{4/s\}$. Let $M_1 = \{3/x, 3/z\}$ and $M_2 = \{5/y, 7/r\}$ be two sub msets of M. Then $R_L(M_1) = \{3/x\}$ and $R_L(M_2) = \{5/y\}$. Therefore $R_L(M_1) \oplus R_L(M_2) = \{3/x, 5/y\}$. But $M_1 \oplus M_2 = \{3/x, 5/y, 3/z, 7/r\}$ and $R_L(M_1 \oplus M_2) = \{3/x, 5/y, 3/z, 7/r\}$. Thus $R_L(M_1 \oplus M_2) \neq R_L(M_1) \oplus R_L(M_2)$.
Also $M_1 \cup M_2 = \{3/x, 5/y, 3/z, 7/r\}$ and $R_L(M_1 \cup M_2) = \{3/x, 5/y, 3/z, 7/r\}$. But $R_L(M_1) \cup R_L(M_2) = \{3/x, 5/y\}$. Therefore $R_L(M_1 \cup M_2) \neq R_L(M_1) \cup R_L(M_2)$.

Proposition 50. *If an mset relation R on M is a reflexive mset relation, then the following conditions hold.*
(i) $R_L(\phi) = \phi$
(ii) $R_L(N) \subseteq N$.

Proof.
(i) Since R is a reflexive mset relation on M, then $x \in {}^m \langle m/x \rangle R$ for all $x \in {}^m M$, also there is no $x \in {}^m M$ such that $\langle m/x \rangle R \subseteq \phi$ hence $R_L(\phi) = \phi$.
(ii) Assume that $x \in {}^m R_L(N)$, then $\langle m/x \rangle R \subseteq N$. Since R is a reflexive mset relation on M, then $x \in {}^m \langle m/x \rangle R$ for all $x \in {}^m M$ and there is no $y \in {}^n M \ominus N$ such that $\langle n/y \rangle R \subseteq N$. Thus $x \in {}^m N$ and so $R_L(N) \subseteq N$.

Following example gives the result (ii) in proposition 50 is not true in general.

Example 51. *Let* $R = \{(3/x, 3/x)/9, (5/y, 5/y)/25, (3/z, 3/z)/9, (7/r, 7/r)/49,$
$(3/x, 5/y)/15, (5/y, 3/z)/15, (3/z, 3/x)/9, (7/r, 3/x)/21\}$ *be any reflexive mset*
relation on a nonempty mset $M = \{3/x, 5/y, 3/z, 7/r\}$. *Then,* $\langle 3/x \rangle R = \{3/x\}$,
$\langle 5/y \rangle R = \{5/y\}, \langle 3/z \rangle R = \{3/z\}$ *and* $\langle 7/r \rangle R = \{3/x, 7/r\}$. *If* $M_1 =$
$\{5/y, 3/z, 7/r\}$, *then* $R_L(M_1) = \{5/y, 3/z\}$, *thus* $R_L(M_1) \neq M_1$.

Proposition 52. *Let* R *be an equivalence mset relation on a nonempty mset*
M, *then the following conditions hold.*
  (i) For every $N \subseteq M$, $N \subseteq R_L(R_U(N))$.
  (ii) For every $N \subseteq M$, $R_U(N) \subseteq R_L(R_U(N))$.

Proof

  (i) Let $x \in {}^m N$, we want to show that $x \in {}^m R_L(R_U(N))$, i.e., $\langle m/x \rangle R \subseteq$
   $R_U(N)$ or $\langle n/y \rangle R \cap N \neq \varnothing$ for all $y \in {}^n \langle m/x \rangle R$. Since R is an equivalence
   mset relation, then $\langle m/x \rangle R = \langle n/y \rangle R$ or $\langle m/x \rangle R \cap \langle n/y \rangle R = \varnothing$ for all
   $x \in {}^m M$, $y \in {}^n M$, then for all $x \in {}^m N$ and $y \in {}^n \langle m/x \rangle R$, $\langle n/y \rangle R \cap N \neq \varnothing$.
   i.e., $y \in {}^n R_U(N)$ for all $y \in {}^n \langle m/x \rangle R$, then $\langle m/x \rangle R \subseteq R_U(N)$, thus
   $x \in {}^m R_L(R_U(N))$ and so $N \subseteq R_L(R_U(N))$.
  (ii) Let $x \in {}^m R_U(N)$, we want to show that $x \in {}^m R_L(R_U(N))$. Since $x \in$
   ${}^m R_U(N)$, then $\langle m/x \rangle R \cap N \neq \varnothing$, also R is an equivalence mset relation
   hence, $\langle n/y \rangle R \cap N \neq \varnothing$ for all $y \in {}^n \langle m/x \rangle R$, then $y \in {}^n R_U(N)$ for all
   $y \in {}^n \langle m/x \rangle R$. i.e., $\langle m/x \rangle R \subseteq R_U(N)$, thus $x \in {}^m R_L(R_U(N))$. Hence
   $R_U(N) \subseteq R_L(R_U(N))$. □

Following example gives the result (i) in proposition 52 is not true in general.

Example 53. *Let* $R = \{(3/x, 3/x)/9, (5/y, 5/y)/25, (3/z, 3/z)/9, (7/r, 7/r)/49,$
$(3/x, 5/y)/15,$ $(5/y, 3/x)/15,$ $(5/y, 3/z)/15,$ $(3/z, 5/y)/15,$ $(3/x, 3/z)/9,$
$(3/z, 3/x)/9\}$ *be an equivalence mset relation on a nonempty mset* $M =$
$\{3/x, 5/y, 3/z, 7/r\}$. *Then* $\langle 3/x \rangle R = \langle 5/y \rangle R = \langle 3/z \rangle R = \{3/x, 5/y, 3/z\}$ *and*
$\langle 7/r \rangle R = \{7/r\}$. *If* $M_1 = \{3/z, 7/r\}$, *then* $R_U(M_1) = \{3/x, 5/y, 3/z, 7/r\}$ *and*
$R_L(R_U(M_1)) = \{3/x, 5/y, 3/z, 7/r\}$, *thus* $M_1 \neq R_L(R_U(M_1))$.

Following example gives the first condition in proposition 52 is not hold if the
relation R is tolerance mset relation.

Example 54. *Let* $R = \{(3/x, 3/x)/9, (5/y, 5/y)/25, (3/z, 3/z)/9, (7/r, 7/r)/49,$
$(3/x, 5/y)/15, (5/y, 3/x)/15, (5/y, 3/z)/15, (3/z, 5/y)/15\}$ *be a tolerance mset re-*
lation on a nonempty mset $M = \{3/x, 5/y, 3/z, 7/r\}$. *Then* $\langle 3/x \rangle R = \{3/x, 5/y\}$,
$\langle 5/y \rangle R = \{5/y\}, \langle 3/z \rangle R = \{5/y, 3/z\}$ *and* $\langle 7/r \rangle R = \{7/r\}$. *If* $M_1 = \{3/x, 3/z\}$,
then $R_U(M_1) = \{3/x, 3/z\}$ *and* $R_L(R_U(M_1)) = \phi$, *thus* $M_1 \neq R_L(R_U(M_1))$.

Theorem 55. *If* R *is an equivalence mset relation on a nonempty mset* M, *then*
$R_U(N) = R_L(R_U(N))$.

Proof. Follows from proposition 50 and proposition 52.

Theorem 56. *If* R *is a irreflexive, symmetric and transitive mset relation on*
a nonempty mset M, *then* $N \subseteq R_L(R_U(N))$ *for all* $N \subseteq M$.

Proof. Let R be an irreflexive, symmetric and transitive mset relation on a nonempty mset M, i.e., $\langle m/x \rangle R = \{m/x\}$ or ϕ for all $x \in {}^m M$ and let $x \in {}^m N$, then there are two cases:

The first is $\langle m/x \rangle R = \phi$ and hence $\langle m/x \rangle R \subseteq R_U(N)$, then $x \in {}^m R_L(R_U(N))$. The second is $\langle m/x \rangle R = \{m/x\}$ and so $\langle m/x \rangle R \cap N \neq \phi$, then $x \in {}^m R_U(N)$, i.e., $\langle m/x \rangle R \subseteq R_U(N)$, hence $x \in {}^m R_L(R_U(N))$. Thus $N \subseteq R_L(R_U(N))$.

The following example gives the theorem 56 is not true in general.

Example 57. *Let* $R = \{(3/x, 5/y)/15, (5/y, 3/x)/15, (5/y, 3/z)/15, (3/z, 5/y)/15,$ $(3/x, 3/z)/9, (3/z, 3/x)/9\}$ *be an irreflexive, symmetric and transitive mset relation on a nonempty mset* $M = \{3/x, 5/y, 3/z, 7/r\}$. *Then* $\langle 3/x \rangle R = \{3/x\}$, $\langle 5/y \rangle R = \{5/y\} \langle 3/z \rangle R = \{3/z\}$ *and* $\langle 7/r \rangle R = \phi$. *If* $M_1 = \{3/x, 5/y\}$, *then* $R_U(M_1) = \{3/x, 5/y\}$ *and* $R_L(R_U(M_1)) = \{3/x, 5/y, 7/r\}$, *thus* $M_1 \neq R_L(R_U(M_1))$.

Proposition 58. *For any mset relation R on a nonempty mset M the following conditions hold.*

   (i) *For every* $N \subseteq M$, $R_U(N) = [R_L(N^c)]^c$.
   (ii) $R_U(\phi) = \phi$.
   (iii) *For any sub msets M_1 and M_2 of M,* $R_U(M_1 \cup M_2) = R_U(M_1) \cap R_U(M_2)$.
   (iv) *If* $M_1 \subseteq M_2$ *then* $R_U(M_1) \subseteq R_U(M_2)$.
   (v) *For any sub msets M_1 and M_2 of M,* $R_U(M_1 \cap M_2) \subseteq R_U(M_1) \cap R_U(M_2)$.
   (vi) *For any sub msets M_1 and M_2 of M,* $R_U(M_1 \oplus M_2) = R_U(M_1) \oplus R_U(M_2)$.
   (vii) *For any sub msets M_1 and M_2 of M,* $R_U(M_1 \ominus M_2) \subseteq R_U(M_1) \ominus R_U(M_2)$.
   (viii) *For every* $N \subseteq M$, $R_U(R_U(N)) \subseteq R_U(N)$.

Proof. The proof is the same as for proposition 48.

Proposition 59. *If an mset relation R on M is a reflexive mset relation, then the following condition hold.*
   (i) $R_U(M) = M$.
   (ii) $N \subseteq R_U(N)$.

Proof. The proof is the same as for proposition 50.

Proposition 60. *Let R be an equivalence mset relation on a nonempty mset M, then the following condition hold.*
   (i) $R_U(R_L(N)) \subseteq N$.
   (ii) $R_U(R_L(N)) \subseteq R_L(N)$.

Proof. The proof is the same as for proposition 52.

Theorem 61. *If R is an equivalence mset relation on a nonempty mset M, then* $R_U(R_L(N)) = R_L(N)$.

Proof. Follows from proposition 59 and proposition 60.

Theorem 62. *If R is a irreflexive, symmetric and transitive mset relation on a nonempty mset M, then $R_U(R_L(N)) \subseteq N$ for all $N \subseteq M$.*

Proof. The proof is the same as for theorem 56.

Theorem 63. *If an mset relation R on a nonempty mset M is a reflexive mset relation, then $R_L(N) \subseteq N \subseteq R_U(N)$.*

Proof. Follows from proposition 50 and proposition 59.

Theorem 64. *If R is any mset relation on a nonempty mset M, then $R_L(M_1^c \cup M_2) \subseteq R_L(M_1)^c \cup R_L(M_2)$ for all $M_1, M_2 \subseteq M$.*

Proof. Let $x \notin^m R_L(M_1)^c \cup R_L(M_2)$, then $x \notin^m R_L(M_1)^c$ and $x \notin^m R_L(M_2)$ hence, $x \in^m R_L(M_1)$ and $x \notin^m R_L(M_2)$, i.e., $\langle m/x \rangle R \subseteq M_1$ and $\langle m/x \rangle R \nsubseteq M_2$, then $\langle m/x \rangle R \nsubseteq M_1^c$ and $\langle m/x \rangle R \nsubseteq M_2$, hence $\langle m/x \rangle R \nsubseteq M_1^c \cup M_2$, thus $x \notin^m R_L(M_1^c \cup M_2)$, i.e., $R_L(M_1^c \cup M_2) \subseteq R_L(M_1)^c \cup R_L(M_2)$.

The following example gives the result in theorem 64 is not true in general.

Example 65. *In Example 49, if $M_1 = \{3/x, 3/z\}$ and $M_2 = \{5/y, 7/r\}$, then $R_L(M_1^c \cup M_2) = \{5/y, 4/s\}$, $R_L(M_2) = \{5/y\}$, $R_L(M_1) = \{3/x\}$ and $R_L(M_1)^c = \{5/y, 3/z, 7/r, 4/s\}$, then $R_L(M_1)^c \cup R_L(M_2) = \{5/y, 3/z, 7/r, 4/s\}$. Thus $R_L(M_1^c \cup M_2) \neq R_L(M_1)^c \cup R_L(M_2)$.*

The following theorem states that a reflexive mset relation is sufficient for the mset approximation operators to be a pair of interior and closure operators on multisets.

Theorem 66. *If R is a reflexive mset relation on M, then the pair of lower and upper mset approximations are a pair of interior and closure operators satisfying Kuratowski axioms of M-topological spaces.*

Proof. The proof follows from definition 44 and propositions 48, 50, 58 and 59.

Remark 67. *Suppose R is a reflexive mset relation on M. The M-topology generated by the lower and upper mset approximation operators is the same as the M-topology τ_3 generated by the post msets as a sub M-base.*

7   Conclusion and Future Work

Pawlak introduced the concept of rough sets which have a wide range of applications in various fields like artificial intelligence, cognitive sciences, knowledge discovery from databases, machine learning expert systems etc. [6,13,15,16,20,24,25]. These types of situations deal with objects and attributes. Many situations may occur where the counts of the objects in the universe of discourse are not single. In information systems we come across situations which involve the concept of a multiset.

This paper begins with a brief survey of the notion of msets introduced by Yager, different types of collection of msets and operations under such collections. It also gives the definition of mset relation, equivalence mset relation and mset function introduced by the authors. After presenting the preliminaries and basic definitions, the authors introduced the notion of M-topological space. Basis, sub basis, closure and interior are defined in the context of msets. M-topologies are obtained by using mset relation. Rough multiset is introduced in terms of lower and upper approximations. Authors used an M-topological concept to investigate Pawlaks rough set theory by replacing its universe by multiset. Authors used the M-base $\{\langle m/x \rangle R : x \in {}^m M\}$ of an M-topology τ on M generated by any mset relation R, to investigate new definitions for the rough multisets. In this case the generalized approximation mset space (M, R) is an M-topological space for any reflexive mset relation R. M-topology τ, constructed from an mset relation on M, may help in formalizing many applications.

An information system is formally defined as a pair $I = (M, A)$ where M is a non empty finite set of objects called the universe of discourse and A is a non empty finite set of attributes. Information multisystems are represented using multisets instead of crisp sets. Multisets are defined as those sets in which an element may have several occurrences as opposed to just one in a crisp set. The very definition of multisets implies the omission of object identifiers. Number of occurrences of each object is denoted in an additional column named C (counts or multiplicity). Formally, an information multisystem can be defined as a triple, $I = (M, A, R)$ where M is an mset of objects, A is the set of attributes, R is an mset relation defined on M. For example, the chemical system can be defined as $S = (M, A, R)$ where M is the mset of all possible molecules, A is an algorithm describing the reaction vessel or domain and how the rules are applied to the molecules inside the vessel and R is the set of "collision" rules, representing the interaction among the molecules. Thus, if M is a collection of molecules (it's form a multiset) and R is a set of reaction rules on M (mset relation) then M-topology τ generated by R is a collection of chemical equations for M (knowledge base for M).

Finally the concept of rough multisets and related properties with the help of lower mset approximation and upper mset approximations are important frameworks for certain types of information multisystems. Chakrabarthy [2,6] introduced two types of bags call IC bags and n^k bags, which is suitable for situations where the counts of the objects in information system are not fixed and are represented in the form of intervals of positive integers and power set of positive integers $(P(N))$. These kinds of problems appear, for instance, during a nuclear fission, when a nucleus (consisting of protons and neutrons) is split into multiple nuclei, each of them with its own number of protons and neutrons. Thus the rough multiset can be associated to IC bags or n^k bags with the help of lower mset approximations and upper mset approximations. This association could be used for certain types of decision analysis problems and could prove useful as mathematical tools for building decision support systems.

Acknowledgement. The authors would like to thank the referees and James F.Peters and Andrzej Skowron, Editor-in-Chief of the Transactions on Rough Sets, for carefully examining the paper and providing very helpful comments and suggestions, the authors are also grateful to Mrs. Dorothy Anita for improving the linguistic quality of the paper.

References

1. Syropoulos, A.: Mathematics of Multisets in Multiset Processing, pp. 347–358. Springer, Heidelberg (2001)
2. Chakrabarty, K.: Bags with Interval Counts. Foundations of Computing and Decision Sciences 25, 23–36 (2000)
3. Chakrabarty, K., Biswas, R., Nanda, S.: Fuzzy Shadows. Fuzzy Sets and Systems 101, 413–421 (1999)
4. Chakrabarty, K., Biswas, R., Nanda, S.: On Yagers theory of bags and fuzzy bags. Computer and Artificial Intelligence 18, 1–17 (1999)
5. Chakrabarty, K., Gedeon, T.: On Bags and Rough Bags. In: Proceedings Fourth Joint Conference on Information Sciences, North Carolina, USA, vol. 1, pp. 60–63 (1998)
6. Chakrabarty, K.: Loan Despi; n^k-bags. International Journal of Intelligent Systems 22, 223–236 (2007)
7. Flapan, E.: When Topology Meets Chemistry. Cambridge University Press, Cambridge (2000)
8. Simmons, G.F.: Introduction to Topology and Modern Analysis International edn., Mc Graw Hill, New York (1963)
9. Girish, K.P., John, S.J.: General Relations between partially ordered multisets and their chains and antichains. Mathematical Communications 14(2), 193–206 (2009)
10. Girish, K.P., John, S.J.: Relations and Functions in Multiset Context. Information Sciences 179, 758–768 (2009)
11. Grzymala-Busse, J.: Learning from examples based on rough multisets. In: Proceedings of the Second International Symposium on Methodologies for Intelligent Systems, Charlotte, North Carolina, USA, pp. 325–332 (1987)
12. Liu, G.: Generalized rough sets over fuzzy lattices. Information Sciences 178, 1651–1662 (2007)
13. Jena, S.P., Ghosh, S.K., Tripathy, B.K.: On the theory of bags and lists. Information Sciences 132, 241–254 (2001)
14. Kelly, J.L.: General topology. Springer, New York (1955); 2nd edn. Van Nostrand, New York (1957)
15. Lashin, E.F., Kozae, A.M., Abo Khadra, A.A., Medhat, T.: Rough set theory for topological spaces. International Journal of Approximate Reasoning 40, 35–43 (2005)
16. Lashin, E.F., Medhat, T.: Topological reduction of information systems. Chaos, Solitons and Fractals 25, 277–286 (2005)
17. Pawlak, Z.: Classification of objects by means of attributes. Polish Academy of Sciences, 429
18. Pawlak, Z.: A treatise on rough sets. Transactions of Rough Sets 4, 1–17 (2005)
19. Pawlak, Z.: Rough sets. International Journal of Computer Sciences 11, 341–356 (1982)
20. Pawlak, Z.: Rough sets and Fuzzy sets. Fuzzy Sets and Sytems 17, 99–102 (1985)

21. Pawlak, Z.: Some issues on rough sets. Transactions of Rough Sets I, 1–58 (2004)
22. Pawlak, Z., Skowron, A.: Rough sets and Boolean reasoning. Information Sciences 177, 41–73 (2007)
23. Pawlak, Z., Skowron, A.: Rough sets: Some extensions. Information Sciences 177, 28–40 (2007)
24. Pawlak, Z., Skowron, A.: Rudiments of rough sets. Information Sciences 177, 3–27 (2007)
25. Peters, J., Pal, S.: Cantor, fuzzy, near and rough sets in image analysis. In: Pal, S., Peters, J. (eds.) Rough Fuzzy Image Analysis: Foundations and Methodologies, pp. 1.1–1.15. CRC Press, Taylor and Francis Group, Boca Raton, U.S.A (2010), ISBN 9778-1-4398-0329-5
26. Polkowski, L.: Rough Sets. Mathematical Foundations. Springer, Berlin (2002)
27. Singh, D., Singh, J.N.: Some combinatorics of multisets. Int. J. Math. Educ. Sci. Tech. 34, 489–499 (2003)
28. Skowron, A.: On topology in Information System. Bull. Polish. Sci. Math. 36, 477–479 (1988)
29. Skowron, A., Peters, J.: Rough-granular computing, Handbook on Granular Computing, ch. 13, pp. 285–328. John Wiley & Sons, Inc., N.Y., USA (2008)
30. Deng, T., Chen, Y., Xu, W., Dai, Q.: A novel approach to fuzzy rough sets based on a fuzzy covering. Information Sciences 177, 2308–2326 (2007)
31. Blizard, W.D.: Multiset Theory. Notre Dame Journal of Logic 30, 36–65 (1989)
32. Willard, S.: General Topology. Addison-Wesley, Reading (1970)
33. Zhu, W.: Generalized rough sets based on relations. Information Sciences 177, 4997–5011 (2007)
34. Zhu, W.: Topological approaches to covering rough sets. Information Sciences 177, 1499–1508 (2007)
35. Yager, R.R.: On the theory of bags. Int. J. General Syst. 13, 23–37 (1986)
36. Yao, Y.Y.: Rough Sets, Neighborhood Systems, and Granular Computing. In: Proceedings of the 1999 IEEE Canadian Conference on Electrical and Computer Engineering, pp. 9–12 (1999)

A Rough Set Exploration of
Facial Similarity Judgements

Daryl H. Hepting[1], Richard Spring[1], and Dominik Ślęzak[2,3]

[1] Department of Computer Science, University of Regina
3737 Wascana Parkway, Regina, SK, S4S 0A2 Canada
dhh@cs.uregina.ca, rspring@gmail.com
[2] Institute of Mathematics, University of Warsaw
Banacha 2, 02-097 Warsaw, Poland
[3] Infobright Inc.
Krzywickiego 34 pok. 219, 02-078 Warsaw, Poland
slezak@infobright.com

Abstract. Facial recognition is routine for most people; yet describing the process of recognition, or describing a face to be recognized reveals a great deal of complexity inherent in the activity. Eyewitness identification remains an important element in judicial proceedings: it is very convincing, yet it is not very accurate.

A study was conducted in which participants were asked to sort a collection of facial photographs into an unrestricted number of piles, based on their individual judgements of facial similarity. Participants then labelled each pile. Three different qualities identified in the photos may have been used by participants in making similarity judgements. Choosing photos with the stipulation that half have the quality and half do not, the composition of each pile made by each participant was analysed. A pile is rated as "quality present" (or missing), if it contained significantly more of "quality present" (or missing) photos. Otherwise, it was rated as "quality undecided". As a result, an information system was obtained with objects corresponding to the participants and attributes corresponding to the pairs of photos. Further, a decision attribute was added for each of the qualities. The decision classes were determined by setting a threshold for the percentage of QU photos. Initially, this threshold was determined by observation. The rough set based attribute reduction methodology was applied to this data in order to build classifiers for each quality. Other thresholds for QU were also considered, based on computational experimentation, in order to improve the accuracy of the classifiers.

This paper describes the initial study, the computational approach which includes an important pre-processing step and development of a useful heuristic, the results from the evaluation, and a list of opportunities for future work. Although different participants created quite different sortings of photos, the rough set analysis detected photos with more general significance. This may lead to a practical test for individual abilities, as well as to inferring what discriminations people use in face recognition.

J.F. Peters et al. (Eds.): Transactions on Rough Sets XIV, LNCS 6600, pp. 81–99, 2011.
© Springer-Verlag Berlin Heidelberg 2011

1   Introduction

Eyewitness identification holds a prominent role in many judicial settings, yet it is generally not accurate. Verbal overshadowing [14] is an effect that can obscure a witness's recollection of face when she is asked to describe the face to create a composite sketch. Alternatively, if the witness is asked to examine a large collection of photos, her memory may become saturated and she may mistakenly judge the current face similar to another she had previously examined (i.e., inaccurate source monitoring) and not to the one she is trying to recall [3].

It is hypothesized that if the presentation of images can be personalized, the eyewitness may have to deal with fewer images, minimizing both of the negative effects discussed. This research examines whether there may be a reliable basis for such personalization by analyzing data from a sorting study of 356 randomly ordered photographs, 178 each of Caucasian and First Nations faces, performed by 25 participants. The study avoided verbalization completely while sorting. Once sorting was complete, each participant was asked to label each of the piles that she had made. Each participant was asked to group the 356 photos into an unconstrained number of piles, according to perceived similarity. As a participant encountered a photo, she could only place that photo without reviewing or disturbing any existing piles. This requirement is similar to sequential eyewitness lineup procedures [15].

Race, along with labels such as "ears" or "lips", were treated as qualities that were a part of the judgement process used by participants. After choosing as many photos as possible, with the stipulation that half had the quality (called QP – for Quality Present) and half did not (QM – Quality Missing), the composition of each pile made by each participant was analysed. A pile was rated as QP (QM), if it contained significantly more of QP (QM) photos. Otherwise, it was rated as QU (Quality Undecided). All photos in the pile were then relabelled in the same manner. The result is that for each participant, all photos were finally classified as either QP, QM, or QU. This inquiry is indirectly related to binary classification – by construction, all photos began as either QP or QM. However, the participants' accuracy in correctly classifying each photo was not of interest, because division of photos into QP and QM was subjective. Rather, participants' classification of photos may help to identify a strategy that is being used in the sorting. For example, some participants may be making use of a quality in their sorting. However, it is not sufficient to say that other participants are not using that quality in their sorting. Further difficulties arise when participants place photos together not because they are similar but because they are not sufficiently different.

The idea of sorting photos is not new. There are also some approaches to reduce the number of photos viewed before finding the target (see e.g. [13]). However, when comparing to the others, the present methodology requires less information (for example, no photo similarity ratings are requested) and leads to a far simpler data layout (see the next section for details). Consequently, when building classification models and considering attribute selection criteria, the goal is to apply very straightforward techniques, which do not make any assumptions about the training data (cf. [11]). Indirectly, each participant made

63,190 pairwise similarity judgements. This quantity of data made it a good candidate for rough set attribute reduction methodology [1]. Yet, analyzing the data for all 63,190 pairs is troublesome for various reasons, including speed. Therefore, a pre-processing step is devised to reduce the number of pairs under consideration to those most likely to contribute to an accurate classifier. This technique may be of general interest for problems that have a distance metric that can be applied between decision classes.

Section 2 describes the study in detail. Section 3 describes a method to analyze the behaviour of each participant, by examining the composition of each pile made, which led to the determination of decision classes used later. Section 4 describes the method for preprocessing pair data. Section 5 describes the rough set analysis [11] of the pair sets selected through our filtering. Section 6 presents ways in which we have tried to improve the accuracy of the classifier, and explores the potential effectiveness of a heuristic derived from the filtering computations. Section 7 presents some conclusions and avenues for future work.

2   Sorting Study of Facial Photographs

The stimulus photo set comprised equal numbers ($n = 178$) of Caucasian and First Nation faces. Cross-race identification of faces is an important topic of ongoing research [12], and this sorting study seeks to contribute to this body of work. It is focused on similarity judgements as a way to understand the way people perceive structure in the stimuli set. It may be that not everyone perceives the same structure. Therefore, if a person's preferred structure could be ascertained easily, it could be used to improve identification accuracy.

Figure 2 shows two photos from the stimulus set. Photographs were of the head and shoulders of each individual in a front facing pose wearing casual clothing. Subjects for these photographs were positioned 5 feet from the camera and 2 feet from the background wall. All distinguishing materials (e.g., glasses, piercing) were removed for the purposes of the photograph. All photographs were edited using Adobe Photoshop 7.0. Photographs for the facial recognition task were cropped to include only the subject's head and shoulders, while the background colour was changed from white to grey. The photographs were laminated on 5 by 4 inch cards. Participants were asked to view photos one at a time and place each photo on a pile with similar photos, without disturbing existing piles. Therefore, not all participants would make the same direct comparison: only if one photo was visible when the other was being placed. The number of piles was not constrained. Within the 25 participants, the number of piles made ranged between 4 and 38.

Participants were recruited through the Computer Science department participant pool at the University of Regina [8], and they received a research credit in exchange for their participation. The participant responses to a demographic questionnaire administered at the start of the study are summarized in Table 1. Notice that none of the participants have a First Nations' ancestor and almost none of them had exposure to First Nations' people.

Table 1. Participants' Modal Response to Demographic Questions (N=25). Note: Education = years post-secondary education; C = Caucasian; FN = First Nations.

Characteristic	Mode (n)	Ancestry/Exposure	Mode (n)
Gender	Female (13)	Ancestor C	Yes (24)
Age	18-25(20)	Ancestor FN	No (25)
Education	1-4 (18)	Ancestor Other	No (21)
Study/Interest	Science (13)	Exposure C	Yes (24)
Computer Familiarity	Medium(14)	Exposure FN	No (23)
		Exposure Other	No (21)

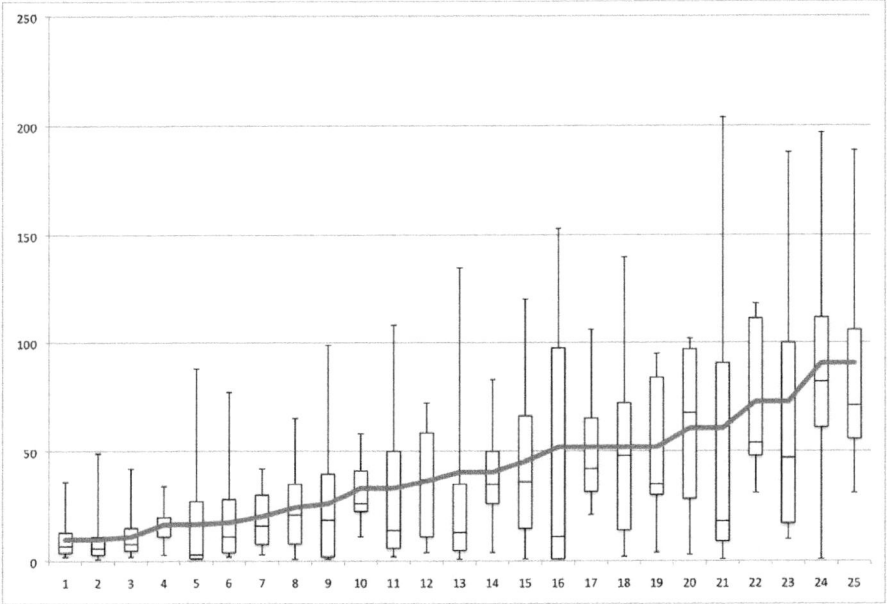

Fig. 1. Box and whiskers plot of pile sizes per participant, sorted in descending order of number of piles made. The line shows the average pile size.

The photographs were sorted in a variety of ways. Figure 1 shows a summary of the piles that were made. It is interesting to note that for many of the participants, regardless of the number of piles they made, some piles contained very few photos.

3   Analysis of Pile Composition

A method was developed to analyze the composition of each pile made by participants, based on a bimodal identification of the photos. This method used knowledge about the problem's specifics and some simple statistical data preparation techniques [2].

Fig. 2. Two photos from the collection which participants were asked to sort according to similarity

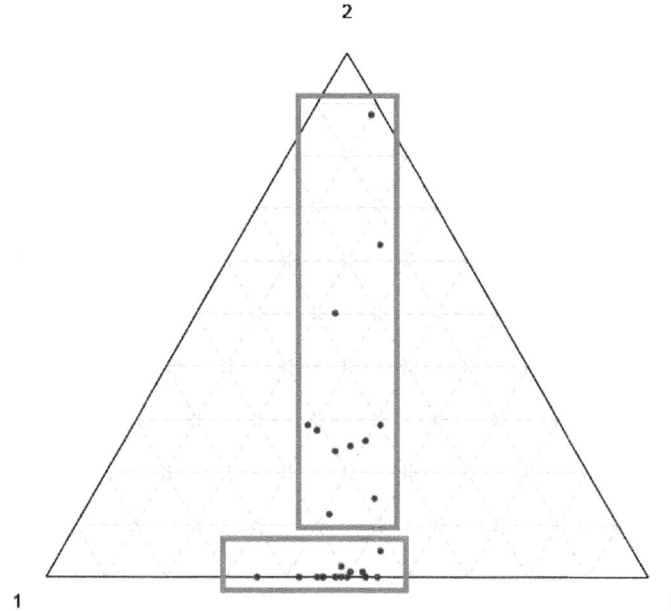

Fig. 3. Distribution of participants based on their classification of photos. Each point reflects the mix of Caucasian, First Nations, and Unidentified photos as rated by a participant. Vertex 1 is Caucasian, Vertex 3 is First Nations, and Vertex 2 is Undecided. A threshold on QU (RU) of 5% was used. (When referring to a particular quality, Q will be replaced by R for Race, E for Ears, or L, for Lips). The lower rectangle identifies participants in D_{RP} (with no or very few photos classified as RU). The upper rectangle identifies participants in D_{RM} (with many photos classified as RU).

Table 2. Labels for facial parts and general facial characteristics. Many labels are used for every picture. The table is organized by percentage of photos available for comparison. The percentage available column indicates the number of photos available for comparison, which is the twice the minimum of the percentage of labelled photos and the percent This analysis only looked at the presence or absence of a label, so "ears" and "lips" were chosen (44%) because these allowed the most photos to be used.

Label	Photos	% Labelled	% Unlabelled	% Available
ears	157	44	56	88
lips	157	44	56	88
eyebrows	217	61	39	78
head/face size	125	35	65	70
facial hair	243	68	32	64
nose	259	73	27	54
forehead	65	18	82	36
jaw/chin	318	89	11	22
cheeks	25	7	93	14
neck	25	7	93	14
teeth	14	4	96	8
hair	356	100	0	0
eyes	356	100	0	0
skin/complexion	356	100	0	0
head/face shape	356	100	0	0

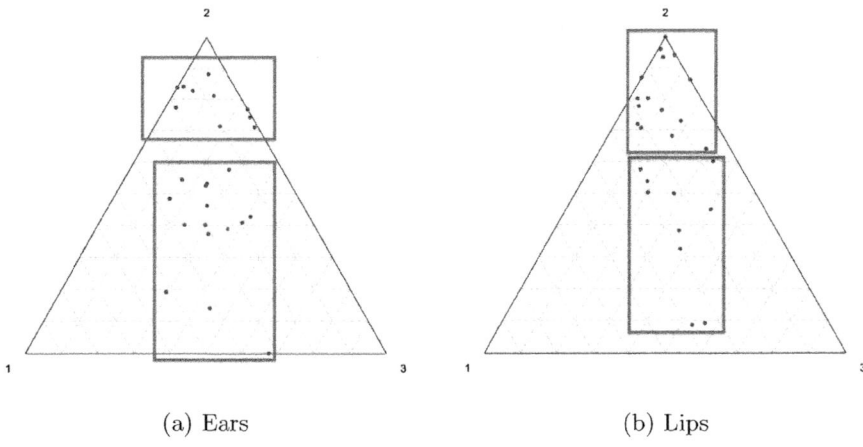

(a) Ears   (b) Lips

Fig. 4. Each point reflects the mix of photos classified by a participant. A point in the center represents an equal mix of photos classified QP, QM, and QU. This figure shows the distribution of participants based on their classification of photos with respect to the "ears"/"lips" label. Many points are located between Vertex 1 and Vertex 2, representing approximately equal numbers of photos classified as QP and QM. If a point is near Vertex 3, that participant classified most of the photos as QU. For the Ears/Lips quality, decision classes were constructed with a threshold of 60% for QU.

The first division of photos was based on the stimulus set to which each belongs. Nominally, this division is on the basis of Race, which may be either "Caucasian" or "First Nations". As many photos as possible are chosen, with the stipulation that half have the quality (called QP – for Quality Present) and half do not (QM – Quality Missing), For the Race (Caucasian/First Nations) quality, we consider all the photos. However, other qualities may involve only a subset of the photos.

The number of Quality Present (QP) and Quality Missing (QM) photos in each pile was expressed as a percentage. The CHITEST function in Microsoft Excel was used to compare the ratio of QP to QM against an expected equal distribution. The pile was classified as QU (for Quality Undecided) if the distribution was not significantly different from equal ($p \geq 0.05$). If the distribution was significantly different from equal ($p < 0.05$), the pile was classified as QP, if QP > QM or as QM if QM > QP. All pictures in that pile were then relabelled according the pile's classification. By analysing in this way all piles made by a participant, we determined the number of photos relabelled as QP, QM, and QU. This data lends itself to display using a trilinear plot.

Figure 3 shows the plot for participants according to the Race quality. By observation, a threshold for QU was set and two decision classes were formed: D_{RP} (QU < 5%, $n = 14$) and D_{RM} (QU $\geq$ 5%, $n = 11$).

Labels associated with the piles were then examined. Table 2 shows the distribution of labels across photos. The labels "ears" and "lips" were chosen because they permitted the largest number of photos to be used.

Because the labelled photos were not by definition matched with an equal number of unlabelled photos, the piles had to be virtually reconstructed. Since 157 photos were labelled in each case, 42 of the 199 unlabelled photos were removed at random. Once the number of labelled photos matched the number of unlabelled photos, the same bimodal comparison method was applied.

Figure 4 shows the distribution of participants according to this method. Two decision classes were formed for each label, D_{QP} (QU < 60%, $n = 10$ for Ears, $n = 15$ for Lips) and D_{QM} (QU $\geq$ 60%, $n = 15$ for Ears, $n = 10$ for Lips).

4   Preprocessing of Pair Data

For each observed quality (Race, Ears, Lips), a threshold for the percentage of QU (Quality Undecided) was set to determine the two decision classes to be used, D_{QP} and D_{QM}. Because a similarity judgement was recorded for each pair of the $\binom{356}{2} = 63,190$ photos, it is possible to compare the judgement of the two decision classes with respect to each pair. In order to reduce the number of pairs under consideration, those which one decision class predominantly rated as similar and the other decision class predominantly rated as dissimilar may be most desirable. This follows an approach similar to the feature extraction/selection phase in knowledge discovery and data mining [2,10]. Figure 5, which shows similarity ratings between the decision classes for the Race quality, indicates that this may be a fruitful course of action.

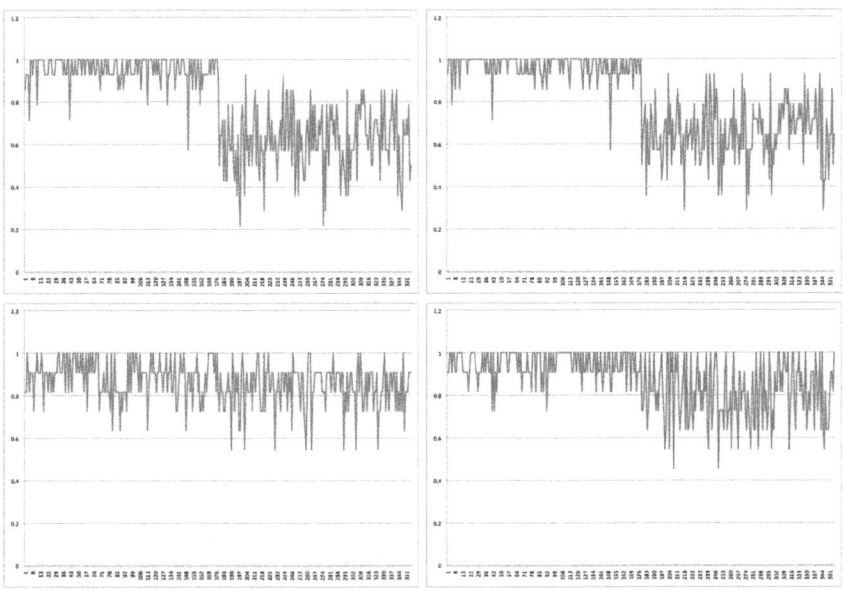

Fig. 5. The photos in the pair from Figure 2, whose similarity was rated differently between D_{RP} and D_{RM}, are compared against all other photos, by decision class. The top graphs show results for the D_{RP} and the bottom graphs show results for the D_{RM}.

Table 3. Pair table for Race quality, QU (RU) threshold = 5%

	Gap (Δ)				
d	1.0	≥ 0.9	≥ 0.8	≥ 0.7	≥ 0.6
0.0	0	0	0	0	0
≤ 0.1		0	0	0	1
≤ 0.2			0	0	7
≤ 0.3				17	82
≤ 0.4					130

Table 4. Pair table for Ears quality, QU (EU) threshold = 60%

	Gap (Δ)				
d	1.0	≥ 0.9	≥ 0.8	≥ 0.7	≥ 0.6
0.0	0	0	2	3	7
≤ 0.1		2	9	31	59
≤ 0.2			22	77	250
≤ 0.3				159	498
≤ 0.4					675

Table 5. Pair table for Lips quality, QU (LU) threshold = 60%

	Gap (Δ)				
d	1.0	≥ 0.9	≥ 0.8	≥ 0.7	≥ 0.6
0.0	0	0	0	0	0
≤ 0.1		0	0	0	2
≤ 0.2			3	14	16
≤ 0.3				24	78
≤ 0.4					166

Table 6. Comparison of the total number of pair table entries with the measured average accuracy of classifiers presented by Hepting et al. [5]

Quality	Table Sum	Avg. Accuracy
Ears	1782	99.2, σ 1.69
Lips	303	94.4, σ 3.86
Race	237	92.38, σ 4.38

The normalized distance (d) for each pair was computed for both decision classes (D_{QP}, D_{QM}). A distance of 0 means that all participants in a decision class have rated the pair as similar, whereas a distance of 1 means that all participants in a decision class have rated the pair as dissimilar. Pairs are grouped on the basis of 2 parameters: $d = min(d_{D_{QP}}, d_{D_{QM}})$ and $\Delta = |d_{D_{QP}} - d_{D_{QM}}|$.

The so-called pair table comprises the numbers of attributes (pairs of photos) for all parameter pairs (d, Δ, by increments of 0.1) for which $d + \Delta \leq 1$, $0 \leq d \leq 0.4$, and $1.0 \geq \Delta \geq 0.6$. Compared to previous work [5,6] that introduced the pairs table, the line for $d = 0$ is added because it is important on its own and should be highlighted separately. On the other hand, the line for $0.4 < d \leq 0.5$ is removed because those pairs are not able to contribute substantially to the discrimination between decision classes. Pairs found at $d = 0, \Delta = 1$ divide the participants exactly into the two decision classes and herein they are referred to as "splitting pairs".

Tables 3, 4, and 5 show the pair tables for each of the qualities considered in this paper. Cells for which $d + \Delta > 1$ are greyed out.

5   Rough Set Attribute Reduction Methodology

The rough set attribute reduction methodology [11] was applied to each of the pair tables in Tables 3, 4, and 5. For each non-zero entry in a pair table, we repeat the following steps 10 times.

1. Preprocessing: Create an input file comprising data for each participant (object) including judgements (attributes) about each pair (0 if similar, 1 if dissimilar) identified by the particular values of d and Δ (always such that $d + \Delta \leq 1$). The decision variable is the value determined in Section 3, which indicates the decision of whether the participant belongs to D_{QP} or D_{QM}. See Table 7 for an example.

2. Preprocessing: Split input file (50/50): Each table was split with approximately 50% of participants in a training set (data from 12 or 13 participants) and 50% of participants in a testing set (data from 13 or 12 participants).

3. Training: Calculate the reducts in training file using genetic algorithms [16] in RSES (Rough Set Exploration System). The genetic algorithms procedure calculates the top N (with $N = 10$) reducts possible for a given analysis, if N reducts could be found.

4. Testing: Use classifiers based on the reducts generated in Training, test the results on the data in the testing set.

5. Classification: Observe the classification accuracy and report the results.

If a classifier with 100% accuracy and 100% coverage is not found within 10 tries, it may still exist. Choosing 12 or 13 of 25 participants for training leads to a possible $\binom{25}{13} = 5,200,300$ combinations and classifiers.

Table 7. 16 pairs selected as input to RSES (Rough Set Exploration System) for classification for the Lips quality ($d = 0.2$, $\Delta = 0.6$)

object	Photograph Pairs																class
	1969a-2094a	4211a-5893a	2660a-8127a	1296a-2811a	3722a-4158a	2094a-6682a	058-149	083-117	1716a-7001a	040-108	1032a-1867a	0011a-7453a	1296a-6682a	5241a-8164a	0079a-6524a	1032a-8831a	
1	1	0	0	1	0	1	0	1	1	0	0	1	0	0	1	0	0
2	0	0	0	0	0	1	0	1	1	0	0	0	1	0	1	0	1
3	0	0	0	0	0	0	0	0	0	0	0	0	0	0	0	0	0
4	0	0	1	0	0	0	0	0	0	0	0	0	0	0	0	0	0
5	1	1	1	1	1	1	1	1	1	1	1	1	1	1	1	1	1
6	1	1	1	1	1	1	1	1	1	1	1	1	1	1	1	1	1
7	1	0	0	0	0	0	0	0	0	1	0	0	1	0	0	0	0
8	0	0	0	0	0	0	0	1	0	0	0	0	0	0	0	0	0
9	0	0	0	1	0	0	0	0	1	1	0	0	0	0	0	0	0
10	1	1	1	1	1	1	1	0	1	1	1	1	1	1	1	1	1
11	0	1	1	1	1	1	0	0	0	0	0	0	1	1	1	0	0
12	0	0	0	0	0	0	0	0	0	1	0	0	0	0	0	0	0
13	0	0	0	0	0	0	0	0	1	0	0	0	0	0	0	0	0
14	1	1	1	1	1	1	1	1	1	1	1	1	1	1	1	1	1
15	1	1	1	1	0	0	0	1	1	1	1	1	1	1	1	1	1
16	0	0	0	0	0	0	0	0	0	0	0	0	0	0	1	0	0
17	0	0	0	0	0	0	0	0	0	0	0	0	0	1	0	0	0
18	0	0	0	0	0	0	0	0	0	0	0	0	0	0	0	0	0
19	1	1	0	0	0	1	0	0	0	0	1	0	0	0	0	1	0
20	1	1	1	1	1	1	0	1	1	1	1	1	1	1	1	1	1
21	0	0	0	0	0	0	0	0	0	0	0	1	0	0	0	0	0
22	0	0	0	0	0	0	1	1	0	0	1	0	0	1	0	0	0
23	0	1	1	0	0	0	0	0	0	0	0	1	0	0	0	1	0
24	1	1	1	1	0	1	1	1	1	1	1	1	1	1	1	1	1
25	1	1	1	1	1	1	1	1	1	1	1	1	1	1	1	1	1

Table 8. Full accuracy and full coverage classifiers for Ears/Lips. In each entry where they were found, the count. Dashes where no such classifiers were found. Most succinct classifier is marked with *, with number of reducts in parentheses.

d	Gap (Δ)				
	1.0	≥ 0.9	≥ 0.8	≥ 0.7	≥ 0.6
0.0	-	-	0	0	0
≤ 0.1		2*(2)	8	7	5
≤ 0.2			3	4	2
≤ 0.3				1	3
≤ 0.4					0

(a) Ears

d	Gap (Δ)				
	1.0	≥ 0.9	≥ 0.8	≥ 0.7	≥ 0.6
0.0	-	-	-	-	-
≤ 0.1		-	-	-	0
≤ 0.2			2*(1)	1	3
≤ 0.3				1	1
≤ 0.4					1

(b) Lips

(a) A participant likely belongs to D_{EM} if she rates either pair as dissimilar.

(b) A participant likely belongs to D_{LM} if she rates both pairs as dissimilar.

Fig. 6. The two pairs of photos in the most succinct classifier for (a) Ears and (b) Lips

Table 9. The most popular pairs, based on all reducts from all full accuracy/full coverage classifiers for the three decision class groups. Each of these pairs occurred at least 2% of the time. Photo names ending in 'a' are Caucasian (C), otherwise First Nations (FN). Notice for Race that these pairs are split evenly between C and FN. For Ears, almost all important pairs are FN while for Lips, almost all important pairs are C. There are no mixed pairs. Also, there is very little overlap between classifiers.

Race	Ears	Lips
087-142	095-106	1296a-6682a
0662a-4919a	146-172	0079a-6524a
004-050	111-121	1296a-2811a
6281a-9265a	037-176	1716a-7001a
059-128	033-121	1032a-8831a
1859a-1907a	058-157	4211a-5893a
039-125	4833a-9948a	083-117
8027a-8922a	038-095	0011a-7453a
050-176	038-068	5241a-8164a
	061-065	2094a-6682a
		1969a-2094a
		040-108
		2660a-8127a

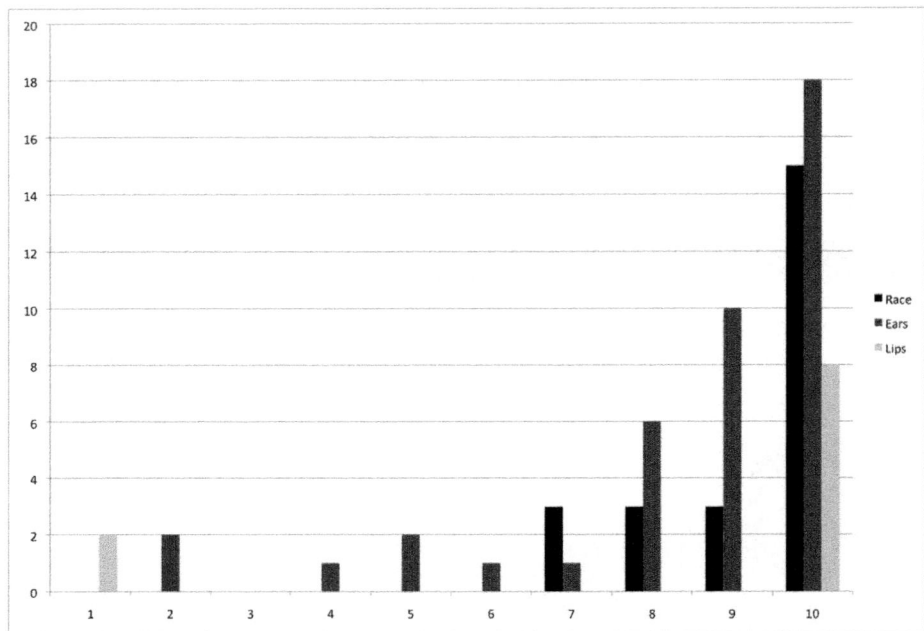

Fig. 7. Distribution of reducts for full accuracy/full coverage classifiers. Length is the horizontal axis. Very succinct classifiers were obtained for qualities Ears and Lips, in comparison with Race.

Figures 6(a) and 6(b) show the pairs of photos upon which the most succinct classifiers (Table 8) are built. Table 9 reports the most frequently occurring pairs of photos.

5.1 Comparison

Figure 7 shows the distribution of reducts amongst full accuracy and full coverage classifiers for the 3 decision class pairs based on observed qualities.

6 Validation

As can be seen in Table 6, the classifiers built for the Ears quality in [5,6] were very accurate, the most accurate of the three qualities considered (Race, Ears, and Lips). Yet, none of the 63,190 pairs split the participants exactly into the decision classes D_{EP_0}/D_{EM_0}.

Could the accuracy of the classifiers be improved by modifying the composition of the decision classes?

In previous work [5,6] the pairs in each cell of the pair table were used as input to one of the rough set classifier construction systems [1,11]. A relationship between the pair tables and the accuracy of classifiers is suggested [7] and requires

Table 10. Pairs with the highest similarity rating with respect to D_{EP_0}, presented for both decision classes

Pair	$D_{EP_0}(n = 15)$	$D_{EM_0}(n = 10)$	Total Similar
1907a-3600a	12	3	15/25
1182a-1391a	14	3	17/25
3905a-9948a	12	5	17/25
050-105	13	6	19/25
089-137	12	7	19/25
4426a-9700a	12	7	19/25

Table 11. Pairs with the highest similarity rating with respect to D_{EM_0}, presented for both decision classes.

Pair	$D_{EM_0}(n = 10)$	$D_{EP_0}(n = 15)$	Total Similar
152-153	10	2	12/25
4833a-9948a	10	2	12/25
023-114	10	4	14/25
105-136	10	5	15/25
1306a-2038a	10	5	15/25
2660a-7669a	10	6	16/25
5774a-9717a	10	6	16/25
060-098	10	7	17/25
1969a-4158a	10	7	17/25
2094a-3722a	10	7	17/25
3669a-3943a	10	7	17/25

further investigation. Table 6 summarizes the pair tables' properties and the observed accuracy of classifiers obtained for particular decision classes. Here, it is used as a heuristic (H) that replaces more thorough classifier training processes. Intuitively, decision classes that more clearly separate the participants will have more differing pairs of photos.

The original decision class pair is denoted by D_{QP_0} and D_{QM_0}, and the refinements considered here are denoted by D_{QP_i}/D_{QM_i}, $i = 1, 2, ...$

Further experimental efforts were undertaken to improve the overall model by not only optimizing the classifier but also by tuning the definitions of the decision classes. In this way, it is hoped to converge on the pairs of photos most significant to participants in their judgements of similarity.

6.1 Small Perturbations

Small adjustments to the composition of the decision classes for Ears may further improve the accuracy of the classifiers obtained. Some pairs are very close to splitting D_{EP_0}/D_{EM_0}, according to Tables 10 and 11. With a preference for an approximately equal split of similarity votes to dissimilarity votes, pairs 152-153 and 4833a-9948a with 12/25 similarity votes each are identified as the best candidates with which to create new decision classes. It seems reasonable to

Table 12. The most popular pairs which appeared in reducts of classifiers with full accuracy and coverage, as well as information which participants rated those pairs as similar (indicated below with an 'S')

Pairs	DEM0 (n=10)										DEP0 (n=15)															
	7	13	14	17	19	20	21	22	24	25	1	2	3	4	5	6	8	9	10	11	12	15	16	18	23
038-068	S		S	S	S	S	S	S	S	S									S						
038-095	S	S		S	S	S	S	S	S	S	S			S											
061-065	S		S	S	S	S	S	S	S	S							S	S							
095-106	S	S	S	S	S	S	S		S	S															
146-172	S	S	S	S		S	S	S	S	S															
111-121	S	S	S	S	S	S		S	S	S						S									
037-176	S	S	S	S	S	S		S	S	S															S
033-121	S	S	S	S	S	S		S	S	S				S											
058-157	S	S	S		S	S	S	S	S	S					S										
4833a-9948a	S	S	S	S	S	S	S	S	S	S								S							S
	10	8	9	9	9	10	7	9	10	10	1	0	0	2	1	1	1	2	1	0	0	0	0	0	2

Table 13. Discrimination values calculated for all participants. Membership in decision classes is also displayed. An 'X' indicates membership in D_{EM}.

Discrim	57	59	62	63	66	71	74	75	76	77	79	81	82	85	85	85	87	89	89	90	92	94	94	95	96
Particip	25	9	20	24	7	14	18	13	17	21	22	10	23	4	11	16	15	8	12	19	2	1	6	5	3
D_{E_0}	X		X	X	X		X	X	X	X										X					
D_{E_6}	X		X	X	X																				

choose modifications which make small changes, but improve discriminability. New decision class pairs are formed based on 152-153 (D_{EP_1}/D_{EM_1}) and 4833a-9948a (D_{EP_2}/D_{EM_2}), first separately and then in combination (D_{EP_3}/D_{EM_3}).

Considering the original decision classes for the quality Ears (D_{EP_0}/D_{EM_0}), all of the most similar pairs have $d = 0$ for D_{EM_0} but $\Delta \neq 1$ for D_{EM_0} (Table 11). Therefore, another reasonable strategy may be to move participants from D_{EP} to D_{EM}, so that $d = 0$ for D_{EP}. Based on Table 10, pairs 1182a-1391a and 1907a-3600a have been chosen for this purpose. The new decision class pairs are denoted as D_{EP_4}/D_{EM_4} and D_{EP_5}/D_{EM_5}, respectively.

6.2 Intersection of Reduct Pair Similarities

Hepting et al. [6] presented a list of popular pairs (see Table 9) in reducts computed as in Section 5 for the quality Ears (with decision classes D_{EP_0}/D_{EM_0}). Each of these popular pairs was examined to determine which participants rated them as similar. The results are presented in Table 12. There are 4 participants who rated all 10 of the popular pairs as similar. The decision class pair D_{EP_6}/D_{EM_6} was constructed to capture this division. Table 14 shows that much better accuracy of classification might be expected for this decision class pair, although only a small amount of participants are classified. This remains to be

Table 14. Comparison of heuristic related to the Ears quality

Decision Classes	Heuristic (H)
D_{EP_0}/D_{EM_0}	1794
D_{EP_1}/D_{EM_1}	278
D_{EP_2}/D_{EM_2}	1261
D_{EP_3}/D_{EM_3}	678
D_{EP_4}/D_{EM_4}	491
D_{EP_5}/D_{EM_5}	80
D_{EP_6}/D_{EM_6}	30736
D_{EP_7}/D_{EM_7}	26166

Table 15. Race quality for various thresholds of QU (RU)

%RU	D_{RP}	D_{RM}	Classified	Heuristic
0	10	15	10	2073
5	14	11	11	237
15	16	9	9	61
25	18	7	7	178
30	22	3	3	10982
50	23	2	2	59173
65	24	1	1	268238
90	25	0	0	-

Table 16. Ears quality for various thresholds of QU (EU)

%EU	D_{EP}	D_{EM}	Classified	Heuristic
0	1	24	1	59058
15	2	23	2	2505
20	3	22	3	420
40	5	20	5	347
45	9	16	9	581
50	11	14	11	1972
55	14	11	11	889
60	15	10	10	1794
75	18	7	7	4202
80	20	5	5	16187
85	24	1	1	268238
90	25	0	0	-

Table 17. Lips quality for various thresholds of QU (LU)

%LU	D_{LP}	D_{LM}	Classified	Heuristic
10	2	23	2	2834
35	3	22	3	420
40	4	21	4	543
45	5	20	5	228
50	7	18	7	126
55	8	17	8	270
60	10	15	10	303
65	11	14	11	270
70	12	13	12	283
75	15	10	10	1028
80	19	6	6	2921
90	21	4	4	18465
95	23	2	2	100647
100	25	0	0	-

verified in further experiments, by applying both rough set and not rough set classification models [1,10].

6.3   Discrimination

To pursue the undecidedness of participants, a measure of the discrimination that each participant applies to the set of photos is computed as a percentage by subtracting the fraction of all possible pairs made by the participant from 100. Consider the two extremes: if a participant made 356 piles (1 photo in each), there would be no pairs and so discrimination would be 100%. If the participant made only 1 pile containing all photos, the participant's sorting could make all possible pairs so discrimination would be 0%. Table 13 presents the discrimination values calculated for all participants. Membership in 2 different decision class pairs is also identified: D_{EP_0}/D_{EM_0} and D_{EP_6}/D_{EM_6}.

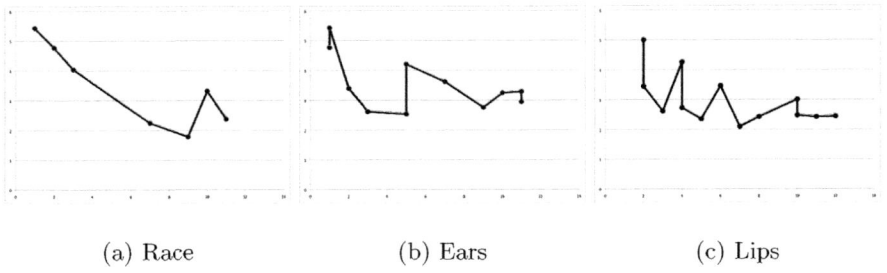

(a) Race   (b) Ears   (c) Lips

Fig. 8. Plots of number of participants classified against log10(heuristic) values. When few participants are classified, the heuristic values are large (at the left of the graphs). Relatively high values of heuristic for which many participants are classified, are hypothesized to represent the best choices for further processing.

Table 18. Pair table for Race quality, QU (RU) threshold = 0% (H = 2073)

	Gap (Δ)				
d	1.0	≥ 0.9	≥ 0.8	≥ 0.7	≥ 0.6
0.0	1	1	2	6	19
≤ 0.1		4	9	57	111
≤ 0.2			11	98	396
≤ 0.3				125	569
≤ 0.4					664

Table 19. Pair table for Ears quality, QU (EU) threshold = 50% (H = 1972)

	Gap (Δ)				
d	1.0	≥ 0.9	≥ 0.8	≥ 0.7	≥ 0.6
0.0	0	0	1	3	3
≤ 0.1		4	7	18	23
≤ 0.2			18	47	94
≤ 0.3				184	646
≤ 0.4					924

Table 20. Pair table for Lips quality, QU (LU) threshold = 75% (H = 1028)

	Gap (Δ)				
d	1.0	≥ 0.9	≥ 0.8	≥ 0.7	≥ 0.6
0.0	0	0	3	6	12
≤ 0.1		0	7	28	45
≤ 0.2			12	52	182
≤ 0.3				72	279
≤ 0.4					330

Table 21. Full accuracy/coverage classifiers for Race quality, with QU (RU) threshold = 0. Most succinct classifier is marked with *, with number of reducts in parentheses

	Gap (Δ)				
d	1.0	≥ 0.9	≥ 0.8	≥ 0.7	≥ 0.6
0.0	0	0	10*(1)	4	3
≤ 0.1		10	10	5	3
≤ 0.2			9	3	3
≤ 0.3				3	1
≤ 0.4					2

From Table 13, participant 9 seems to be a good candidate to add to D_{EM}, even though Table 12 does not agree. This configuration, as described, is considered as the decision class pair D_{EP_7}/D_{EM_7}.

Table 14 compares the heuristic (sum of pair table entries), H, for each of the decision class pairs considered thus far. It is somewhat surprising that only D_{EP_6}/D_{EM_6} and D_{EP_7}/D_{EM_7}, which classify very few participants, have heuristic values greater than D_{EP_0}/D_{EM_0}.

6.4   QU Threshold

From this small example, it appears that the manipulation of decision classes may be harder than first imagined. It seems that the parameterization in terms of QP, QM, and QU has captured some important aspect of this problem. Therefore, as a final examination of means to improve the accuracy of the classifiers, different values are explored for the threshold on QU%. In Tables 15 to 17, the decision classes D_{QP} and D_{QM} are laid out in increments of 5%. Figure 8 shows the relationship between heuristic and the number of participants classified for each quality. It may be easy to have a very high heuristic value while only classifying a small fraction of the participants.

Further examining the Race quality decision classes described in Table 18, which is the highest heuristic among the qualities considered (Tables 19 and 20), leads to the RSES results shown in Table 21.

7   Conclusions and Future Work

Cross-race identification of faces is an important topic of ongoing research [9,12], and this sorting study seeks to contribute to this body of work. The focus of this work has been on the labelling of similarity judgements as a way to understand the way people perceive structure in the stimuli set.

This work has demonstrated that rough set can help in accuracy and clarity of the results. This work also lends support to the pairs table technique as a broadly applicable method. While results reported in Table 21, for example, explicitly list only the frequency of full accuracy and full coverage classifiers, the overall average accuracy and coverage are very high. This would seem to indicate that there are definite patterns in the data, waiting to be discovered with the appropriate approach.

Table 22. The distribution of the 25 participants by decision class combinations, based on Tables 3, 4, and 5

Decision Class Combination	Members
D_{RM}, D_{EM}, D_{LM}	3
D_{RM}, D_{EM}, D_{LP}	3
D_{RM}, D_{EP}, D_{LM}	2
D_{RM}, D_{EP}, D_{LP}	7
D_{RP}, D_{EM}, D_{LM}	5
D_{RP}, D_{EM}, D_{LP}	3
D_{RP}, D_{EP}, D_{LM}	0
D_{RP}, D_{EP}, D_{LP}	2

Succinct tests for each quality have been identified. Further studies with more participants will help to validate these tests with more participants. It may also be fruitful to examine the interactions between the decision class pairs (Table 22, see also Table 9).

It is not clear on what basis the popular photos are differentiated by participants Therefore, repertory grid techniques [4] will be used to gain further insight into the decision process that may be at work.

The amount of overlap of attributes in reducts is particularly interesting. More than simply counting occurrences, a more detailed analysis of the correlations between attributes is also warranted.

This data has been explored with RSES. However, there are more advanced rough set approaches and tools, with their own parameters, that can be used in this situation. It may be indeed useful to experiment with ensembles of reducts, approximate reducts, and so on, both within and outside of RSES [10,17].

These experiments based on the Race, Ears, and Lips qualities have been an important step. However, there are a total of 16,777,235 binary decision classes available on this data. There may be other, better, classifiers available. A wide ranging numerical experiment is warranted (Table 9).

Acknowledgements. The authors thank Timothy Maciag and Katherine Arbuthnott. Daryl Hepting also thanks Jennifer Hepting. He was supported by the Natural Sciences and Engineering Research Council (NSERC) of Canada. Dominik Ślęzak was supported by the grants N N516 368334 and N N516 077837 from the Ministry of Science and Higher Education of the Republic of Poland.

References

1. Bazan, J.G., Szczuka, M.: The Rough Set Exploration System. In: Peters, J.F., Skowron, A. (eds.) Transactions on Rough Sets III. LNCS, vol. 3400, pp. 37–56. Springer, Heidelberg (2005)
2. Berthold, M., Hand, D.J.: Intelligent Data Analysis. Springer, Heidelberg (1999)
3. Dysart, J., Lindsay, R., Hammond, R., Dupuis, P.: Mug shot exposure prior to lineup identification: Interference, transference, and commitment effects. Journal of Applied Psychology 86(6), 1280–1284 (2002)
4. Gaines, B.R., Shaw, M.L.G.: Knowledge acquisition tools based on personal construct psychology. The Knowledge Engineering Review 8, 49–85 (1993)
5. Hepting, D., Maciag, T., Spring, R., Arbuthnott, K., Ślęzak, D.: A rough sets approach for personalized support of face recognition. Rough Sets, Fuzzy Sets, Data Mining and Granular Computing, 201–208 (2009)
6. Hepting, D., Spring, R., Maciag, T., Arbuthnott, K., Ślęzak, D.: Classification of facial photograph sorting performance based on verbal descriptions. In: Rough Sets and Current Trends in Computing, pp. 570–579 (2010)
7. Hepting, D.H., Ślęzak, D., Spring, R.: Validation of Classifiers for Facial Photograph Sorting Performance. In: Kim, T.H., Pal, S.K., Grosky, W.I., Pissinou, N., Shih, T.K., Ślęzak, D. (eds.) SIP/MulGraB 2010. CCIS, vol. 123, pp. 184–195. Springer, Heidelberg (2010)

8. Hepting, D.: Ethics and usability testing in computer science education. ACM SIGCSE Bulletin 38(2), 76–80 (2006)
9. Jackiw, L.B., Arbuthnott, K.D., Pfeifer, J.E., Marcon, J.L., Meissner, C.A.: Examining the cross-race effect in lineup identification using caucasian and first nations samples. Canadian Journal of Behavioural Science/Revue Canadienne des Sciences du Comportement 40(1), 52–57 (2008)
10. Kuncheva, L.: Combining Pattern Classifiers: Methods and Algorithms. Wiley, Chichester (2004)
11. Pawlak, Z.: Rough set approach to knowledge-based decision support. European Journal of Operational Research 99, 48–57 (1997)
12. Platz, S., Hosch, H.: Cross-racial/ethnic eyewitness identification: A field study 1. Journal of Applied Social Psychology (January 1988),
http://www3.interscience.wiley.com/journal/119451843/abstract
13. Pryke, S., Lindsay, R., Pozzulo, J.: Sorting mug shots: methodological issues. Applied Cognitive Psychology 14(1), 81–96 (2000)
14. Schooler, J.W., Ohlsson, S., Brooks, K.: Thoughts beyond words: when language overshadows insight. Journal of Experimental Psychology: General 122, 166–183 (1993)
15. Steblay, N., Dysart, J., Fulero, S., Lindsay, R.: Eyewitness accuracy rates in sequential and simultaneous lineup presentations: A meta-analytic comparison. Law and Human Behavior 25(5), 459–473 (2001)
16. Wróblewski, J.: Genetic algorithms in decomposition and classification problems. In: Polkowski, L., Skowron, A. (eds.) Rough Sets in Knowledge Discovery 2: Applications, Case Studies and Software Systems, pp. 471–487. Physica-Verlag, Heidelberg (1998)
17. Wróblewski, J.: Adaptive aspects of combining approximation spaces. In: Pal, S.K., Polkowski, L., Skowron, A. (eds.) Rough–Neural Computing. Techniques for Computing for Words, pp. 139–156. Springer, Heidelberg (2004)

Evolutionary Tolerance-Based Gene Selection in Gene Expression Data

Na Jiao

Department of Information Science and Technology
East China University of Political Science and Law
Shanghai 201620, P.R. China
zdx.jn@163.com

Abstract. Gene selection is to select the most informative genes from the whole gene set, which is a key step of the discriminant analysis of microarray data. Rough set theory is an efficient mathematical tool for further reducing redundancy. The main limitation of traditional rough set theory is the lack of effective methods for dealing with real-valued data. However, gene expression data sets are always continuous. This has been addressed by employing discretization methods, which may result in information loss. This paper investigates one approach combining feature ranking together with features selection based on tolerance rough set theory. Moreover, this paper explores the other method which can utilize the information contained within the boundary region to improve classification accuracy in gene expression data. Compared with gene selection algorithm based on rough set theory, the proposed methods are more effective for selecting high discriminative genes in cancer classification.

Keywords: microarray data, gene selection, feature ranking, tolerance rough set theory, cancer classification.

1 Introduction

DNA microarray is a technology to measure the expression levels of thousands of genes, which is quite suitable for comparing the gene expression levels in tissues under different conditions, such as healthy versus diseased. Gene expression data has two major characteristics. (1) There are only very few samples of each type of cancers (usually from several to several tens). (2) There are a large number of genes (several thousands). In other words, the data is high dimensional.

Discriminant analysis of microarray data has been widely studied to assist diagnosis. Because lots of genes in the original gene set are irrelevant or even redundant for specific discriminant problem, gene selection is usually introduced to preprocess the original gene set for further analysis.

Extracting and analyzing information from gene expression data is helpful for reducing dimensionality, removing irrelevant data, improving learning accuracy, and enhancing output comprehensibility. There are two basic categories of feature selection algorithms, namely filter and wrapper models [1]. Filter methods

J.F. Peters et al. (Eds.): Transactions on Rough Sets XIV, LNCS 6600, pp. 100–118, 2011.
© Springer-Verlag Berlin Heidelberg 2011

are essentially data pre-processing or data filtering methods. Filter methods select feature subsets independently of any learning algorithm and rely on various measures of the general characteristics of the training data [2]. Simple methods based on statistical tests (T-test, F-test) have been shown to be effective [3,4]. The idea of these methods is that features are ranked and the top ones or those that satisfy a certain criterion are selected. More sophisticated methods were also developed [5]. They also have the virtue of being easily and very efficiently computed. In filter methods, the characteristics in the feature selection are uncorrelated to those of the learning methods, therefore they have better generalization property. In wrapper methods, feature selection is "wrapped" around a learning method: the usefulness of a feature is directly judged by the estimated accuracy of the learning method. One can often obtain a set with a very small number of features, which can give high accuracy, because the characteristics of the features match well with those of the learning method. Wrapper methods typically require extensive computation to search the best features [6].

Features using existing feature selection such as filter and wrapper have redundancy because genes have similar scores in similar pathways. Rough set theory can be used to eliminate such redundancy. Rough set theory [7-21], proposed by Pawlak in 1982, is widely applied in many fields of data mining such as classification and feature selection. Many researchers propose various feature selection approaches to apply rough set theory into gene selection [22-24]. Rough set theory is employed to generate reducts, which represent the minimal sets of non-redundant features capable of discerning between all objects, in a multiobjective Framework[22]. It is a rough feature selection algorithm using redundancy reduction for effective handling of high-dimensional microarray gene expression data. An initial pre-processing for redundancy reduction has been done, which aids faster convergence along the search space. Moreover, a reduction in the rows (object pairs) of the decision table was made by restricting comparisons only between objects belonging to different classes-giving the decision table. The rough set theory-based method requires the data to be of categorical type. [23] proceeds by pre-processing gene expression data. The continuous microarray gene expression values discretize using equal width method into different intervals. Positive region-based approach works on significance of attributes and its contribution for classification. The algorithm takes whole attribute set and finds the short reducts. The main contribution of [24] is the definitions on relevance of genes and interaction of genes using rough set, and the method call RMIMR (Rough Maximum Interaction-Maximum Relevance) is proposed. However, traditional rough set theory-based methods are restricted to the requirement that all data must be discrete. Existing methods [25] are to discretize the data sets and replace original data values with crisp values. This is often inadequate, as degrees of objects to the descretized values are not considered. Discretization ignores their discrimination. This may cause information loss. A better choice to solve the problem may be the use of tolerance rough set theory [26,27].

This paper presents two gene selection method based on tolerance rough set theory. By using tolerance relations, the strict requirement of complete equivalence can be relaxed, and a more flexible approach to subset selection can be developed. The first proposed method is comprised two steps. In step 1, we rank all genes with the T-test and select the most promising genes. In step 2, we apply tolerance rough set theory-based method to the selected genes in step 1. A distance metric is introduced into the second proposed algorithm, which can examine the uncertain information contained in the boundary region of tolerance rough set theory. The experimental results demonstrate that the proposed algorithms are more effective than gene selection approach based on rough set theory for achieving good classification performance.

The rest of this paper is organized as follows. Section 2 describes the relevant primary concept of rough set theory, gene expression data and T-test. Section 3 proposes tolerance rough set theory-based gene selection method. Section 4 introduces tolerance-based gene selection via an evaluation measure within the boundary region. A simple example is presented in Section 5. Section 6 presents experimental results on two benchmark gene expression data sets. Finally, Section 7 gives a summary and future research.

2   Preliminaries

In this section, we briefly introduce the basic concepts of rough set theory and T-test.

2.1   Rough Set Theory

There is a classificatory feature in gene expression data sets. We can formalize the gene expression data set into a decision system.

Definition 1. Decision table.
A decision table is defined as $T = \langle U, C \cup D, V, f \rangle$, where U is a non-empty finite set of objects; C is a set of all condition features (also called conditional attributes) and D is a set of decision features (also called decision attributes); $V = \bigcup_{a \in C \cup D} V_a$, V_a is a set of feature values of feature a; and $f : U \times (C \cup D) \to V$ is an information function for every $x \in U$, $a \in C \cup D$.

For any $B \subseteq C \cup D$, an equivalence (indiscernibility) relation induced by B on U is defined as Definition 2.

Definition 2. Equivalence relation.

$$IND(B) = \{(x, y) \in U \times U | \forall b \in B, b(x) = b(y)\}. \tag{1}$$

where $b(x) = f(x, b)(x \in U, b \in C \cup D)$.

The family of all equivalence classes of $IND(B)$, i.e., the partition induced by B, is given in Definition 3.

Definition 3. Partition.

$$U/IND(B) = \{[x]_B \,|\, x \in U\},\tag{2}$$

where $[x]_B$ is the equivalence class containing x. All the elements in $[x]_B$ are equivalent (indiscernible) with respect to B. Equivalence classes are elementary sets in rough set theory.

For any $X \subseteq U$ and $B \subseteq C$, X could be approximated by the lower and upper approximations.

Definition 4. Lower approximation and upper approximation.

$$\underline{B}X = \{x \,|\, [x]_B \subseteq X\},\tag{3}$$

$$\overline{B}X = \{x \,|\, [x]_B \cap X \neq \emptyset\}.\tag{4}$$

Empty set is also the elementary set of the approximation space.

Let $B \subseteq C$, the positive, negative and boundary regions of the partition $U/IND(D)$ with respect to B are defined as Definition 5.

Definition 5. Positive, negative and boundary regions.

$$POS_B(D) = \cup_{X \in U/IND(D)} \underline{B}X,\tag{5}$$

$$NEG_B(D) = U - \cup_{X \in U/IND(D)} \overline{B}X,\tag{6}$$

$$BND_B(D) = \cup_{X \in U/IND(D)} \overline{B}X - \cup_{X \in U/IND(D)} \underline{B}X.\tag{7}$$

The positive region is the set of all samples that can be certainly classified as belonging to the blocks of $U/IND(D)$ using B. The negative region is the set of all samples which can not be classified as belonging to the blocks of $U/IND(D)$ using B. The boundary region is the difference between the upper and lower approximation.

By employing the definition of the positive region it is possible to calculate the rough set degree of dependency of a set of features D on B.

Definition 6. Degree of dependency of feature.

$$\gamma_B(D) = |POS_B(D)| \,/\, |U|.\tag{8}$$

2.2   Microarray and Gene Expression Data

DNA is the basic material that makes up human chromosomes, which stands for deoxyribonucleic acid. DNA microarrays measure the expression of a gene in a cell by measuring the amount of fluorescently labeled mRNA (messenger ribonucleic acid) present for that gene. Microarrays (gene arrays or gene chips) are considered a breakthrough technology in biology, aiding in more accurate diagnosis, prognosis, treatment planning and facilitating the quantitative study of thousands of genes simultaneously from a single sample of cells [22,28,29].

Here is the work procession for a DNA microarray. The nucleotide sequences for a few thousand genes are printed on a glass slide by a robotic printing device. A target sample and a reference sample are labeled with the fluorescence levels (using red and green dyes), and each are hybridized with the DNA on the slide. Through fluoroscopy, the log (red/green) intensities of RNA hybridizing at each site are measured. The result is a few thousand numbers, measuring the expression level of each gene in the target relative to the reference sample. Positive values show higher expression in the target versus the reference, and vice versa for negative values.

A gene expression dataset gathers together the expression values from a series of DNA microarray experiments, with each column representing an experiment. Therefore, there are several thousand rows representing individual genes, and tens of columns representing samples: in the particular example of Figure 1 there are 6830 genes (rows) and 64 samples (columns), although for clarity only a random sample of 100 rows are shown. The figure indicates the data set as a heat map, ranging from negative (green) to positive (red). The samples are 64 cancer tumors from different patients.

2.3   T-test

Feature subset selection is an important step to narrowing down the feature number prior to data mining. T-test [30,31] is widely used. We assume that there are two classes of samples in a gene expression data set.

Definition 7. T-test.

The T-value for gene a is expressed by:

$$t(a) = \frac{\mu_1 - \mu_2}{\sqrt{\sigma_1^2/n_1 + \sigma_2^2/n_2}}, \tag{9}$$

where μ_i and σ_i are the mean and the standard deviation of the expression levels of gene a for $i = 1, 2$. When there are multiple classes of samples, the T-value is typically computed for one class versus all the other classes. The top genes ranked by T-value can be selected for data mining. Feature set so obtained has certain redundancy because genes in similar pathways probably all have very similar score. If several pathways involved in perturbation but one has main influence it is possible to describe this pathway with fewer genes, therefore feature selection based on rough set theory is used to minimize the feature set.

2.4   Gene Selection Algorithm Based on Rough Set Theory

Gene selection algorithm based on rough set theory (GSRS) for gene expression data is composed of T-test and feature selection based on rough set theory [21]. After feature ranking, top ranked genes are selected to form the feature set. The values of all continuous features are discretized. Rough set theory-based feature

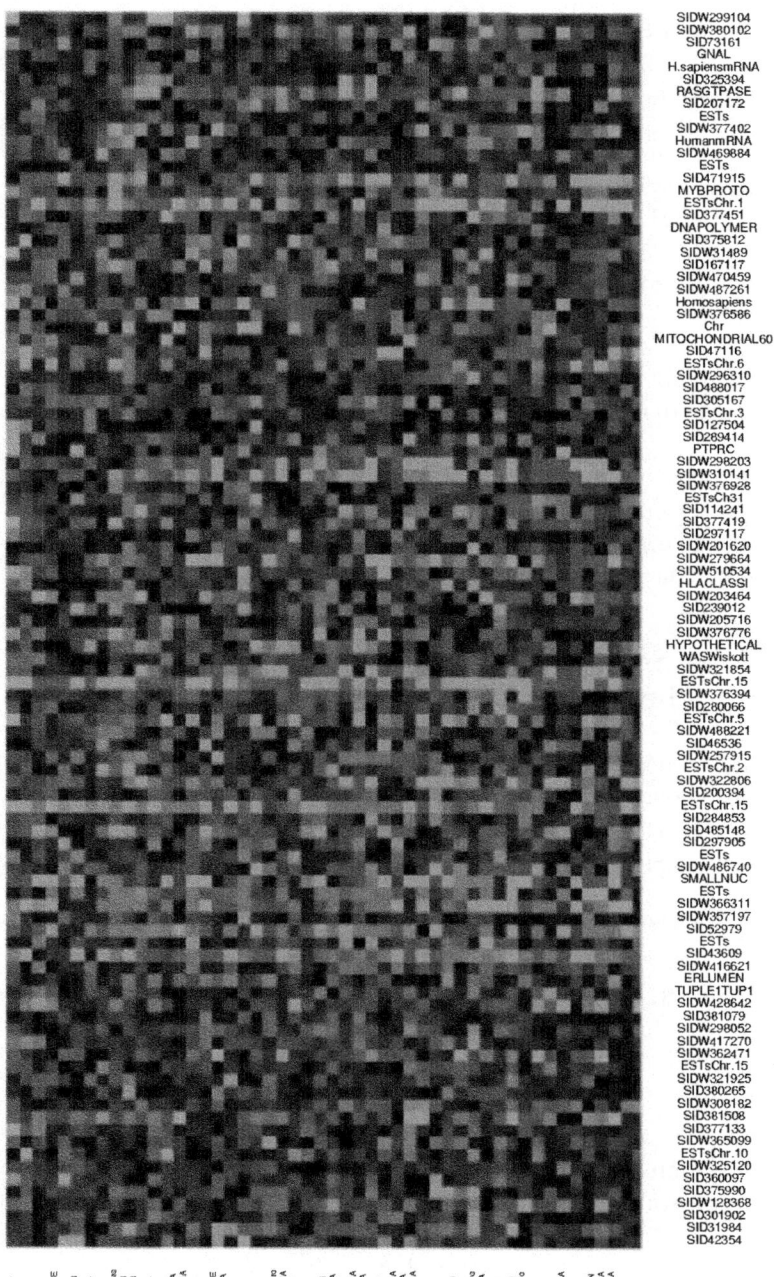

Fig. 1. DNA microarray data

selection method starts with the full set and consecutively deletes one feature at a time until we obtain a reduction.

Algorithm 1. Gene selection algorithm based on rough set theory (GSRS)

    (1) Calculate T-value of each gene, select top ranked n genes to form the feature set C

    (2) Discretize the feature set C

    (3) Set $P = C$

    (4) **do**

    (5)   **for each** $a \in P$

    (6)   **if** $\gamma_{P-\{a\}}(D) == \gamma_C(D)$

    (7) $P = P - \{a\}$

    (8) **until** $\gamma_{(P-\{a\})}(D) < \gamma_{(C)}(D)$

    (9) **return** P

We number the conditional features in P and then begin the loop. The order of the selection conditional features is on the basis of serial number. The loop continues to evaluate in the above manner by deleting conditional features, until the dependency value of the current reduct is less than that of the dataset.

3 Gene Selection Algorithm Based on Tolerance Rough Set Theory

In order to deal with real-valued data, we employ a similarity relation. This allows a relaxation in the way equivalence classes are considered.

3.1 Similarity Measures

In this approach, suitable similarity measure, given in [10,13], is described in Definition 8.

Definition 8. Similarity measure.

$$S_a(x, y) = 1 - \frac{|a(x) - a(y)|}{|a_{\max} - a_{\min}|}, \tag{10}$$

where $a(x) = f(x, a)$ $(x \in U, a \in C \cup D)$, and $a_{\max}$ and $a_{\min}$ denote the maximum and minimum values respectively for feature a. When considering more than one feature, the defined similarities must be combined to provide a measure of the overall similarity of objects. For a subset of features, B, the overall similarity measure is defined as Definition 9.

Definition 9. Overall similarity measure.

$$S_{B,\tau} = \left\{ (x, y) \,\Big|\, \frac{\sum\limits_{a \in B} S_a(x, y)}{|B|} \geq \tau \right\} \tag{11}$$

where τ is a global similarity threshold; it determines the required level of similarity for inclusion within tolerance classes. This framework allows for the specific case of traditional rough set theory by defining a suitable similarity measure and threshold ($\tau = 1$). From this, for any $B \subseteq C \cup D$, $0 < \tau \leq 1$, the so-called tolerance classes that are generated by a given similarity relation for an object are defined as Definition 10.

Definition 10. Similarity relation.

$$S_{B,\tau}(x) = \{y|x, y \in U, (x, y) \in S_{B,\tau}\}. \tag{12}$$

For any $X \subseteq U$, $B \subseteq C$ and $0 < \tau \leq 1$, lower and upper approximations are then defined in a similar way to traditional rough set theory.

Definition 11. Modified lower approximation and upper approximation.

$$\underline{B_\tau}X = \{x|S_{B,\tau}(x) \subseteq X\}, \tag{13}$$

$$\overline{B_\tau}X = \{x|S_{B,\tau}(x) \cap X \neq \emptyset\}. \tag{14}$$

The tuple $\langle \underline{B_\tau}X, \overline{B_\tau}X \rangle$ is called a tolerance-based rough set. Let $B \subseteq C$ and $0 < \tau \leq 1$, the positive, negative, boundary regions and the dependency function can be defined as follows.

Definition 12. Modified positive, negative and boundary regions.

$$POS_{B,\tau}(D) = \cup_{X \in U/S_{D,\tau}} \underline{B_\tau}X. \tag{15}$$

$$NEG_{B,\tau}(D) = U - \cup_{X \in U/S_{D,\tau}} \overline{B_\tau}X \tag{16}$$

$$BND_{B,\tau}(D) = \cup_{X \in U/S_{D,\tau}} \overline{B_\tau}X - \cup_{X \in U/S_{D,\tau}} \underline{B_\tau}X \tag{17}$$

Definition 13. Modified degree of dependency of feature.

$$\gamma_{B,\tau}(D) = |POS_{B,\tau}(D)| / |U|. \tag{18}$$

From these definitions, a feature selection method can be formulated that uses the tolerance-based degree of dependency, $\gamma_{B,\tau}(D)$, to gauge the significance of feature subsets.

3.2 Tolerance Rough Set Theory-Based Gene Selection Method

Gene selection algorithm based on tolerance rough set theory for gene expression data combines feature ranking together with feature selection based on tolerance rough set theory. Similarly, T-test can eliminate such redundant genes. T-test is used to feature ranking as the first step and select top ranked n genes to form the feature set. Tolerance rough set theory-based feature selection method can judge every feature and delete the features that are superfluous.

Algorithm 2. Gene selection algorithm based on tolerance rough set theory (GSTRS)

      (1) Calculate T-value of each gene, select top ranked n genes to form the
           feature set C
      (2) Set $P = C$
      (3) **do**
      (4)   **for each** $a \in P$
      (5)   **if** $\gamma_{P-\{a\},\tau}(D) == \gamma_{C,\tau}(D)$
      (6) $P = P - \{a\}$
      (7) **until** $\gamma_{P-\{a\},\tau}(D) < \gamma_{C,\tau}(D)$
      (8) **return** P

For each $a \in P$, the order of the selection conditional features is the same as GSRS. The stopping criteria are automatically defined through the use of the dependency measure when the deletion of further features results in a decrease in dependency. The time complexity of GSTRS is $O\left(|C|^2 * |U|^2\right)$.

4   Tolerance-Based Gene Selection via an Evaluation Measure within the Boundary Region

The information contained in the boundary region is uncertain, but it may be useful for the discovery of reducts. When only the positive region is considered and the boundary region is ignored for reducts, the useful information contained within the uncertain region is lost. [13-15] use a distance metric to determine the proximity of objects in the boundary region to those in the positive region.

4.1   Distance Metric

Considering computational complexity, the mean of all object attribute values in the already defined tolerance lower approximation is calculated.

    For any $X \subseteq U$, $B \subseteq C$ and $0 < \tau \leq 1$, the mean of modified lower approximation is defined as Definition 14.

Definition 14. Mean of modified lower approximation.

$$\underline{B_\tau}X_M = \left\{ \frac{\sum_{o \in \underline{B_\tau}X} a(o)}{|\underline{B_\tau}X|} | \forall a \in B \right\} \tag{19}$$

According to the mean of modified lower approximation, the distance function for the proximity of objects in the boundary region is formulated as Definition 15.

Definition 15. Distance function for the proximity of objects in the boundary region.

$$DIS_{B,\tau}(D) = \sum_{\mu \in BND_{B,\tau}(D)} \theta_B\left(\underline{B_\tau}X_M, \mu\right), \tag{20}$$

where $\theta_B\left(B_\tau X_M, \mu\right) = \sqrt{\sum_{a\in B} f_a\left(\underline{B_\tau}X_M, \mu\right)^2}$ is a function of Euclidean distance $(f_a\left(x, y\right) = a\left(x\right) - a\left(y\right))$. Euclidean distance is employed in the approach documented here. Euclidean distance is widely used as the distance measure. Euclidean distance is applied to measure the similarity between genes because it is effective to reflect functional similarity such as positive and negative correlation, interdependency as well as closeness in values. This function is the sum of all objects in the boundary region from the mean of modified lower approximation. It is employed to search for the best feature subset. A measure of the attribute can be obtained when an attribute is removed from the set of considered conditional attributes. The less the change in DIS is, the less important the attribute is.

4.2   Gene Selection Based on Distance Function-Assisted Tolerance Rough Set Theory

Gene selection algorithm based on tolerance rough set theory via an evaluation measure in the boundary region (GSTRSB) is similar to GSTRS algorithm. Top ranked genes are selected to form the feature set. GSTRSB can judge every feature and delete redundant features. In each loop, there may be more than one superfluous candidate. Although the deletion of these features does not result in a decrease in dependency, the minimum impact on the distance function for the proximity of objects in the boundary region from the mean of modified lower approximation can guide to discover better feature subset. All the superfluous candidates are examined by the value of DIS, the best selected superfluous feature is the one that results in the minimum difference on DIS.

Algorithm 3. Gene selection algorithm based on tolerance rough set theory via an evaluation measure in the boundary region (GSTRSB)
 (1) Calculate T-value of each gene, select top ranked n genes to form the
   feature set C
 (2) Set $P = C, T = P, best = \emptyset$
 (3) **do**
 (4) $P = T$
 (5) $T = P - \{k\}, \exists k \in P$
 (6) $best = k$
 (7) **for each** $a \in P$
 (8)   **if** $\left|DIS_{P-\{a\},\tau}\left(D\right) - DIS_{P,\tau}\left(D\right)\right| < \left|DIS_{T,\tau}\left(D\right) - DIS_{P,\tau}\left(D\right)\right|$
 (9) $T = P - \{a\}$
 (10) $best = a$
 (11) **until** $\gamma_{P-\{best\},\tau}\left(D\right) < \gamma_{C,\tau}\left(D\right)$
 (12) **return** P

  The order of the selection $a \in P$ is the same as GSTRS. The termination criteria is tolerance rough set dependency when the deletion of any single features results in a decrease in dependency.
  In the GSTRSB approach, we need consider objects in the boundary region and this inevitably adds to the computation. All of those objects in the lower

approximation are also taken into account when calculating a modified lower approximation object for each concept. At this lower level the additional factors include: the calculation of the modified lower approximation mean, the calculation of the upper approximation, and the calculation of the distances of objects in the boundary regions from the modified lower approximation mean. From a high level point-of-view the GSTRSB has a time complexity of $O\left(\left(|C|^2 + |C|\right) * |U|^2\right)$ $\approx O\left(|C|^2 * |U|^2\right)$, where $|N|$ is the number of objects and $|C|$ is the number of features.

5   A Simple Example

To illustrate the operation of feature selection algorithm based on tolerance rough set theory, it is applied to a simple example dataset in Table 1, which contains three real-valued conditional features and a crisp-valued decision feature. Set $\tau = 0.8$. $C = \{a, b, c\}$. $D = \{d\}$.

Table 1. Example dataset

Objects	a	b	c	d
1	0.3	0.4	0.2	R
2	0.3	1	0.6	A
3	0.4	0.3	0.4	R
4	0.9	0.4	0.7	R
5	0.9	0.7	0.7	A
6	1	0.4	0.7	A

The following tolerance classes are generated:

$U/S_{D,\tau} = \{\{1,3,4\}, \{2,5,6\}\}$,
$U/S_{C,\tau} = \{\{1\}, \{2\}, \{3\}, \{5\}, \{4,6\}\}$,
$U/S_{C-\{a\},\tau} = U/S_{\{b,c\},\tau} = \{\{1\}, \{2\}, \{3\}, \{5\}, \{4,6\}\}$,
$U/S_{C-\{b\},\tau} = U/S_{\{a,c\},\tau} = \{\{1\}, \{2\}, \{3\}, \{5\}, \{4,6\}\}$,
$U/S_{C-\{c\},\tau} = U/S_{\{a,b\},\tau} = \{\{1,3\}, \{4,6\}, \{2\}, \{5\}\}$,
$U/S_{C-\{a,b\},\tau} = U/S_{\{c\},\tau} = \{\{1\}, \{2\}, \{3\}, \{4,5,6\}\}$,
$U/S_{C-\{a,c\},\tau} = U/S_{\{b\},\tau} = \{\{1,3,4,6\}, \{2\}, \{5\}\}$,
$U/S_{C-\{b,c\},\tau} = U/S_{\{a\},\tau} = \{\{1,2,3\}, \{4,5,6\}\}$.

The reduct for GSTRS:
Considering feature set , the lower approximations of the decision classes are calculated as follows:

$\underline{C_\tau}\{1,3,4\} = \underline{\{a,b,c\}_\tau}\{1,3,4\} = \{x|S_{\{a,b,c\},\tau}(x) \subseteq \{1,3,4\}\} = \{1,3\}$,
$\underline{C_\tau}\{2,5,6\} = \underline{\{a,b,c\}_\tau}\{2,5,6\} = \{x|S_{\{a,b,c\},\tau}(x) \subseteq \{2,5,6\}\} = \{2,5\}$.

Hence, the positive region can be constructed:

$$POS_{C,\tau}(D) = \cup_{X \in U/S_{D,\tau}} \underline{C_\tau} X = \underline{C_\tau} \{1,3,4\} \cup \underline{C_\tau} \{2,5,6\} = \{1,2,3,5\}.$$

The resulting degree of dependency is:

$$\gamma_{C,\tau}(D) = \frac{|POS_{C,\tau}(D)|}{|U|} = \frac{|\{1,2,3,5\}|}{|\{1,2,3,4,5,6\}|} = \frac{4}{6}.$$

For feature set $C - \{a\}$, the corresponding dependency degree is:

$$\gamma_{C-\{a\},\tau}(D) = \frac{|POS_{C-\{a\},\tau}(D)|}{|U|} = \frac{|\{1,2,3,5\}|}{|\{1,2,3,4,5,6\}|} = \frac{4}{6},$$
$$\gamma_{C-\{a\},\tau}(D) = \gamma_{\{b,c\},\tau}(D) = \gamma_{C,\tau}(D) = \frac{4}{6}.$$

Feature a is deleted from feature set C. Similarly, the dependency degree of feature set $\{b,c\} - \{b\}$ is:

$$\gamma_{\{b,c\}-\{b\},\tau}(D) = \frac{|POS_{\{b,c\}-\{b\},\tau}(D)|}{|U|} = \frac{|\{1,2,3\}|}{|\{1,2,3,4,5,6\}|} = \frac{3}{6},$$
$$\gamma_{\{b,c\}-\{b\},\tau}(D) = \frac{3}{6} < \gamma_{C,\tau}(D) = \frac{4}{6}.$$

The reduct for the GSTRS algorithm is $\{b,c\}$.

The reduct for GSTRSB:
The positive and boundary regions can be constructed:

$$POS_{C,\tau}(D) = \cup_{X \in U/S_{D,\tau}} \underline{C_\tau} X = \underline{C_\tau} \{1,3,4\} \cup \underline{C_\tau} \{2,5,6\} = \{1,2,3,5\}.$$
$$BND_{C,\tau}(D) = \cup_{X \in U/S_{D,\tau}} \overline{C_\tau} X - \cup_{X \in U/S_{D,\tau}} \underline{C_\tau} X = \overline{C_\tau} \{1,3,4\} \cup \overline{C_\tau} \{2,5,6\} -$$
$$\{1,2,3,5\} = \{1,2,3,4,5,6\} - \{1,2,3,5\} = \{4,6\}.$$

In order to calculate the distances of the boundary objects from the lower approximation, it is necessary to achieve the mean of modified lower approximation.

$$\underline{C_\tau} X_M = \left\{ \frac{\sum_{o \in \underline{C_\tau} X} a(o)}{|\underline{C_\tau} X|} | \forall a \in C \right\}$$
$$= \left\{ \frac{\sum a(1),a(2),a(3),a(5)}{|4|} : \frac{\sum b(1),b(2),b(3),b(5)}{|4|} : \frac{\sum c(1),c(2),c(3),c(5)}{|4|} \right\}$$
$$= \left\{ \frac{0.3+0.3+0.4+0.9}{|4|} : \frac{0.4+1+0.3+0.7}{|4|} : \frac{0.2+0.6+0.4+0.7}{|4|} \right\}$$
$$= \{0.475 : 0.6 : 0.475\}$$

The distances of all of the objects in the boundary region from the mean of modified lower approximation are:

$$obj4 \sqrt{f_a(\underline{B_\tau} X_M, 4)^2 + f_b(\underline{B_\tau} X_M, 4)^2 + f_c(\underline{B_\tau} X_M, 4)^2}$$
$$= \sqrt{(0.475 - 0.9)^2 + (0.6 - 0.4)^2 + (0.475 - 0.7)^2} = 0.52,$$
$$obj6 \sqrt{f_a(\underline{B_\tau} X_M, 6)^2 + f_b(\underline{B_\tau} X_M, 6)^2 + f_c(\underline{B_\tau} X_M, 6)^2}$$
$$= \sqrt{(0.475 - 1)^2 + (0.6 - 0.4)^2 + (0.475 - 0.7)^2} = 0.605.$$

$$DISC_{C,\tau}(D) = \sum_{\mu \in BND_{C,\tau}(D)} \theta_C\left(C_\tau X_M, \mu\right) = 0.52 + 0.605 = 1.125.$$

Similarly, $DISC_{C-\{a\},\tau}(D) = 1.66$, $DISC_{C-\{b\},\tau}(D) = 1.05$, $DISC_{C-\{c\},\tau}(D) = 0.95$.

$$\left|DISC_{C-\{a\},\tau}(D) - DISC_{C,\tau}(D)\right| = |1.66 - 1.125| = 0.535,$$
$$\left|DISC_{C-\{b\},\tau}(D) - DISC_{C,\tau}(D)\right| = |1.05 - 1.125| = 0.075,$$
$$\left|DISC_{C-\{c\},\tau}(D) - DISC_{C,\tau}(D)\right| = |0.95 - 1.125| = 0.175.$$
$best = b.$

The resulting degree of dependency is:

$$\gamma_{C,\tau}(D) = \frac{|POS_{C,\tau}(D)|}{|U|} = \frac{|\{1,2,3,5\}|}{|\{1,2,3,4,5,6\}|} = \frac{4}{6}.$$

For feature set $C - \{b\}$, the corresponding dependency degree is:

$$\gamma_{C-\{b\},\tau}(D) = \frac{|POS_{C-\{b\},\tau}(D)|}{|U|} = \frac{|\{1,2,3,5\}|}{|\{1,2,3,4,5,6\}|} = \frac{4}{6},$$
$$\gamma_{C-\{b\},\tau}(D) = \gamma_{\{a,c\},\tau}(D) = \gamma_{C,\tau}(D) = \frac{4}{6}.$$

Feature b is deleted from feature set C.

$$DIS_{\{a,c\}-\{a\},\tau}(D) = 1.2,$$
$$DIS_{\{a,c\}-\{c\},\tau}(D) = 0,$$
$$\left|DIS_{\{a,c\}-\{a\},\tau}(D) - DIS_{\{a,c\},\tau}(D)\right| = |1.2 - 1.05| = 0.15,$$
$$\left|DIS_{\{a,c\}-\{c\},\tau}(D) - DIS_{\{a,c\},\tau}(D)\right| = |0 - 1.05| = 1.05,$$
$best = a.$

Similarly, the dependency degree of feature set $\{a,c\} - \{a\}$ is:

$$\gamma_{\{a,c\}-\{a\},\tau}(D) = \frac{|POS_{\{a,c\}-\{a\},\tau}(D)|}{|U|} = \frac{|\{1,2,3\}|}{|\{1,2,3,4,5,6\}|} = \frac{3}{6},$$
$$\gamma_{\{a,c\}-\{a\},\tau}(D) = \frac{3}{6} < \gamma_{C,\tau}(D) = \frac{4}{6}.$$

Therefore, the GSTRSB algorithm terminates and outputs the reduct $\{a,c\}$.

6   Experiments

To evaluate the performance of the proposed algorithms, we applied them to four benchmark gene expression data sets: Lymphoma data set (http://llmpp.nih.gov /lymphoma), Liver cancer data set (http://genome-www.stanford.edu/hcc/), Colon cancer data set (http://microarray.princeton.edu/oncology) and Leukemia data set (http://www.broad.mit.edu/MPR). The Lymphoma data set is a collection of 96 samples. There are 42 B-cell and 54 Other type samples having 4026 genes. The Liver cancer data set is a collection of gene expression measurements from 156 samples and 1648 genes. There are 82 cases of HCCs and 74 cases of nontumor livers. The Colon cancer data set consists of 62 samples and 2000 genes, and the samples are composed of 40 tumor biopsies collected from tumors and 22 normal biopsies collected from the healthy part of the colons of the same patient. The Leukemia data set consists of 72 samples and 7129 genes,

Table 2. Details of the microarray data

Data sets	Genes	Classes	Samples
Lymphoma	4026	B-cell	42
		Other type	54
Liver cancer	1648	HCCs	82
		nontumor livers	74
Leukemia	7129	ALL	47
		AML	25
Colon cancer	2000	Colon cancer	40
		Normal	22

including 25 cases of acute myeloblastic leukemia (AML) and 47 cases of acute lymphoblastic leukemia (ALL). The data sets are shown in Table 2.

GSRS, GSTRS and GSTRSB are run on the four data sets. Firstly, T-test is employed as a filter on Lymphoma, Liver cancer, Leukemia and Colon cancer. The top ranked 50 genes are selected. When there are missing values in data sets, these values are filled with mean values for continuous features and majority values for nominal features [32]. As four data sets are real-valued, for GSRS algorithm, discretization of every feature of the four data sets is Equal Frequency per Interval [25]. For GSTRS and GSTRSB algorithms, set τ =0.85, 0.9 and 0.95. A range of three tolerance values (0.85-0.95 in intervals of 0.05) were employed when considering the datasets. It should be considered that the ideal tolerance value for any given dataset can only be optimised for that dataset by repeated experimentation. The range of values chosen is an attempt to demonstrate the ideal tolerance threshold for each dataset without exhaustive experimentation. The reduction results and running time are listed in Table 3, 4 and 5 respectively.

Two factors need to be considered for comparing GSRS, GSTRS and GSTRSB. One is the number of selected genes. Table 3, Table 4 and Table 5 show that GSRS, GSTRS and GSTRSB can select a better small set of discriminative genes on four gene expression data sets. The number of selected features by GSRS is generally more than that by GSTRS and GSTRSB. GSTRSB can find the number of selected features that are equal and sometimes smaller than that by GSTRS. The results illustrate that tolerance rough set theory methods can avoid information loss and GSTRSB can find information contained in the boundary region. At the

Table 3. Reduction results and running time (τ =0.85)

Data sets	Reduction results			Running time(S)		
	GSRS	GSTRS	GSTRSB	GSRS	GSTRS	GSTRSB
Lymphoma	5	5	5	65.02	68.62	81.32
Liver cancer	4	4	4	23.12	28.49	36.74
Leukemia	6	6	6	114.18	131.43	165.23
Coloncancer	4	4	4	25.66	26.74	28.97

same time, by Table 3,4,5, we can also observe that GSRS algorithm has a higher computational performance as compared to GSTRS and GSTRSB algorithms. Running time is not our considered factor since our objective is not to improve the computational efficiency.

The other considered factor is classification accuracy of the selected genes of four data sets. Two classifers, C5.0 and KNN, are respectively adopted. The powerful KNN classifier was employed. In the present study, the C5.0 has been implemented for performance evaluation of the proposed methods. These are briefly outlined below.

KNN computes minimum distance from instances or prototypes.

C5.0 creates decision trees by choosing the most informative features. Leaf nodes represent classifications while branches show the related features leading to the classifications.

Table 4. Reduction results and running time (τ =0.9)

Data sets	Reduction results			Running time(S)		
	GSRS	GSTRS	GSTRSB	GSRS	GSTRS	GSTRSB
Lymphoma	7	7	7	66.31	70.45	80.56
Liver cancer	6	6	6	25.06	27.53	37.82
Leukemia	6	6	5	110.25	143.22	178.73
Coloncancer	6	5	5	24.79	28.67	30.11

Table 5. Reduction results and running time (τ =0.95)

Data sets	Reduction results			Running time(S)		
	GSRS	GSTRS	GSTRSB	GSRS	GSTRS	GSTRSB
Lymphoma	9	8	8	63.15	64.83	77.21
Liver cancer	7	6	6	24.33	31.17	39.44
Leukemia	8	8	6	116.32	138.34	171.11
Coloncancer	5	5	4	26.42	27.31	30.26

As there are a relatively small number of samples, leave-one-out accuracy is adopted. The results are shown in Table 6-11 respectively. For different values of k, the results of KNN are also presented in the table 6, 8, 10.

Experimental results show the selected genes by tolerance rough set theory-based approach have higher classification accuracy than the selected genes by rough set theory-based method. GSTRSB outperforms GSTRS in some of the examples. The reason may be that tolerance rough set theory method can retain the information hidden in the data and improve classification accuracy. Moreover, more optimal results may be obtained by the discovery of information within the boundary region.

Table 6. Classification accuracy using KNN classifier (τ =0.85)

Data sets	KNN								
	k=3			k=5			k=7		
	GSRS	GSTRS	GSTRSB	GSRS	GSTRS	GSTRSB	GSRS	GSTRS	GSTRSB
Lymphoma	90.2%	92.4%	93.1%	90.2%	92.4%	93.1%	92.2%	93.5%	94.8%
Liver cancer	85.1%	89.3%	89.3%	85.1%	89.3%	89.3%	85.1%	89.3%	89.3%
Leukemia	78.5%	82.4%	82.4%	77.4%	79.3%	79.3%	77.4%	79.3%	79.3%
Coloncancer	72.3%	73.6%	75.5%	74.2%	78.8%	79.7%	70.2%	72.5%	73.4%

Table 7. Classification accuracy using C5.0 classifier (τ =0.85)

Data sets	C5.0		
	GSRS	GSTRS	GSTRSB
Lymphoma	92.6%	93.8%	95.5%
Liver cancer	84.6%	87.7%	87.7%
Leukemia	78.5%	80.1%	81.4%
Coloncancer	75.7%	78.3%	78.3%

Table 8. Classification accuracy using KNN classifier (τ =0.9)

Data sets	KNN								
	k=3			k=5			k=7		
	GSRS	GSTRS	GSTRSB	GSRS	GSTRS	GSTRSB	GSRS	GSTRS	GSTRSB
Lymphoma	93.5%	94.8%	94.8%	93.5%	94.8%	94.8%	95.5%	96.2%	96.2%
Liver cancer	89.6%	92.5%	92.5%	89.6%	92.5%	92.5%	89.6%	92.5%	92.5%
Leukemia	86.3%	88.9%	89.3%	82.7%	85.2%	87.6%	80.1%	85.8%	86.4%
Coloncancer	73.7%	74.3%	78.2%	79.4%	81.3%	82.9%	79.4%	81.3%	82.9%

Table 9. Classification accuracy using C5.0 classifier (τ =0.9)

Data sets	C5.0		
	GSRS	GSTRS	GSTRSB
Lymphoma	95.2%	97.4%	97.4%
Liver cancer	91.3%	94.3%	94.3%
Leukemia	86.1%	88.2%	87.1%
Coloncancer	81.4%	83.6%	85.5%

Table 10. Classification accuracy using KNN classifier (τ =0.95)

Data sets	KNN								
	k=3			k=5			k=7		
	GSRS	GSTRS	GSTRSB	GSRS	GSTRS	GSTRSB	GSRS	GSTRS	GSTRSB
Lymphoma	86.5%	88.3%	87.7%	88.4%	90.3%	89.1%	90.8%	93.3%	93.9%
Liver cancer	83.7%	87.3%	87.3%	83.7%	87.3%	87.3%	83.7%	87.3%	87.3%
Leukemia	76.3%	78.5%	80.6%	74.6%	76.8%	75.4%	72.2%	74.3%	75.1%
Coloncancer	69.5%	70.8%	72.6%	72.5%	74.3%	75.8%	73.4%	75.8%	76.9%

Table 11. Classification accuracy using C5.0 classifier (τ =0.95)

Data sets	C5.0		
	GSRS	GSTRS	GSTRSB
Lymphoma	89.9%	91.1%	93.6%
Liver cancer	82.2%	85.6%	85.6%
Leukemia	75.3%	77.4%	76.3%
Coloncancer	73.3%	74.1%	74.9%

7   Conclusions

In this paper, we address gene selection of tolerance rough set theory. By constructing an example, we show how the technique works. This paper extends the research of traditional rough set theory and establishes one direction for seeking efficient algorithms for gene expression data. Our methods are applied to the gene selection of cancer classification. Experimental results show their validity. The further work is to investigate other application areas of the described methods.

References

1. Ding, C., Peng, H.C.: Minimum Redundancy Feature Selection from Microarray Gene Expression Data. In: 2nd IEEE Computer Society Bioinformatics Conference (CSB 2003), pp. 523–529 (2003)
2. Tibshirani, R., Hastie, T., Narashiman, B., Chu, G.: Diagnosis of Multiple Cancer Types by Shrunken Centroids of Gene Expression. Nat'l Academy of Sciences, 6567–6572 (2002)
3. Dudoit, S., Fridlyand, J., Speed, T.: Comparison of discrimination methods for the classification of tumors using gene expression data. Tech. Report 576, Dept of Statistics, UC Berkeley (2000)
4. Model, F., Adorján, P., Olek, A., Piepenbrock, C.: Feature selection for DNA methylation based cancer classification. Bioinformatics 17, 157–164 (2001)
5. Ben-Dor, A., Bruhn, L., Friedman, N., Nachman, I., Schummer, M., Yakhini, Z.: Tissue classification with gene expression profiles. J. Comput. Biol. 7, 559–584 (2000)

6. Kohavi, R., John, G.H.: Wrappers for Feature Subset Selection. Artificial Intelligence, 273–324 (1997)
7. Pawlak, Z.: Rough Sets. International Journal of Information Computer Science 11(5), 341–356 (1982)
8. Greco, S., Inuiguchi, M., Slowinski, R.: Fuzzy Rough Sets and Multiple-Premise Gradual Decision Rules. International Journal of Approximate Reasoning 41(2), 179–211 (2006)
9. Dubois, D., Prade, H.: Putting Rough Sets and Fuzzy Sets Together. In: Intelligent Decision Support, pp. 203–232 (1992)
10. Jensen, R., Shen, Q.: Tolerance-Based and Fuzzy-Rough Feature Selection. In: Proceedings of the 16th International Conference on Fuzzy Systems (FUZZ- IEEE 2007), pp. 877–882 (2007)
11. Yang, M., Yang, P.: A Novel Condensing Tree Structure for Rough Set Feature Selection. Neurocomputing 71, 1092–1100 (2008)
12. Hu, Q.H., Liu, J.F., Yu, D.: Mixed Feature Selection Based on Granulation and Approximation. Knowledge-Based Systems 21, 294–304 (2008)
13. Parthaláin, N.M., Shen, Q.: Exploring The Boundary Region of Tolerance Rough Sets for Feature Selection. Pattern Recognition 42, 655–667 (2009)
14. Parthaláin, N.M., Shen, Q., Jensen, R.: Distance measure assisted rough set feature selection. In: Proceedings of the 16th International Conference on Fuzzy Systems (FUZZ-IEEE 2007), pp. 1084–1089 (2007)
15. Slezak, D.: Various approaches to reasoning with frequency based decision reducts: a survey. In: Polkowski, L., Tsumoto, S., Lin, T.Y. (eds.) Rough Set Methods and Applications, pp. 235–285. Physica-Verlag, Heidelberg (2000)
16. Yao, Y.Y., Zhao, Y.: Discernibility Matrix Simplification for Constructing Attribute Reducts. Information Sciences 179, 867–882 (2009)
17. Miao, D.Q., Wang, J.: Information-Based Algorithm for Reduction of Knowledge. In: IEEE International Conference on Intelligent Processing Systems, pp. 1155–1158 (1997)
18. Yang, X.B., Xie, J., Song, X.N., Yang, J.Y.: Credible Rules in Incomplete Decision System Based on Descriptors. Knowledge-Based Systems 22, 8–17 (2009)
19. Shen, Q., Chouchoulas, A.: A Rough-Fuzzy Approach for Generating Classification Rules. Pattern Recognition 35, 2425–2438 (2002)
20. Qian, Y.H., Dang, C.Y., Liang, J.Y., Zhang, H.Y., Ma, J.M.: On The Evaluation of The Decision Performance of An Incomplete Decision Table. Data & Knowledge Engineering 65, 373–400 (2008)
21. Wang, G.Y.: Rough Set Theory and Knowledge Acquisition. Xi'an Jiaotong University Press, Xi'an (2001)
22. Banerjee, M., Mitra, S., Banka, H.: Evolutinary-Rough Feature Selection in Gene Expression Data. IEEE Transaction on Systems, Man, and Cyberneticd, Part C: Application and Reviews 37, 622–632 (2007)
23. Momin, B.F., Mitra, S., Datta Gupta, R.: Reduct Generation and Classification of Gene Expression Data. In: Proceeding of First International Conference on Hybrid Information Technology (ICHICT 2006), New York, pp. 699–708 (2006)
24. Li, D.F., Zhang, W.: Gene Selection Using Rough Set Theory. In: Wang, G.-Y., Peters, J.F., Skowron, A., Yao, Y. (eds.) RSKT 2006. LNCS (LNAI), vol. 4062, pp. 778–785. Springer, Heidelberg (2006)
25. Grzymala-Busse, J.W.: Discretization of Numerical Attributes. In: Klosgen, W., Zytkow, J. (eds.) Handbook of Data Mining and Knowledge Discovery, pp. 218–225. Oxford University Press, Oxford (2002)

26. Skowron, A., Stepaniuk, J.: Tolerance approximation spaces. Fundamenta Informaticae 27, 245–253 (1996)
27. Slowinski, R., Vanderpooten, D.: A generalized definition of rough approximations based on similarity. IEEE Trans. on Knowl. and Data Eng. 12, 331–336 (2000)
28. Special issue on bioinformatics. IEEE Comput. 35(7) (2002)
29. Hastie, T., Tibshirani, R., Friedman, J.: The elements of statistical learning: data mining, Inference, and Prediction. Springer, Heidelberg (2001)
30. Golub, T.R., Slonim, D.K., Tamayo, P., Huard, C., Gaasenbeek, M., Mesirov, J.P., Coller, H., Loh, M.L., Downing, J.R., Caligiuri, M.A., Bloomfield, C.D., Lander, E.S.: Molecular Classification of Cancer: Class Discovery and Class Prediction by Gene Expression Monitoring. Science 286, 531–537 (1999)
31. Wang, L.P., Feng, C., Xie, X.: Accurate Cancer Classification Using Expressions of Very Few Genes. IEEE/ACM Transactions on Computational Biology and Bioinformatics 4, 40–53 (2007)
32. Grzymala-Busse, J.W., Grzymala-Busse, W.J.: Handling Missing Attribute Values. In: Maimon, O., Rokach, L. (eds.) Handbook of Data Mining and Knowledge Discovery, pp. 37–57 (2005)

New Approach in Defining Rough Approximations

E.K.R. Nagarajan and D. Umadevi⋆

Centre for Research and Post Graduate Studies in Mathematics,
Ayya Nadar Janaki Ammal College (Autonomous),
Sivakasi - 626 124, Tamil Nadu, India
nagarajanekr@yahoo.com, umavasanthy@yahoo.com

Abstract. In this paper, we discuss some algebraic structures of the set of all lower($B^{\blacktriangledown}$) and upper($B^{\blacktriangle}$) approximations defined through the basic map φ in the general setting of complete atomic Boolean lattice B. In fact, we prove that if φ is extensive and closed, then $B^{\blacktriangledown}$ and $B^{\blacktriangle}$ are algebraic completely distributive lattices. A representation theorem for algebraic completely distributive lattices in terms of $B^{\blacktriangledown}$ under the existence of an extensive and closed map φ is given. We also prove that if φ is extensive and symmetric, then $B^{\blacktriangledown}$ and $B^{\blacktriangle}$ are complete ortholattices. A representation theorem for complete ortholattices in terms of $B^{\blacktriangle}$ under the existence of an extensive and symmetric map φ is shown. Further, we define a map $\langle\varphi\rangle$ induced from the basic map φ and study the properties of the rough approximations defined with respect to $\langle\varphi\rangle$ under various conditions on the basic map φ.

Keywords: Rough approximations, complete atomic Boolean lattice, linearly ordered set, algebraic completely distributive lattice, complete ortholattice.

1   Introduction

Rough set theory proposed by Pawlak [22] is an approach to handle vague concepts. The two basic philosophical assumptions by Pawlak in rough set theory are that to each object of the considered universe there is associated a set of information describing their attributes and classification of the observed objects based on their attribute properties is the fundamental source of our knowledge about them. Information about the observed objects are stored in the form of an information table represented by an information system. An information system is a pair (U, A), where U is the universe (set of objects) and A is the set of all attributes describing the properties of the objects. Each attribute $a \in A$ is a map $a : U \rightarrow V_a$, where each V_a consists of values of the attribute a. Objects in the universe are organized into disjoint classes in such a way that each class contains objects that are indiscernible by means of same set of their attribute properties.

⋆ Research work of the authors was supported by University Grants Commission under the Major Research Project No. F.No.-33- 100/2007 (SR).

These classes are called elementary sets which form the basic knowledge granules of the universe. Knowledge obtained through the classification is represented by an indiscernibility relation. That is, to every subset B of A, there is associated an indiscernibility relation I_B as follows:

$$I_B = \{(x, y) \in U \times U / \; for \; all \; a \in B, a(x) = a(y)\}$$

where $a(x)$ is the attribute value of the object x with respect to the attribute a. Usually the indiscernibility relation is an equivalence relation and their equivalence classes are the basic knowledge granules. Indiscernibility relation generated in this way is the mathematical basis for rough set theory. Hence the information about the objects in the considered universe represented by an information system is analyzed by means of an approximation space consisting of the universe and an indiscernibility relation. Generally an approximation space is a pair (U, E), where E is an equivalence relation on a non - empty set U and E is assumed to be an indiscernibility relation derived from an information system.

When the information about the objects of U is complete, E becomes an identity relation. Then the equivalence classes reduces to singletons. That is, each object in U can be certainly and uniquely characterized. When the information about the objects of U is totally insufficient, E becomes $U \times U$. That is, we cannot distinguish objects. In between these two cases, when the information about the objects of U is incomplete, there exist some distinct objects which cannot be distinguished with the available knowledge. Thus they cannot be uniquely characterized. Pawlak's idea is to roughly approximate those vague concepts by two precise concepts based on certainty and possibility. These rough approximations are usually called lower and upper approximations respectively. In rough set theory, concepts are represented by subsets of the universe. The lower and upper approximations of a subset X of U is defined as follows:

$$X^{\blacktriangledown} = \{x \in U / [x]_E \subseteq X\} \; and$$
$$X^{\blacktriangle} = \{x \in U / [x]_E \cap X \neq \phi\} \qquad (1)$$

where $[x]_E$ is an equivalence class containing x. Two subsets X and Y in an approximation space (U, E) are said to be roughly equal ($\equiv$) if their corresponding lower and upper approximations are equal. This rough equality is an equivalence relation on subsets of U. The equivalence classes $\{[X]_{\equiv} / X \subseteq U\}$ of the rough equality relation are called rough sets. Each element in the same rough set looks the same, when observed through the knowledge given by an indiscernibility relation E.

Classical sets are characterized by characteristic functions and fuzzy sets are characterized by membership functions. Similarly rough sets are characterized by their lower and upper approximations. Iwinski [12] simply represented the rough set $[X]_{\equiv}$ by their lower and upper approximations as an ordered pair $(\underline{X}, \overline{X})$. The collection of all rough sets of an approximation space is called a rough sets system. He used an ordering relation defined componentwise on the rough sets system of an approximation space (U, E) and obtained a lattice structure. Subsequentely, many researchers have discussed the structure of rough sets system and obtained many representation theorems for it. One can find those structures in the following literature [3,4,5,10,12,20,21,23,24].

Information systems may be non - deterministic or with some attribute values missing for more complex real time problems. In such cases, the indiscernibility relation need not be an equivalence relation. Indiscernibility relation may vary with the values of the attributes in an information system. Different types of information systems and their indiscernibility relations were discussed in detail in [18,19]. To analyze such type of information systems, the notion of generalized approximation space (U, R) was introduced by relaxing equivalence relation to an arbitrary binary relation R and many authors have studied about it in [25,27,28,29,30,31]. In the generalized approximation space (U, R), the lower and upper approximations are usually defined as follows : for any $X \subseteq U$,

$$X^{\blacktriangledown} = \{x \in U/R(x) \subseteq X\} \ and$$
$$X^{\blacktriangle} = \{x \in U/R(x) \cap X \neq \phi\} \tag{2}$$

where $R(x) = \{y \in U/xRy\}$.

A research in this direction is extended by considering the lower and upper approximations as rough approximation operators and defining those operators in abstract algebraic structures by various approximable substructures as a generalization of Pawlak's rough approximation operators. Some of the literature in this direction are [6,7,8,9,13,14]. In this paper, we consider one such abstract generalized rough approximation operators defined by Jarvinen which is suitable for all the rough approximation operators based on binary relations.

In [13], to study the rough approximations based on indiscernibility relations which are not necessarily reflexive, symmetric and transitive, Jarvinen defined in a lattice theoretical setting two maps which mimic the rough approximation operators. In a complete atomic Boolean lattice $(B, \leq)$, the rough approximations $x^{\blacktriangledown}$ and $x^{\blacktriangle}$ for any $x \in B$, are defined using a basic map $\varphi : \mathcal{A}(B) \to B$, where $\mathcal{A}(B)$ is the set of all atoms of B. He denoted the set of all lower and upper approximations of elements of the complete atomic Boolean lattice B as $B^{\blacktriangledown}$ and $B^{\blacktriangle}$ respectively. He showed that the sets $B^{\blacktriangledown}$ and $B^{\blacktriangle}$ together with the ordering $\leq$ induced from the complete atomic Boolean lattice $(B, \leq)$ form complete lattices. The ordered sets $(B^{\blacktriangledown}, \leq)$ and $(B^{\blacktriangle}, \leq)$ are distributive lattices if φ is extensive and closed. $(B^{\blacktriangledown}, \leq)$ and $(B^{\blacktriangle}, \leq)$ are complete atomic Boolean lattices if the map φ is symmetric and closed and are equal complete atomic Boolean lattices in the sense $B^{\blacktriangledown} = B^{\blacktriangle}$ if the map φ is extensive, symmetric and closed. Further the example 3.16 of [13] shows that $(B^{\blacktriangledown}, \leq)$ and $(B^{\blacktriangle}, \leq)$ are not necessarily distributive if φ is extensive and symmetric. He also extended this work in his paper [14]. This paper is an extended and revised version of the presentation [16] made by the authors in the Twelfth International Conference on Rough Sets, Fuzzy Sets, Data Mining and Granular Computing 2009 at IIT Delhi, New Delhi.

The organization of this paper is as follows: The following section contains some basic definitions and results from lattice theory and from papers [13,14]. In section 3, we give a necessary and sufficient condition for the ordered sets

$(B^{\blacktriangledown}, \leq)$ and $(B^{\blacktriangle}, \leq)$ to be linearly ordered. Then we define the join($\sqcup$) and the meet($\sqcap$) operations on the ordered sets $(B^{\blacktriangledown}, \leq)$ and $(B^{\blacktriangle}, \leq)$. This algebraic approach helps us to define some additional binary and unary operations in $B^{\blacktriangledown}$ and $B^{\blacktriangle}$ under various conditions on φ. As $(B^{\blacktriangledown}, \leq)$ and $(B^{\blacktriangle}, \leq)$ are dually isomorphic, we discuss those operations only for the algebraic structure $(B^{\blacktriangle}, \sqcup, \sqcap)$. Two algebraic structures of $B^{\blacktriangledown}$ and $B^{\blacktriangle}$ under two different conditions on the basic map φ and their representations in terms of $B^{\blacktriangledown}$ and $B^{\blacktriangle}$ respectively are newly included in this extended version. Further in section 4, in order to lift the definition of rough approximations defined in [1,2] to the general setting of complete atomic Boolean lattice, we define a new map $\langle \varphi \rangle$ induced from the basic map φ and study the properties of the rough approximation operators defined with respect to the map $\langle \varphi \rangle$ under various conditions on the basic map φ. A study with an example to compare the considered two basic knowledge granules in an information system is newly included in section 4. Finally, conclusion of the paper is given in section 5.

2   Preliminaries

An ordered set $(P, \leq)$ is linearly ordered if for every $x, y \in P$, either $x \leq y$ or $y \leq x$. A map $f : P \to P$ is said to be extensive, if $x \leq f(x)$, for all $x \in P$. The map f is order - preserving if $x \leq y$ implies $f(x) \leq f(y)$. The map $f : P \to P$ is said to be idempotent if $f(f(x)) = f(x)$. A closure operator on an ordered set P is an idempotent, extensive and order-preserving self-map. A self-map f is called topological closure operator (also called Kuratowski closure operator) on a complete lattice L if it is idempotent, extensive and satisfies $f(0) = 0$ and $f(a \vee b) = f(a) \vee f(b)$ for all $a, b \in L$. Further, if f is a closure operator and a complete join-morphism on a complete lattice L, then f is called Alexandrov closure operator on L. The dual of Alexandrov closure operator is Alexandrov interior operator

Definition 1. *A bounded lattice* $(L, \vee, \wedge, 0, 1)$ *is said to be a complemented lattice if for each element* $x \in L$, *there exists an element* $y \in L$ *such that* $x \vee y = 1$, $x \wedge y = 0$. *The complement of* x *is denoted by* x'.

Definition 2. *A bounded lattice* $(L, \vee, \wedge, 0, 1)$ *is said to be an ortholattice if there exists a unary operation* $' : L \to L$ *satisfying the following conditions*

1. $x \vee x' = 1$, $x \wedge x' = 0$
2. $x \leq y \Rightarrow y' \leq x'$
3. $x'' = x$.

Definition 3. *A lattice* $(L, \vee, \wedge)$ *is distributive if one of the following laws holds for any* $x, y, z \in L$

$$x \wedge (y \vee z) = (x \wedge y) \vee (x \wedge z)$$
$$x \vee (y \wedge z) = (x \vee y) \wedge (x \vee z)$$

Definition 4. *A complemented distributive lattice is a Boolean algebra.*

Definition 5. *An algebra* $(L, \vee, \wedge, \rightarrow, \leftarrow, 0, 1)$ *is a bi - Heyting algebra if* $(L, \vee, \wedge, 0, 1)$ *is a bounded distributive lattice and is closed under* $\rightarrow, \leftarrow$, *where for each* $x, y \in L$, $x \rightarrow y$ *is the greatest element* $a \in L$, *such that* $x \wedge a \leq y$ *and* $x \leftarrow y$ *is the least element* $b \in L$, *such that* $x \leq y \vee b$.

In the above definition, $x \rightarrow y$ is called the relative pseudocomplement of x with respect to y and $x \leftarrow y$ is called the dual relative pseudocomplement of x with respect to y. $x \rightarrow 0$ is the pseudocomplement of x in L and $1 \leftarrow x$ is the dual pseudocomplement of x in L.

Definition 6. *A lattice L is completely distributive if it satisfies the following conditions :*

$$\bigwedge_{i \in I} \left(\bigvee_{j \in J_i} a_{ij} \right) = \bigvee_{f \in \prod_{i \in I} J_i} \left(\bigwedge_{i \in I} a_{if(i)} \right)$$

$$\bigvee_{i \in I} \left(\bigwedge_{j \in J_i} a_{ij} \right) = \bigwedge_{f \in \prod_{i \in I} J_i} \left(\bigvee_{i \in I} a_{if(i)} \right)$$

where I and J_i are non - empty index sets and $a_{ij} \in L$.

Completely distributive lattices are always complete. In a completely distributive lattice $(L, \vee, \wedge, 0, 1)$, the relative pseudocomplement is defined by $x \rightarrow y = \vee \{a \in L / x \wedge a \leq y\}$ and dually we can define dual relative pseudocomplement. Hence completely distributive lattices are bi - Heyting algebras.

Definition 7. *Let L be a complete lattice and a be an element of L. Then a is called compact iff $a \leq \vee X$, for some $X \subseteq L$ implies $a \leq \vee X_1$, for some finite $X_1 \subseteq X$. A complete lattice L is called algebraic iff every element of L is the join of compact elements.*

Result 8. *Let L be a lattice. Then the following are equivalent :*

1. *L is isomorphic to a complete ring of sets;*
2. *L is algebraic and completely distributive;*
3. *L is distributive and doubly algebraic (that is, both L and L^{∂} are algebraic);*
4. *L is algebraic, distributive and every element of L is a join of completely join irreducible elements of L.*

For further definitions and results in lattice theory the readers are asked to refer [11,15,26]. Let us recall some definitions and results given in [13,14]. Throughout this paper, B is a complete atomic Boolean lattice $(B, \leq)$ and $\mathcal{A}(B)$ is the set of atoms of B.

Definition 9. *[13] The map $\varphi : \mathcal{A}(B) \longrightarrow B$ is said to be*

1. *extensive, if $x \leq \varphi(x)$, for all $x \in \mathcal{A}(B)$.*
2. *symmetric, if $x \leq \varphi(y) \Longrightarrow y \leq \varphi(x)$, for all $x, y \in \mathcal{A}(B)$.*

3. *closed, if* $x \leq \varphi(y) \implies \varphi(x) \leq \varphi(y)$, *for all* $x, y \in \mathcal{A}(B)$.

For a map $\varphi : \mathcal{A}(B) \longrightarrow B$ there exists a unique map $\psi : \mathcal{A}(B) \longrightarrow B$ satisfying the condition:

$$a \leq \varphi(b) \iff b \leq \psi(a)$$

Definition 10. *[14] Let* $\varphi : \mathcal{A}(B) \longrightarrow B$ *be a map. For any element* $x \in B$,

$$
\begin{aligned}
x^{\blacktriangledown} &= \bigvee \{a \in \mathcal{A}(B)/\varphi(a) \leq x\} \\
x^{\blacktriangle} &= \bigvee \{a \in \mathcal{A}(B)/\varphi(a) \wedge x \neq 0\} \\
x^{\triangledown} &= \bigvee \{a \in \mathcal{A}(B)/\psi(a) \leq x\} \\
x^{\triangle} &= \bigvee \{a \in \mathcal{A}(B)/\psi(a) \wedge x \neq 0\}
\end{aligned}
\tag{3}
$$

For any binary relation R on a set U if we define

$$\varphi : U \to \wp(U) \text{ by } \varphi(x) = R(x) \text{ and}$$

$$\psi : U \to \wp(U) \text{ by } \psi(x) = R^{-1}(x), \text{ for all } x \in U$$

Then the following observations are obvious:

i. R is reflexive $\Leftrightarrow$ φ is extensive $\Leftrightarrow$ ψ is extensive;
ii. R is symmetric $\Leftrightarrow$ φ is symmetric $\Leftrightarrow$ ψ is symmetric;
iii. R is transitive $\Leftrightarrow$ φ is closed $\Leftrightarrow$ ψ is closed.

Result 11. *[14] Let B be a complete atomic Boolean lattice and $\varphi : \mathcal{A}(B) \longrightarrow B$ be a map. Then for all $a \in \mathcal{A}(B)$ and $x \in B$ the following hold:*

i. $a \leq x^{\blacktriangledown} \Leftrightarrow \varphi(a) \leq x$ *and* $a \leq x^{\blacktriangle} \Leftrightarrow \varphi(a) \wedge x \neq 0$
ii. $a \leq x^{\triangledown} \Leftrightarrow \psi(a) \leq x$ *and* $a \leq x^{\triangle} \Leftrightarrow \psi(a) \wedge x \neq 0$
iii. $a^{\blacktriangle} = \psi(a)$ *and* $a^{\triangle} = \varphi(a)$.
iv. $x^{\triangle} = \bigvee \{\varphi(a)/a \in \mathcal{A}(B) \text{ and } a \leq x\}$
v. $x^{\blacktriangle} = \bigvee \{\psi(a)/a \in \mathcal{A}(B) \text{ and } a \leq x\}$

For any $S \subseteq B$, let $S^{\blacktriangledown} = \{x^{\blacktriangledown}/x \in S\}$ and $S^{\blacktriangle} = \{x^{\blacktriangle}/x \in S\}$.

3   Some Algebraic Structures of Rough Approximations

Let us denote the set $\{\varphi(a)/a \in \mathcal{A}(B)\}$ by $\varphi(\mathcal{A}(B))$.

Theorem 12. *Let B be a complete atomic Boolean lattice and $\varphi : \mathcal{A}(B) \longrightarrow B$ be a map. Then $(B^{\blacktriangle}, \leq)$ is linearly ordered if, and only if, $(\varphi(\mathcal{A}(B)), \leq)$ is linearly ordered.*

Proof. Assume that $(\varphi(\mathcal{A}(B)), \leq)$ is linearly ordered. Let $x, y \in B^{\blacktriangle}$. Then $x = u^{\blacktriangle}, y = v^{\blacktriangle}$, for some $u, v \in B$. Suppose x and y are not comparable in $B^{\blacktriangle}$. Then there exist $a, b \in \mathcal{A}(B)$ such that $a \leq x = u^{\blacktriangle}$ and $a \not\leq y = v^{\blacktriangle}, b \leq y = v^{\blacktriangle}$ and $b \not\leq x = u^{\blacktriangle}$. Then, we have $\varphi(a) \wedge u \neq 0$ and $\varphi(a) \wedge v = 0, \varphi(b) \wedge v \neq 0$ and $\varphi(b) \wedge u = 0$. Since $(\varphi(\mathcal{A}(B)), \leq)$ is linearly ordered, we have $\varphi(a) \leq \varphi(b)$ or $\varphi(b) \leq \varphi(a)$. If $\varphi(a) \leq \varphi(b)$, then $\varphi(a) \wedge u \neq 0$ implies $\varphi(b) \wedge u \neq 0$. Then $b \leq u^{\blacktriangle}$, which is a contradiction to the assumption. Similar contradiction also occurs when $\varphi(b) \leq \varphi(a)$. Hence $(B^{\blacktriangle}, \leq)$ is linearly ordered.

Conversely, assume that $(B^{\blacktriangle}, \leq)$ is linearly ordered. Suppose $(\varphi(\mathcal{A}(B)), \leq)$ is not linearly ordered. Then there exist $a, b \in \mathcal{A}(B)$ such that neither $\varphi(a) \leq \varphi(b)$ nor $\varphi(b) \leq \varphi(a)$. Then there exist $c, d \in \mathcal{A}(B)$ such that $c \leq \varphi(a)$ and $c \not\leq \varphi(b), d \leq \varphi(b)$ and $d \not\leq \varphi(a)$. Then $c \leq \varphi(a)$ and $c \not\leq \varphi(b)$ implies $\varphi(a) \wedge c \neq 0$ and $\varphi(b) \wedge c = 0$. Then, we have $a \leq c^{\blacktriangle}$ and $b \not\leq c^{\blacktriangle}$. Similarly, $d \leq \varphi(b)$ and $d \not\leq \varphi(a)$ implies $b \leq d^{\blacktriangle}$ and $a \not\leq d^{\blacktriangle}$. Thus there exist $a, b \in \mathcal{A}(B)$ such that $a \leq c^{\blacktriangle}$ and $a \not\leq d^{\blacktriangle}, b \leq d^{\blacktriangle}$ and $b \not\leq c^{\blacktriangle}$. This implies there exist $c^{\blacktriangle}, d^{\blacktriangle} \in B^{\blacktriangle}$ such that $c^{\blacktriangle} \not\leq d^{\blacktriangle}$ and $d^{\blacktriangle} \not\leq c^{\blacktriangle}$, which is a contradiction to the assumption. Hence $(\varphi(\mathcal{A}(B)), \leq)$ is linearly ordered.

Since $(B^{\blacktriangle}, \leq)$ is dually order isomorphic to $(B^{\blacktriangledown}, \leq)$, we have the following corollary.

Corollary 13. *Let B be a complete atomic Boolean lattice and $\varphi : \mathcal{A}(B) \longrightarrow B$ be a map. Then $(B^{\blacktriangledown}, \leq)$ is linearly ordered if, and only if, $(\varphi(\mathcal{A}(B)), \leq)$ is linearly ordered.*

Remark 14. *Let us consider a map $\varphi : U \longrightarrow \wp(U)$ defined by $\varphi(x) = R(x)$ for all $x \in U$, where $R(x) = \{y \in U / xRy\}$. If R is transitive and connected(for every $x, y \in U$, either $y \in R(x)$ or $x \in R(y)$), then for every $x, y \in U$, either $R(x) \subseteq R(y)$ or $R(y) \subseteq R(x)$. Therefore, by the above theorem $(\wp(U)^{\blacktriangle}, \subseteq)$ and $(\wp(U)^{\blacktriangledown}, \subseteq)$ are linearly ordered sets.*

The join $(\sqcup)$ and the meet $(\sqcap)$ operations in the complete lattice $(B^{\blacktriangle}, \leq)$ are as follows: For any $S^{\blacktriangle} \subseteq B^{\blacktriangle}$

$$\sqcup S^{\blacktriangle} = \left(\bigvee_{x \in S} x \right)^{\blacktriangle}$$

$$\sqcap S^{\blacktriangle} = \bigvee \{a^{\blacktriangle} / a^{\blacktriangle} \leq \bigwedge_{x \in S} x^{\blacktriangle}\}$$

The join $(\sqcup)$ and the meet $(\sqcap)$ operations in the complete lattice $(B^{\blacktriangledown}, \leq)$ are as follows: For any $S^{\blacktriangledown} \subseteq B^{\blacktriangledown}$

$$\sqcup S^{\blacktriangledown} = \bigwedge \{a^{\blacktriangledown} / \bigvee_{x \in S} x^{\blacktriangledown} \leq a^{\blacktriangledown}\}$$

$$\sqcap S^{\blacktriangledown} = \left(\bigwedge_{x \in S} x \right)^{\blacktriangledown}$$

where $\bigvee$ and $\bigwedge$ are the join and meet operations in B respectively.

Lemma 15. *If $\varphi : \mathcal{A}(B) \longrightarrow B$ is an extensive and closed map, then the following properties hold for any $x, y \in B$:*

i. $x^{\blacktriangledown} \leq x \leq x^{\blacktriangle}; x^{\triangledown} \leq x \leq x^{\triangle}$

ii. $x^{\blacktriangledown\blacktriangledown} = x^{\blacktriangledown}; x^{\blacktriangle\blacktriangle} = x^{\blacktriangle}; x^{\triangledown\triangledown} = x^{\triangledown}; x^{\triangle\triangle} = x^{\triangle}$

iii. $x^{\blacktriangle\triangledown} = x^{\blacktriangle}; x^{\triangledown\blacktriangle} = x^{\triangledown}; x^{\triangle\blacktriangledown} = x^{\triangle}; x^{\blacktriangledown\triangle} = x^{\blacktriangledown}$

iv. $(x^{\blacktriangle} \wedge y^{\blacktriangle})^{\blacktriangle} = x^{\blacktriangle} \wedge y^{\blacktriangle}; (x^{\blacktriangledown} \vee y^{\blacktriangledown})^{\blacktriangledown} = x^{\blacktriangledown} \vee y^{\blacktriangledown}$

Proof. For the proof of (i), (ii) and (iii), see [14].
iv. Since φ is extensive, we have $x^{\blacktriangle} \wedge y^{\blacktriangle} \leq (x^{\blacktriangle} \wedge y^{\blacktriangle})^{\blacktriangle}$. Also $x^{\blacktriangle} \wedge y^{\blacktriangle} \leq x^{\blacktriangle}, y^{\blacktriangle}$ implies $(x^{\blacktriangle} \wedge y^{\blacktriangle})^{\blacktriangle} \leq x^{\blacktriangle\blacktriangle}, y^{\blacktriangle\blacktriangle}$. By (ii), we have $(x^{\blacktriangle} \wedge y^{\blacktriangle})^{\blacktriangle} \leq x^{\blacktriangle}, y^{\blacktriangle}$. This implies $(x^{\blacktriangle} \wedge y^{\blacktriangle})^{\blacktriangle} \leq x^{\blacktriangle} \wedge y^{\blacktriangle}$. Hence $(x^{\blacktriangle} \wedge y^{\blacktriangle})^{\blacktriangle} = x^{\blacktriangle} \wedge y^{\blacktriangle}$. Similarly, we can prove $(x^{\blacktriangledown} \vee y^{\blacktriangledown})^{\blacktriangledown} = x^{\blacktriangledown} \vee y^{\blacktriangledown}$.

Lemma 16. *If $\varphi : \mathcal{A}(B) \longrightarrow B$ is an extensive and closed map, then $\sqcap S^{\blacktriangle} = \wedge S^{\blacktriangle}$ and $\sqcup S^{\blacktriangle} = \vee S^{\blacktriangle}$.*

Proof. $\sqcap S^{\blacktriangle}$ is the largest element $c^{\blacktriangle} \in B^{\blacktriangle}$ such that $c^{\blacktriangle} \leq \wedge S^{\blacktriangle}$. This implies $c^{\blacktriangle} = c^{\blacktriangle\blacktriangle} \leq (\wedge S^{\blacktriangle})^{\blacktriangle}$. By lemma 15(iv), we have $(\wedge S^{\blacktriangle})^{\blacktriangle} = \wedge S^{\blacktriangle}$. Thus $\wedge S^{\blacktriangle}$ is the largest element $c^{\blacktriangle} \in B^{\blacktriangle}$ such that $c^{\blacktriangle} \leq \wedge S^{\blacktriangle}$. Hence $\sqcap S^{\blacktriangle} = \wedge S^{\blacktriangle}$. Dually, we can show that $\sqcup S^{\blacktriangle} = \vee S^{\blacktriangle}$.

Now, $(B^{\blacktriangle}, \sqcup, \sqcap)$ is a sublattice of the complete atomic Boolean lattice $(B, \vee, \wedge, ', 0, 1)$. Since φ is extensive, by proposition 3.8 of [13], we have 0 and 1 are the least and the greatest elements of $B^{\blacktriangle}$ respectively. Therefore, $(B^{\blacktriangle}, \sqcup, \sqcap, 0, 1)$ is a completely distributive lattice. Also, we have $\{a^{\blacktriangle}/a \in \mathcal{A}(B)\}$ are the non - zero join - irreducible elements and hence are the compact elements of $(B^{\blacktriangle}, \sqcup, \sqcap, 0, 1)$ and for any $x \in B$, $x^{\blacktriangle} = \bigvee_{a \leq x} \psi(a) = \bigvee_{a \leq x} a^{\blacktriangle}$. So, every element in $B^{\blacktriangle}$ is the join of compact elements below it. Therefore $B^{\blacktriangle}$ is compactly generated. Hence $(B^{\blacktriangle}, \sqcup, \sqcap, 0, 1)$ is an algebraic completely distributive lattice.

Since $(B^{\blacktriangledown}, \leq)$ and $(B^{\blacktriangle}, \leq)$ are dually isomorphic, $(B^{\blacktriangledown}, \sqcup, \sqcap, 0, 1)$ is also an algebraic completely distributive lattice and $\{a^{\triangle}/a \in \mathcal{A}(B)\}$ are its non - zero join - irreducible elements.

Theorem 17. *Given any algebraic completely distributive lattice D, there exist a complete atomic Boolean lattice B and an extensive, closed map $\varphi : \mathcal{A}(B) \to B$ such that $D \cong B^{\blacktriangledown}$.*

Proof. Let D be any algebraic completely distributive lattice. Let J be the set of all non - zero join irreducible elements in D. Let $B = \wp(J)$. Then B is the required complete atomic Boolean lattice with $\mathcal{A}(B) = \{\{a\}/a \in J\}$. Then $f : D \to B$ defined by $f(x) = \{a \in J/a \leq_D x\}$ is an embedding of D into B. Let $\varphi : \mathcal{A}(B) \to B$ be defined by

$$\varphi(\{a\}) = \wedge_B \{f(x)/x \in J \text{ and } a \in f(x)\}$$

for all $\{a\} \in \mathcal{A}(B)$. Since for each $a \in J$, $f(a)$ is the least element in B such that $a \in f(a)$, $\varphi(\{a\}) = f(a)$, for all $\{a\} \in \mathcal{A}(B)$. Therefore φ is extensive. Let

$\{a\}, \{b\} \in \mathcal{A}(B)$ be such that $\{a\} \leq_B \varphi(\{b\})$. Then $a \in f(b)$ and $b \in J$ imply $\varphi(\{a\}) \leq_B \varphi(\{b\})$. Therefore, φ is closed. Hence φ is the required extensive and closed map. Since $B^{\blacktriangledown} = B^{\triangle}$, we first show that there exists an isomorphism between D and $B^{\triangle}$. Then the result follows. Let $g : D \to B^{\triangle}$ be defined by

$$g(x) = \vee_{B\{a\} \leq_B f(x)} \varphi(\{a\})$$

By result 11(iv), we have every element of $B^{\triangle}$ is of the form $x^{\triangle} = \vee_{B\{a\} \leq_B x} \varphi(\{a\})$, for some $x \in B$. For each $x \in D$, $f(x) \in B$ and hence $g(x) = \vee_{B\{a\} \leq_B f(x)} \varphi(\{a\}) \in B^{\triangle}$. Now, $g(x) = \vee_{B\{a\} \leq_B f(x)} \varphi(\{a\}) = \vee_{B\{a\} \leq_B f(x)} f(a) = f(\vee_{D\{a\} \leq_B f(x)} a) = f(x)$, for, $f(x)$ contains set of join irreducible elements in D which are less than x. As f is a one - one homomorphism, g is also a one - one homomorphism. Now to prove g is onto, let $z \in B^{\triangle}$. Then $z = \vee_{B\{a\} \leq_B x} \varphi(\{a\})$, for some $x \in B$. We have, $z = \vee_{B\{a\} \leq_B x} f(a) = f(\vee_{D\{a\} \leq_B x} a)$. But $\vee_D \{a \in J / \{a\} \leq_B x\}$ is some $y \in D$. Thus, $z = f(y) = g(y)$, for some $y \in D$. Therefore g is onto. Hence $D \cong B^{\triangle}$.

Let us define two more binary operations $\to$ and $\leftarrow$ in $B^{\blacktriangle}$ as follows:

$$x^{\blacktriangle} \to y^{\blacktriangle} = (x^{\blacktriangle\prime} \vee y^{\blacktriangle})^{\triangledown}$$

$$x^{\blacktriangle} \leftarrow y^{\blacktriangle} = (x^{\blacktriangle} \wedge y^{\blacktriangle\prime})^{\blacktriangle}$$

where $'$ is the complementation in B.

Lemma 18. *If $\varphi : \mathcal{A}(B) \longrightarrow B$ is an extensive and closed map, then for any $x^{\blacktriangle}, y^{\blacktriangle}, z^{\blacktriangle} \in B^{\blacktriangle}$,*

i. $x^{\blacktriangle} \wedge z^{\blacktriangle} \leq y^{\blacktriangle} \Longleftrightarrow z^{\blacktriangle} \leq x^{\blacktriangle} \to y^{\blacktriangle}$,
ii. $x^{\blacktriangle} \leq z^{\blacktriangle} \vee y^{\blacktriangle} \Longleftrightarrow x^{\blacktriangle} \leftarrow y^{\blacktriangle} \leq z^{\blacktriangle}$.

Proof. Let $x^{\blacktriangle}, y^{\blacktriangle} \in B^{\blacktriangle}$.
i. Suppose there exists $z^{\blacktriangle} \in B^{\blacktriangle}$ such that $x^{\blacktriangle} \wedge z^{\blacktriangle} \leq y^{\blacktriangle}$

$$\Rightarrow x^{\blacktriangle\prime} \vee (x^{\blacktriangle} \wedge z^{\blacktriangle}) \leq x^{\blacktriangle\prime} \vee y^{\blacktriangle}$$
$$\Rightarrow (x^{\blacktriangle\prime} \vee x^{\blacktriangle}) \wedge (x^{\blacktriangle\prime} \vee z^{\blacktriangle}) \leq x^{\blacktriangle\prime} \vee y^{\blacktriangle}$$
$$\Rightarrow (x^{\blacktriangle\prime} \vee z^{\blacktriangle}) \leq x^{\blacktriangle\prime} \vee y^{\blacktriangle}$$
$$\Rightarrow z^{\blacktriangle} \leq x^{\blacktriangle\prime} \vee y^{\blacktriangle}$$
$$\Rightarrow z^{\blacktriangle} = z^{\blacktriangle\triangledown} \leq (x^{\blacktriangle\prime} \vee y^{\blacktriangle})^{\triangledown}$$
$$\Rightarrow z^{\blacktriangle} \leq x^{\blacktriangle} \to y^{\blacktriangle}$$

Conversely, suppose there exists $z^{\blacktriangle} \in B^{\blacktriangle}$ such that $z^{\blacktriangle} \leq x^{\blacktriangle} \to y^{\blacktriangle}$. Then $x^{\blacktriangle} \wedge (x^{\blacktriangle\prime} \vee y^{\blacktriangle})^{\triangledown} \leq x^{\blacktriangle} \wedge (x^{\blacktriangle\prime} \vee y^{\blacktriangle}) = (x^{\blacktriangle} \wedge x^{\blacktriangle\prime}) \vee (x^{\blacktriangle} \wedge y^{\blacktriangle}) = (x^{\blacktriangle} \wedge y^{\blacktriangle}) \leq y^{\blacktriangle}$. Thus $x^{\blacktriangle} \wedge (x^{\blacktriangle} \to y^{\blacktriangle}) \leq y^{\blacktriangle}$. Since $z^{\blacktriangle} \leq x^{\blacktriangle} \to y^{\blacktriangle}$, $x^{\blacktriangle} \wedge z^{\blacktriangle} \leq x^{\blacktriangle} \wedge (x^{\blacktriangle} \to y^{\blacktriangle}) \leq y^{\blacktriangle}$.
ii. The proof is similar to that of (i).

Proposition 19. *If $\varphi : \mathcal{A}(B) \longrightarrow B$ is an extensive and closed map, then $(B^{\blacktriangle}, \sqcup, \sqcap, \to, \leftarrow, 0, 1)$ is a bi-Heyting algebra.*

Proof. We have $(B^{\blacktriangle}, \sqcup, \sqcap, 0, 1)$ is a completely distributive lattice. From lemma 18, we have $x^{\blacktriangle} \to y^{\blacktriangle}$ is the relative psuedocomplement of $x^{\blacktriangle}$ with respect to $y^{\blacktriangle}$ and $x^{\blacktriangle} \leftarrow y^{\blacktriangle}$ is the dual relative pseudocomplement of $x^{\blacktriangle}$ with respect to $y^{\blacktriangle}$. Hence $(B^{\blacktriangle}, \sqcup, \sqcap, \to, \leftarrow, 0, 1)$ is a bi-Heyting algebra.

Since $(B^{\blacktriangledown}, \leq)$ and $(B^{\blacktriangle}, \leq)$ are dually isomorphic, $(B^{\blacktriangledown}, \sqcup, \sqcap, \to, \leftarrow, 0, 1)$ is also a bi - Heyting algebra in which $x^{\blacktriangledown} \to y^{\blacktriangledown} = (x^{\blacktriangledown\prime} \vee y^{\blacktriangledown})^{\blacktriangledown}$ and $x^{\blacktriangledown} \leftarrow y^{\blacktriangledown} = (x^{\blacktriangledown} \wedge y^{\blacktriangledown\prime})^{\triangle}$.

Let us define another unary operation $^{\perp}$ in $B^{\blacktriangle}$ as follows:

$$(x^{\blacktriangle})^{\perp} = x^{\triangle\prime\blacktriangle} \tag{4}$$

where $'$ is the complementation in B.

Lemma 20. *If* $\varphi : \mathcal{A}(B) \longrightarrow B$ *is an extensive and symmetric map, then* $(B^{\blacktriangle}, \sqcup, \sqcap, ^{\perp}, 0, 1)$ *is a complete complemented lattice. Moreover, the complementation* $^{\perp}$ *is defined by* $(x^{\blacktriangle})^{\perp} = x^{\blacktriangle\prime\blacktriangle}$, *where* $'$ *is the complementation in the complete atomic Boolean lattice* B.

Proof. It is enough to show that for every $x^{\blacktriangle} \in B^{\blacktriangle}$, $(x^{\blacktriangle})^{\perp}$ is the complement of $x^{\blacktriangle}$. Since φ is symmetric, $x^{\triangle} = x^{\blacktriangle}$ and $a^{\blacktriangle} = \varphi(a)$, for all $a \in \mathcal{A}(B)$. Then $(x^{\blacktriangle})^{\perp} = x^{\blacktriangle\prime\blacktriangle}$. We have to show that $(x^{\blacktriangle})^{\perp} \sqcup x^{\blacktriangle} = 1$ and $(x^{\blacktriangle})^{\perp} \sqcap x^{\blacktriangle} = 0$. We have $(x^{\blacktriangle})^{\perp} \sqcup x^{\blacktriangle} = (x^{\blacktriangle\prime} \vee x)^{\blacktriangle} \leq 1$. Suppose $(x^{\blacktriangle\prime} \vee x)^{\blacktriangle} < 1$. Then there exists $a \in \mathcal{A}(B)$ such that $a \leq 1$ and $a \not\leq (x^{\blacktriangle\prime} \vee x)^{\blacktriangle}$. Then by result 11(i), we have $\varphi(a) \wedge (x^{\blacktriangle\prime} \vee x) = 0$. This implies $(\varphi(a) \wedge x^{\blacktriangle\prime}) \vee (\varphi(a) \wedge x) = 0$. Then $\varphi(a) \wedge x^{\blacktriangle\prime} = 0$ and $\varphi(a) \wedge x = 0$. Since φ is extensive, $a \leq \varphi(a)$. So, we have $a \wedge x^{\blacktriangle\prime} \leq \varphi(a) \wedge x^{\blacktriangle\prime} = 0$ which implies $a \leq x^{\blacktriangle}$. $\varphi(a) \wedge x = 0$ implies $a \not\leq x^{\blacktriangle}$, by result 11(i). Now, we have $a \leq x^{\blacktriangle}$ and $a \not\leq x^{\blacktriangle}$ which is absurd. Thus, $(x^{\blacktriangle})^{\perp} \sqcup x^{\blacktriangle} = 1$.

Next we have to show that $(x^{\blacktriangle})^{\perp} \sqcap x^{\blacktriangle} = 0$. Suppose not. Then there exists $a \in \mathcal{A}(B)$ such that $a^{\blacktriangle} \leq x^{\blacktriangle\prime\blacktriangle} \wedge x^{\blacktriangle}$. This implies $\varphi(a) \leq x^{\blacktriangle\prime\blacktriangle}$ and $\varphi(a) \leq x^{\blacktriangle}$. By result 11(i), $\varphi(a) \leq x^{\blacktriangle\prime\blacktriangle}$ implies $a \leq x^{\blacktriangle\prime\blacktriangle\blacktriangledown} = x^{\blacktriangle\blacktriangledown\blacktriangle\prime}$. Since φ is symmetric, $x^{\blacktriangle\blacktriangledown\blacktriangle} = x^{\blacktriangle}$. Therefore, we have $a \leq x^{\blacktriangle\prime}$. Since φ is extensive, $\varphi(a) \leq x^{\blacktriangle}$ implies $a \leq x^{\blacktriangle}$. Thus, we have $a \leq x^{\blacktriangle\prime}$ and $a \leq x^{\blacktriangle}$ which implies $a \leq x^{\blacktriangle\prime} \wedge x^{\blacktriangle} = 0$ which is a contradiction to $a \in \mathcal{A}(B)$. Thus $(x^{\blacktriangle})^{\perp} \sqcap x^{\blacktriangle} = 0$. Hence $(B^{\blacktriangle}, \sqcup, \sqcap, ^{\perp}, 0, 1)$ is a complete complemented lattice.

Lemma 21. *If* $\varphi : \mathcal{A}(B) \longrightarrow B$ *is an extensive and symmetric map, then the following properties hold for every* $x, y \in B^{\blacktriangle}$

i. $x^{\perp\perp} = x$
ii. $x \leq y \Rightarrow y^{\perp} \leq x^{\perp}$

Proof. Let $x, y \in B^{\blacktriangle}$. Then $x = u^{\blacktriangle}, y = v^{\blacktriangle}$, for some $u, v \in B$.
i. $x^{\perp\perp} = (u^{\blacktriangle})^{\perp\perp} = u^{\blacktriangle\prime\blacktriangle\prime\blacktriangle} = u^{\blacktriangle\blacktriangledown\blacktriangle\prime\prime} = u^{\blacktriangle} = x$.
ii. Suppose $x \leq y$. Then $u^{\blacktriangle} \leq v^{\blacktriangle} \Rightarrow v^{\blacktriangle\prime} \leq u^{\blacktriangle\prime} \Rightarrow v^{\blacktriangle\prime\blacktriangle} \leq u^{\blacktriangle\prime\blacktriangle} \Rightarrow y^{\perp} \leq x^{\perp}$.

Theorem 22. *Let* B *be a complete atomic Boolean algebra and* $\varphi : \mathcal{A}(B) \longrightarrow B$ *be an extensive and symmetric map, then* $(B^{\blacktriangle}, \sqcup, \sqcap, ^{\perp}, 0, 1)$ *is a complete ortholattice.*

Proof. From lemma 20 and lemma 21, $(B^{\blacktriangle}, \sqcup, \sqcap, ^{\perp}, 0, 1)$ is a complete ortholattice.

Since φ is symmetric, $(B^{\blacktriangle}, \leq) \cong (B^{\blacktriangledown}, \leq)$. Hence $(B^{\blacktriangledown}, \sqcup, \sqcap, ^{\perp}, 0, 1)$ is also a complete ortholattice, where the orthcomplementation $^{\perp}$ in $B^{\blacktriangledown}$ is given by $(x^{\blacktriangledown})^{\perp} = x^{\blacktriangledown' \blacktriangledown}$.

Example 23. *Consider a complete atomic Boolean lattice B in fig. 1. Let the map $\varphi : \mathcal{A}(B) \longrightarrow B$ be defined as:*

$$\varphi(a) = e, \varphi(b) = d', \varphi(c) = a', \varphi(d) = e'$$

Then φ is extensive and symmetric and $B^{\blacktriangle} = \{0, e, a', d', e', 1\}$. The structure of $B^{\blacktriangle}$ is shown in fig. 2.

Theorem 24. *Given any complete ortholattice L, there exist a complete atomic Boolean lattice B and an extensive, symmetric map $\varphi : \mathcal{A}(B) \to B$ such that $L \cong B^{\blacktriangle}$.*

Proof. Let L be any complete ortholattice. Consider $S = L - \{0\}$. Let $B = \wp(S)$. Then B is the required complete atomic Boolean lattice with $\mathcal{A}(B) = \{\{a\}/a \in S\}$. Let $\varphi : \mathcal{A}(B) \to B$ be defined by

$$\varphi(\{a\}) = \{x \in S/a \not\leq_L x'\}$$

where $'$ is the orthocomplementation in the complete ortholattice L. Since $a \not\leq_L a'$, we have $\{a\} \leq_B \varphi(\{a\})$, for all $\{a\} \in \mathcal{A}(B)$. Therefore φ is extensive. Let $\{a\}, \{b\} \in \mathcal{A}(B)$ such that $\{a\} \leq_B \varphi(\{b\})$. Then $b \not\leq_L a'$ implies $a \not\leq_L b'$. Then $\{b\} \leq_B \varphi(\{a\})$. Therefore, φ is symmetric. Hence φ is the required extensive and symmetric map. Let $g : L \to B^{\blacktriangle}$ be defined by $g(x) = \bigvee_{B\{a\} \leq_B X} \varphi(\{a\})$, where $\vee_L X = x$. By proposition 3.10 of [13], we have every element of $B^{\blacktriangle}$ is of the form $X^{\blacktriangle} = \bigvee_{B\{a\} \leq_B X} \varphi(\{a\})$, for $X \in B$. This g is the required isomorphism.

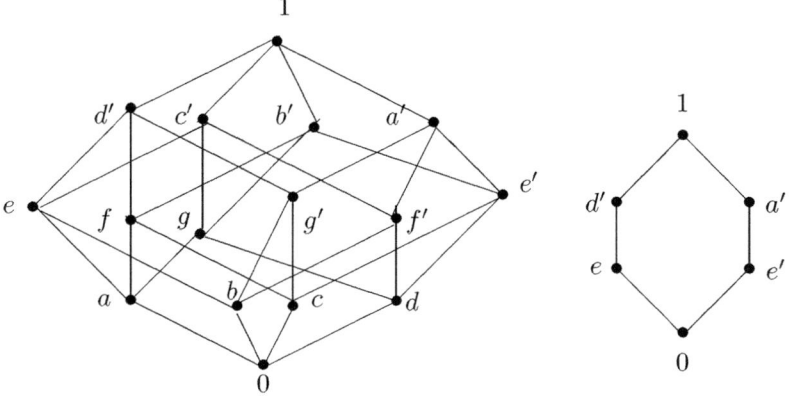

Fig. 1. Complete atomic Boolean lattice B

Fig. 2. Structure of $B^{\blacktriangle}$

Theorem 25. *Let* $\varphi : \mathcal{A}(B) \longrightarrow B$ *be an extensive and closed map. Then* $(B^{\blacktriangle}, \sqcup, \sqcap, ^{\perp}, 0, 1)$ *is a Boolean algebra if, and only if,* φ *is symmetric.*

Proof. Let φ be an extensive and closed map. Then we know that $(B^{\blacktriangle}, \sqcup, \sqcap, 0, 1)$ is a bounded distributive lattice. If φ is symmetric, then by lemma 20, $^{\perp}$ is the complementation in $B^{\blacktriangle}$. Therefore $(B^{\blacktriangle}, \sqcup, \sqcap, ^{\perp}, 0, 1)$ is a Boolean algebra.

Conversely, suppose $(B^{\blacktriangle}, \sqcup, \sqcap, ^{\perp}, 0, 1)$ is a Boolean algebra. Then $^{\perp}$ is a complementation in $B^{\blacktriangle}$ and so for all $x \in B$, we have $(x^{\blacktriangle})^{\perp} \sqcap x^{\blacktriangle} = 0$. Also, we have $a^{\blacktriangle} = \psi(a)$. Then $(x^{\blacktriangle})^{\perp} \sqcap x^{\blacktriangle} = 0$ implies there doesnot exists $a \in \mathcal{A}(B)$, such that $\psi(a) \leq x^{\triangle \prime \blacktriangle} \wedge x^{\blacktriangle}$. Therefore $(x^{\triangle \prime \blacktriangle} \wedge x^{\blacktriangle})^{\triangledown} = 0$. This implies $x^{\triangle \prime \blacktriangle \triangledown} \wedge x^{\blacktriangle \triangledown} = 0$, which implies $x^{\triangle \blacktriangledown \triangle \prime} \wedge x^{\blacktriangle \triangledown} = 0$. By lemma 15 (ii) and (iii), we have $x^{\triangle \blacktriangledown \triangle} = x^{\triangle}$ which implies $x^{\triangle \prime} \wedge x^{\blacktriangle} = 0$. Thus $x^{\blacktriangle} \leq x^{\triangle}$, for all $x \in B$. We have to show that φ is symmetric. Suppose for some $a, b \in \mathcal{A}(B), a \leq \varphi(b)$ and $b \not\leq \varphi(a)$. $a \leq \varphi(b)$ implies $a \wedge \varphi(b) \neq 0$, which implies $b \leq a^{\blacktriangle}$. And $b \not\leq \varphi(a)$ implies $a \not\leq \psi(b)$, which implies $a \wedge \psi(b) = 0$. Then $b \not\leq a^{\triangle}$. Thus $a^{\blacktriangle} \not\leq a^{\triangle}$ which is a contradiction to $x^{\blacktriangle} \leq x^{\triangle}$, for all $x \in B$. Therefore for all $a, b \in \mathcal{A}(B), a \leq \varphi(b)$ and $b \leq \varphi(a)$. Hence φ is symmetric.

4   New Approach in Defining the Rough Approximations

In the generalized approximation space (U, R), $\{R(x)/x \in U\}$ is the usual basic knowledge granules. The reflexive property of the binary relation in the generalized approximation space cannot be relaxed, since an object cannot be distinguisable with itself based on their attribute properties. In [1,2], Allam *et al.* introduced a new definition of lower and upper approximation operators based on binary relation with $\{\langle x \rangle R/x \in U\}$ as basic knowledge granules in the generalized approximation space. The lower and upper approximations proposed in [1] is as follows:

$$\underline{X} = \{x \in U/\langle x \rangle R \subseteq X\} \; and \qquad (5)$$
$$\overline{X} = \{x \in U/\langle x \rangle R \cap X \neq \phi\}$$

where $\langle x \rangle R = \bigcap_{x \in R(y)} R(y)$.

Let us compare the two basic knowledge granules using an information system in the following example.

Example 26. *Consider a many - valued information system given in table 1. It can be analyzed by an indiscernibility relation* R:

$$(x_1, x_2) \in R \iff a(x_1) \cap a(x_2) \neq \phi \qquad (6)$$

The usual basic knowledge granules obtained through the indiscernibility relation (6) regarding the degrees of the candidates are

- $R(x_1) = \{x_1, x_3, x_5\}$
- $R(x_2) = \{x_2, x_3, x_4, x_5\}$

- $R(x_3) = \{x_1, x_2, x_3, x_5\}$
- $R(x_4) = \{x_2, x_4\}$
- $R(x_5) = \{x_1, x_2, x_3, x_5\}$

The basic knowledge granules $\langle x \rangle R$ obtained through the indiscernibility relation (6) are

- $\langle x_1 \rangle R = \{x_1, x_3, x_5\}$
- $\langle x_2 \rangle R = \{x_2\}$
- $\langle x_3 \rangle R = \{x_3, x_5\}$
- $\langle x_4 \rangle R = \{x_2, x_4\}$
- $\langle x_5 \rangle R = \{x_3, x_5\}$

Here $\langle x \rangle R$ contains objects that are more similar to x based on their attribute values than $R(x)$.

In the above example, the basic knowledge granules $\langle x \rangle R$ are finer than $R(x)$. Also, the rough approximation operators in (5) coincides with the standard rough approximation operators in the classical approximation space and some generalized approximation spaces [17].

Jarvinen's definition of lower and upper approximation in (3) is a most suitable general setting for the rough approximation operators based on binary relation, so the rough approximation operators in (5) can be lifted to this general setting. In order to lift the definition to the general setting of complete atomic Boolean lattice, we define a new map $\langle \varphi \rangle$ induced from the basic map φ as follows: for all $a \in \mathcal{A}(B)$,

$$\langle \varphi \rangle(a) = \bigwedge_{a \leq \varphi(b)} \varphi(b)$$

Now, we can define the lower and upper approximations on B with respect to the induced map $\langle \varphi \rangle$.

Definition 27. *Let B be a complete atomic Boolean lattice and $\varphi : \mathcal{A}(B) \longrightarrow B$ be a map. For any $x \in B$, we define*

$$x^{\curlyvee} = \bigvee \{a \in \mathcal{A}(B)/\langle \varphi \rangle(a) \leq x\} \ and$$

$$x^{\curlywedge} = \bigvee \{a \in \mathcal{A}(B)/\langle \varphi \rangle(a) \wedge x \neq 0\}$$

Table 1. Information about the candidates from the applications for a Job

Candidates	Gender	Degrees
x_1	Female	$\{B.Sc., M.Sc., \}$
x_2	Female	$\{B.A., M.A., Ph.D., \}$
x_3	Male	$\{B.Sc., M.Sc., Ph.D., \}$
x_4	Male	$\{B.A., M.A., \}$
x_5	Female	$\{B.Sc., M.Sc., Ph.D., \}$

The elements $x^{\curlyvee}$ and $x^{\curlywedge}$ are the lower and upper approximations of x with respect to $\langle\varphi\rangle$ respectively.

According to lemma 3.3, 3.4 and proposition 3.5 in [13], the following results are true for any map φ. Since $\langle\varphi\rangle$ is also a map, the same results hold for the lower and upper approximations of $x \in B$ with respect to $\langle\varphi\rangle$. Therefore the proofs of the following lemma and proposition are omitted.

Lemma 28. *Let B be a complete atomic Boolean lattice and $\varphi : \mathcal{A}(B) \longrightarrow B$ be a map. Then for elements $x, y \in B$ and $a \in \mathcal{A}(B)$ the following hold:*

i. $a \leq x^{\curlywedge} \iff \langle\varphi\rangle(a) \wedge x \neq 0$
ii. $a \leq x^{\curlyvee} \iff \langle\varphi\rangle(a) \leq x$
iii. $0^{\curlywedge} = 0$ and $1^{\curlyvee} = 1$
iv. $x \leq y \Longrightarrow x^{\curlyvee} \leq y^{\curlyvee}$ and $x^{\curlywedge} \leq y^{\curlywedge}$

For any $S \subseteq B$, denote by $S^{\curlyvee} = \{x^{\curlyvee}/x \in S\}$ and $S^{\curlywedge} = \{x^{\curlywedge}/x \in S\}$.

Proposition 29. *Let B be a complete atomic Boolean lattice and $\varphi : \mathcal{A}(B) \longrightarrow B$ be a map. Then the following hold:*

i. The maps $^{\curlyvee} : B \longrightarrow B$ and $^{\curlywedge} : B \longrightarrow B$ are mutually dual.
ii. For all $S \subseteq B, (\bigvee S)^{\curlywedge} = \bigvee S^{\curlywedge}$ and $(\bigwedge S)^{\curlyvee} = \bigwedge S^{\curlyvee}$
iii. $(B^{\curlyvee}, \leq)$ is a complete lattice with least element 0 and greatest element $1^{\curlyvee}$.
iv. $(B^{\curlywedge}, \leq)$ is a complete lattice with least element $0^{\curlywedge}$ and greatest element 1.

Proposition 30. *Let B be a complete atomic Boolean lattice and $\varphi : \mathcal{A}(B) \longrightarrow B$ be a map. Then the following hold for every $x \in B$,*

i. $x^{\curlywedge\blacktriangle} = x^{\blacktriangle}$
ii. $x^{\curlyvee\blacktriangledown} = x^{\blacktriangledown}$

Proof. i. Let $a \in \mathcal{A}(B)$ be such that $a \leq x^{\curlywedge\blacktriangle}$. Then, we have $\varphi(a) \wedge x^{\curlywedge} \neq 0$. Then there exists $b \in \mathcal{A}(B)$ such that $b \leq \varphi(a)$ and $b \leq x^{\curlywedge}$ which imply $\langle\varphi\rangle(b) \wedge x \neq 0$. Since $b \leq \varphi(a)$, we have $\langle\varphi\rangle(b) \leq \varphi(a)$. Then $\varphi(a) \wedge x \neq 0$ implies $a \leq x^{\blacktriangle}$. This implies $\vee\{a \in \mathcal{A}(B)/a \leq x^{\curlywedge\blacktriangle}\} \leq \vee\{a \in \mathcal{A}(B)/a \leq x^{\blacktriangle}\}$ which implies $x^{\curlywedge\blacktriangle} \leq x^{\blacktriangle}$. Let $a \in \mathcal{A}(B)$ be such that $a \leq x^{\blacktriangle}$. Then we have $\varphi(a) \wedge x \neq 0$. Then there exists $b \in \mathcal{A}(B)$ such that $b \leq \varphi(a)$ and $b \leq x$. Since there exists $b \in \mathcal{A}(B)$ such that $b \leq \varphi(a)$, we have $b \leq \langle\varphi\rangle(b)$. Also, we have $b \leq x$ which implies $\langle\varphi\rangle(b) \wedge x \neq 0$. Then we have $b \leq x^{\curlywedge}$. So, $b \leq \varphi(a)$ and $b \leq x^{\curlywedge}$ imply $\varphi(a) \wedge x^{\curlywedge} \neq 0$. Then, we have $a \leq x^{\curlywedge\blacktriangle}$. Thus, $\vee\{a \in \mathcal{A}(B)/a \leq x^{\blacktriangle}\} \leq \vee\{a \in \mathcal{A}(B)/a \leq x^{\curlywedge\blacktriangle}\}$ which implies $x^{\blacktriangle} \leq x^{\curlywedge\blacktriangle}$. Hence $x^{\curlywedge\blacktriangle} = x^{\blacktriangle}$, for all $x \in B$.

ii) By (i), we have $x^{\curlywedge\blacktriangle} = x^{\blacktriangle}$, for all $x \in B$. Thus for x', we have $x'^{\curlywedge\blacktriangle} = x'^{\blacktriangle}$. By the duality of maps $^{\curlyvee}, ^{\curlywedge}$ and $^{\blacktriangledown}, ^{\blacktriangle}$, we have $x^{\curlyvee\blacktriangledown} = x^{\blacktriangledown}$.

The following lemma shows that $\langle\varphi\rangle$ is always closed for any map φ.

Lemma 31. *Let B be a complete atomic Boolean lattice and $\varphi : \mathcal{A}(B) \longrightarrow B$ be a map. Then the induced map $\langle\varphi\rangle$ is always closed.*

Proof. Let $a, b \in \mathcal{A}(B)$ be such that $a \leq \langle \varphi \rangle(b)$. Then by definition of $\langle \varphi \rangle$, we have

$$a \leq \varphi(c), \text{ for all } c \in \mathcal{A}(B) \text{ such that } b \leq \varphi(c) \qquad (*)$$

Let $x \in \mathcal{A}(B)$ be such that $x \leq \langle \varphi \rangle(a)$. Then by definition of $\langle \varphi \rangle$, we have

$$x \leq \varphi(d), \text{ for all } d \in \mathcal{A}(B) \text{ such that } a \leq \varphi(d) \qquad (**)$$

Let $c \in \mathcal{A}(B)$ be such that $b \leq \varphi(c)$. This implies $a \leq \varphi(c)$ (by $(*)$) which implies $x \leq \varphi(c)$ (by $(**)$). Therefore $x \leq \varphi(c)$, for all $c \in \mathcal{A}(B)$ such that $b \leq \varphi(c)$. Then $x \leq \bigwedge_{b \leq \varphi(c)} \varphi(c)$ implies $x \leq \langle \varphi \rangle(b)$. Thus $\{x \in \mathcal{A}(B)/x \leq \langle \varphi \rangle(a)\} \subseteq \{x \in \mathcal{A}(B)/x \leq \langle \varphi \rangle(b)\}$ implies $\bigvee\{x \in \mathcal{A}(B)/x \leq \langle \varphi \rangle(a)\} \leq \bigvee\{x \in \mathcal{A}(B)/x \leq \langle \varphi \rangle(b)\}$ which implies $\langle \varphi \rangle(a) \leq \langle \varphi \rangle(b)$. Thus $a \leq \langle \varphi \rangle(b)$ implies $\langle \varphi \rangle(a) \leq \langle \varphi \rangle(b)$, for all $a, b \in \mathcal{A}(B)$. Hence the map $\langle \varphi \rangle$ is closed.

Extensiveness of φ implies the extensiveness of $\langle \varphi \rangle$. The following example shows that, there exists a map φ which is not extensive, but their induced map $\langle \varphi \rangle$ is extensive.

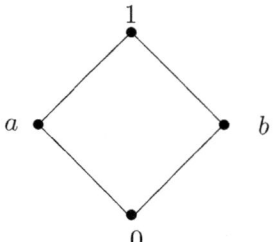

Fig. 3. Complete atomic Boolean lattice B

Example 32. *Consider a complete atomic Boolean lattice B in fig. 3. Let the map $\varphi : \mathcal{A}(B) \longrightarrow B$ be defined by*

$$\varphi(a) = b, \varphi(b) = a$$

Now, $\langle \varphi \rangle(a) = \varphi(b) = a$, $\langle \varphi \rangle(b) = \varphi(a) = b$. We have, $x \leq \langle \varphi \rangle(x)$, for all $x \in \mathcal{A}(B)$. Hence the map $\langle \varphi \rangle$ is extensive, though φ is not extensive. Hence for the extensiveness of the map $\langle \varphi \rangle$, a weaker condition, than the extensive condition on φ is sufficient.

Definition 33. *Let B be a complete atomic Boolean lattice. A map $\varphi : \mathcal{A}(B) \longrightarrow B$ is said to be a cover if*

$$\bigvee_{a \in \mathcal{A}(B)} \varphi(a) = 1$$

Lemma 34. *Let B be a complete atomic Boolean lattice and $\varphi : \mathcal{A}(B) \longrightarrow B$ be a map. Then the following are equivalent:*

i. φ is a cover;
ii. for every $a \in \mathcal{A}(B)$, there exists $b \in \mathcal{A}(B)$ such that $a \leq \varphi(b)$;
iii. $\langle \varphi \rangle$ is extensive.

Proof. (i) $\Longrightarrow$ (ii) Let $a \in \mathcal{A}(B)$. Since φ is a cover, $a = a \wedge 1 = a \wedge (\bigvee\{\varphi(b)/b \in \mathcal{A}(B)\}$. Then there exists $b \in \mathcal{A}(B)$ such that $a \wedge \varphi(b) = a$. This implies there exists $b \in \mathcal{A}(B)$ such that $a \leq \varphi(b)$.

(ii) $\Longrightarrow$ (iii) Let $a \in \mathcal{A}(B)$. Then by assumption, there exists $b \in \mathcal{A}(B)$ such that $a \leq \varphi(b)$. This implies $a \leq \bigwedge_{a \leq \varphi(c)} \varphi(c)$ which implies $a \leq \langle \varphi \rangle(a)$. Thus $a \leq \langle \varphi \rangle(a)$, for all $a \in \mathcal{A}(B)$. Hence $\langle \varphi \rangle$ is extensive.

(iii) $\Longrightarrow$ (i) Suppose $\langle \varphi \rangle$ is extensive. Then $a \leq \langle \varphi \rangle(a)$, for all $a \in \mathcal{A}(B)$. By definition of $\langle \varphi \rangle$, we have for any $a \in \mathcal{A}(B)$, $a \leq \varphi(b)$, for some $b \in A(B)$. Then we have $\{a \in A(B)\} \subseteq \{\varphi(a)/a \in \mathcal{A}(B)\}$ implies $1 = \vee\{a \in \mathcal{A}(B)\} \leq \vee\{\varphi(a)/a \in \mathcal{A}(B)\}$ which implies $\vee\{\varphi(a)/a \in \mathcal{A}(B)\} = 1$. Hence φ is a cover. $\blacksquare$

Proposition 35. *Let B be a complete atomic Boolean lattice and $\varphi : \mathcal{A}(B) \longrightarrow B$ be a map. Then the following are equivalent:*

i. φ is a cover;
ii. $x^{\curlyvee} \leq x$, for all $x \in B$;
iii. $x \leq x^{\curlywedge}$, for all $x \in B$.

Proof. The proof follows from the above lemma and proposition 4.2 in [14]. $\blacksquare$

Proposition 36. *Let B be a complete atomic Boolean lattice and $\varphi : \mathcal{A}(B) \longrightarrow B$ be a map. Then the following are equivalent:*

i. φ is a cover;
ii. $x^{\curlyvee\curlyvee} = x^{\curlyvee}$, for all $x \in B$;
iii. $x^{\curlywedge\curlywedge} = x^{\curlywedge}$, for all $x \in B$.

Proof. The proof follows by using lemma 4.6 and lemma 4.9 in proposition 4.4 of [14]. $\blacksquare$

Corollary 37. *Let B be a complete atomic Boolean lattice and $\varphi : \mathcal{A}(B) \longrightarrow B$ be a map. Then the following are equivalent:*

i. φ is a cover;
ii. $x \longrightarrow x^{\curlyvee}$ is a Alexandrov interior operator;
iii. $x \longrightarrow x^{\curlywedge}$ is a Alexandrov closure operator;

Theorem 38. *Let B be a complete atomic Boolean lattice and $\varphi : \mathcal{A}(B) \longrightarrow B$ be a map. Then*

$$x^{\blacktriangledown} \leq x^{\curlyvee} \leq x \leq x^{\curlywedge} \leq x^{\blacktriangle}$$

holds for all $x \in B$ if, and only if, φ is extensive.

Proof. Let $a \in \mathcal{A}(B)$ be such that $a \leq x^{\blacktriangledown}$. Then we have $\varphi(a) \leq x$. Since $a \leq \varphi(a), \bigwedge_{a \leq \varphi(c)} \varphi(c) \leq \varphi(a) \leq x$. Then $\langle \varphi \rangle(a) \leq x$ implies $a \leq x^{\curlyvee}$. Thus, we have $x^{\blacktriangledown} \leq x^{\curlyvee}$, for all $x \in B$. By duality of the maps $^{\blacktriangledown}, ^{\blacktriangle}$ and $^{\curlyvee}, ^{\curlywedge}$, we have $x^{\curlywedge} \leq x^{\blacktriangle}$. Combining proposition 4.10 with this we have $x^{\blacktriangledown} \leq x^{\curlyvee} \leq x \leq x^{\curlywedge} \leq x^{\blacktriangle}$, for all $x \in B$. Other part is obvious. $\blacksquare$

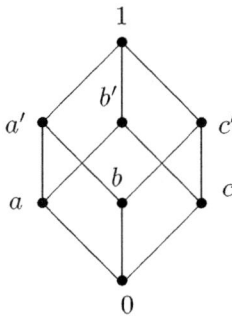

Fig. 4. Complete atomic Boolean lattice B

Example 39. *Consider a complete atomic Boolean lattice B in fig. 4. Let the map $\varphi : \mathcal{A}(B) \longrightarrow B$ be defined by*

$$\varphi(a) = c', \varphi(b) = b' \ and \ \varphi(c) = b'$$

Then obviously φ is a cover but not extensive, for $b \not\leq \varphi(b)$.

We have $\langle\varphi\rangle(a) = a$, $\langle\varphi\rangle(b) = c'$ and $\langle\varphi\rangle(c) = b'$. Since $a \in \mathcal{A}(B)$ is the only atom such that $\varphi(a) \wedge b \neq 0$, $b^{\blacktriangle} = a$. Similarly b is the only atom satisfying the condition $\langle\varphi\rangle(b) \wedge b \neq 0$ which implies $b^{\curlywedge} = b$. Hence $b^{\curlywedge} \not\leq b^{\blacktriangle}$. Also, we have $(b')^{\blacktriangledown} \not\leq (b')^{\curlyvee}$. Thus the chain of inequality in the above theorem does not hold if φ is merely a cover.

Lemma 40. *Let B be a complete atomic Boolean lattice and $\varphi : \mathcal{A}(B) \longrightarrow B$ be a map. Then the following hold:*

i. If φ is extensive and closed, then $\langle\varphi\rangle = \varphi$.
ii. If φ is symmetric and closed, then $\langle\varphi\rangle = \varphi$.

Proof. i) Assume φ is extensive and closed. Then $\langle\varphi\rangle(a) \leq \varphi(b)$, for all $b \in \mathcal{A}(B)$ such that $a \leq \varphi(b)$. Since $a \leq \varphi(a)$ for all $a \in \mathcal{A}(B)$, we have $\langle\varphi\rangle(a) \leq \varphi(a)$, for all $a \in \mathcal{A}(B)$. So, $\langle\varphi\rangle \leq \varphi$. Let $a \in \mathcal{A}(B)$. Let $b \in \mathcal{A}(B)$ be such that $b \leq \varphi(a)$. If there exists $c \in \mathcal{A}(B)$ such that $a \leq \varphi(c)$, then $\varphi(a) \leq \varphi(c)$ (since φ is closed). Therefore $b \leq \varphi(a)$ implies $b \leq \varphi(c)$, for all $c \in \mathcal{A}(B)$ such that $a \leq \varphi(c)$. This shows that $b \leq \bigwedge_{a \leq \varphi(c)} \varphi(c)$ which implies $b \leq \langle\varphi\rangle(a)$. Hence $\varphi = \langle\varphi\rangle$.

ii) Assume φ is symmetric and closed. Let $a \in \mathcal{A}(B)$.
Case 1: $\langle\varphi\rangle(a) = 0$
Then $\bigwedge_{a \leq \varphi(b)} \varphi(b) = 0$ if and only if $a \leq \varphi(b)$, for all $b \in \mathcal{A}(B)$ if and only if $b \leq \varphi(a)$, for all $b \in \mathcal{A}(B)$ (Since φ is symmetric) if and only if $\varphi(a) = 0$. Therefore, we have $\langle\varphi\rangle(a) = 0$ if and only if $\varphi(a) = 0$. Hence $\langle\varphi\rangle = \varphi$ if $\langle\varphi\rangle(a) = 0$.
Case 2: $\langle\varphi\rangle(a) \neq 0$.
Let $b \in \mathcal{A}(B)$ such that $b \leq \langle\varphi\rangle(a)$. Then by definition of $\langle\varphi\rangle$, we have

$$b \leq \varphi(c), \ for \ all \ c \in \mathcal{A}(B) \ such \ that \ a \leq \varphi(c) \qquad (*)$$

Since $\langle\varphi\rangle(a) \neq 0$, $a \leq \varphi(d)$, for some $d \in \mathcal{A}(B)$. Let $d \in \mathcal{A}(B)$ be such that $a \leq \varphi(d)$. Since φ is symmetric, $a \leq \varphi(d)$ implies $d \leq \varphi(a)$. Since φ is closed, we have $\varphi(a) \leq \varphi(d)$ and $\varphi(d) \leq \varphi(a)$. Therefore, $\varphi(a) = \varphi(d)$. Thus, we have $a \leq \varphi(a)$ which implies $b \leq \varphi(a)$ (by(*)). Hence $\langle\varphi\rangle(a) \leq \varphi(a)$. Let $b \in \mathcal{A}(B)$ be such that $b \leq \varphi(a)$. Since $\langle\varphi\rangle(a) \neq 0$, there exists $c \in \mathcal{A}(B)$ such that $a \leq \varphi(c)$. Since φ is closed, $a \leq \varphi(c)$ implies $\varphi(a) \leq \varphi(c)$. So, $b \leq \varphi(c)$, for all $c \in \mathcal{A}(B)$ such that $a \leq \varphi(c)$. Therefore, we have $b \leq \bigwedge_{a \leq \varphi(c)} \varphi(c) = \langle\varphi\rangle(a)$. Thus $\varphi(a) \leq \langle\varphi\rangle(a)$, for all $a \in \mathcal{A}(B)$. Then $\varphi(a) = \langle\varphi\rangle(a)$, for all $a \in \mathcal{A}(B)$. Therefore in the second case also we have $\langle\varphi\rangle = \varphi$.

Remark 41. *If $\varphi : \mathcal{A}(B) \longrightarrow B$ is extensive and closed or symmetric and closed, then the rough approximations of the elements of B with respect to $\langle\varphi\rangle$ and φ are equal.*

The following proposition gives the necessary and sufficient condition for a cover map φ to be equal to its induced map $\langle\varphi\rangle$.

Proposition 42. *Let B be a complete atomic Boolean lattice and $\varphi : \mathcal{A}(B) \longrightarrow B$ be a cover. Then $\langle\varphi\rangle = \varphi$ if, and only if, φ is extensive and closed.*

Proof. If φ is extensive and closed, then by lemma 40 $\langle\varphi\rangle = \varphi$. Conversely, suppose the map φ is a cover and $\langle\varphi\rangle = \varphi$. Then from lemma 34, we have φ is extensive. Also, $\langle\varphi\rangle$ is always closed. This implies φ is also closed. Hence φ is extensive and closed.

5   Conclusion

In this paper, the definition of rough approximation operators given by Jarvinen [13,14] in the general setting of complete atomic Boolean lattices is considered. The ordered sets $(B^{\blacktriangledown}, \leq)$ and $(B^{\blacktriangle}, \leq)$ are considered as algebraic structures by defining their equivalent join and meet operations. Then we prove that if φ is extensive and closed, then $B^{\blacktriangledown}$ and $B^{\blacktriangle}$ are algebraic completely distributive lattices. We prove a theorem for representation of algebraic completely distributive lattices in terms of $B^{\blacktriangledown}$. That is, given any algebraic completely distributive lattice D, there exists a complete atomic Boolean lattice B and an extensive, closed map φ such that $D \cong B^{\blacktriangledown}$. In the case of φ is extensive and symmetric, we prove that $B^{\blacktriangledown}$ and $B^{\blacktriangle}$ are complete ortholattices. Then we prove that, given any complete ortholattice L, there exists a complete atomic Boolean lattice B and an extensive, symmetric map φ such that $L \cong B^{\blacktriangle}$.

In section 4, the definition of rough approximation operators proposed in [1,2] was lifted to the general setting of complete atomic Boolean lattice by defining a new map $\langle\varphi\rangle$ induced from the basic map φ. We investigate some of their properties in this general setting and compare it with rough approximations defined with respect to the basic map φ.

From the example 26 and theorem 38, we conclude that in the generalized approximation space (U, R), the rough approximations $\underline{X}$ and $\overline{X}$ are the better approximates of the set X than the rough approximations $X^{\blacktriangledown}$ and $X^{\blacktriangle}$. So, we may call the rough approximations $\underline{X}$ and $\overline{X}$ with high degree of certainty and possibility as pre - lower and pre - upper approximations of X respectively.

References

1. Allam, A.A., Bakeir, M.Y., Abo-Tabl, E.A.: New approach for Basic Rough set Concepts. In: Slezak, D., Wang, G., Szczuka, M., Düntsch, I., Yao, Y. (eds.) RSFDGrC 2005. LNCS (LNAI), vol. 3641, pp. 64–73. Springer, Heidelberg (2005)
2. Allam, A.A., Bakeir, M.Y., Abo-Tabl, E.A.: Some Methods for Generating Topologies by Relations. Bull. Malays. Math. Sci. Soc. 31(1), 35–45 (2008)
3. Banerjee, M., Chakraborty, M.K.: Rough sets through Algebraic Logic. Fundamenta Informaticae 28, 211–221 (1996)
4. Banerjee, M., Chakraborty, M.K.: Algebras from Rough Sets. In: Sankar, K.P., Polkowski, L., Skowron, A. (eds.) Rough – Neural Computing, pp. 157–185. Springer, Heidelberg (1998)
5. Comer, S.D.: An algebraic approach to the approximation of information. Fundamenta Informaticae 14, 492–502 (1991)
6. Catteneo, G.: Generalized rough sets (Pre - clusivity fuzzy - inutitionistic BZ lattices). Studia Logica 58, 47–77 (1997)
7. Catteneo, G.: Abstract Approximation Spaces for Rough Theories. In: Polkowski, L., skowron, A. (eds.) Rough Sets in Knowledg Discovery I, pp. 59–98. Physica Verlag, Heidelberg (1998)
8. Catteneo, G., Ciucci, D.: Some Methodological Remarks About Categorical Equivalences in the Abstract Approach to Roughness - Part I. In: Wang, G.-Y., Peters, J.F., Skowron, A., Yao, Y. (eds.) RSKT 2006. LNCS (LNAI), vol. 4062, pp. 277–283. Springer, Heidelberg (2006)
9. Degang, C., Wenxiu, Z., Yeung, D., Tsang, E.C.C.: Rough approximations on a complete completely distributive lattice with applications to generalized rough sets. Information Sciences 176, 1829–1848 (2006)
10. Gehrke, M., Walker, E.: On the Structure of Rough Sets. Bulletin of the Polish Academy of Sciences Mathematics 40, 235–245 (1992)
11. Gratzer, G.: General Lattice Theory, 2nd edn. Birkhäuser Verlag, Boston (1998)
12. Iwinski, T.B.: Algebraic Approach to Rough Sets. Bulletin of the Polish Academy of Sciences Mathematics 35, 673–683 (1987)
13. Jarvinen, J.: On the Structure of Rough Approximations. Fundamenta Informaticae 53, 135–153 (2002)
14. Jarvinen, J., Kondo, M., Kortelainen, J.: Modal-Like Operators in Boolean Lattices, Galois Connections and Fixed Points. Fundamenta Informaticae 76, 129–145 (2007)
15. Kalmbach, G.: Orthomodular Lattices. Academic Press Inc., London (1983)
16. Nagarajan, E.K.R., Umadevi, D.: New Approach in Defining Rough Approximations. In: Sakai, H., Chakraborty, M.K., Hassanien, A.E., Ślęzak, D., Zhu, W. (eds.) RSFDGrC 2009. LNCS, vol. 5908, pp. 85–92. Springer, Heidelberg (2009)
17. Nagarajan, E.K.R., Umadevi, D.: A Note on New Approach to Rough sets. Submitted to Transactions on Rough Sets
18. Orlowska, E.: Logic of Nondeterministic Information. Studia Logica XLIV 44, 93–102 (1985)

19. Orlowska, E.: Introduction: What you always wanted to know about rough sets. In: Orlowska, E. (ed.) Incomplete Information: Rough Set Analysis, pp. 1–20. Physica-Verlag, Heidelberg (1998)
20. Pagliani, P.: Rough Sets and Nelson algebras. Fundamenta Informaticae 27, 205–219 (1996)
21. Pagliani, P., Chakraborty, M.K.: A Geometry of approximation Rough Set theory: Logic, Algebra and Topology of Conceptual Patterns. Springer, Heidelberg (2008)
22. Pawlak, Z.: Rough Sets. International Journal of Computer and Information Sciences 5, 341–356 (1982)
23. Polkowski, L.: Rough Sets: Mathematical Foundations. Physica-Verlag, Heidelberg (2002)
24. Pomykala, J., Pomykala, J.A.: The Stone algebra of rough sets. Bulletin of Polish Academy of Sciences Mathematics 36, 495–512 (1988)
25. Slowinski, R., Vanderpooten, D.: A generalized definition of rough approximations based on similarity. IEEE Transactions on Knowledge and Data Engineering 12(2), 331–336 (2000)
26. Szasz, G.: Introduction to Lattice Theory. Academic Press Inc.(London) Ltd., London (1963)
27. Yao, Y.Y.: Two views of the theory of rough sets in finite universes. International Journal of Approximation Reasoning 15, 291–317 (1996)
28. Yao, Y.Y.: Relational Interpretations of Neighborhood Operators and Rough Set Approximation Operators. Information Sciences 111(1-4), 239–259 (1998)
29. Yao, Y.Y.: Constructive and algebraic methods of the theory of rough sets. Information Sciences 109(1-4), 21–47 (1998)
30. Yao, Y.Y.: On generalizing Pawlak approximation operators. In: Polkowski, L., Skowron, A. (eds.) RSCTC 1998. LNCS (LNAI), vol. 1424, pp. 298–307. Springer, Heidelberg (1998)
31. Zhu, W.: Generalized rough sets based on relations. Information Sciences 177(22), 4997–5011 (2007)

Tolerance Rough Set Theory Based Data Summarization for Clustering Large Datasets

Bidyut Kr. Patra and Sukumar Nandi

Department of Computer Science & Engineering, Indian Institute of Technology
Guwahati,
Guwahati, Assam , India, PIN-781039
{bidyut,sukumar}@iitg.ernet.in

Abstract. Finding clusters in large datasets is an interesting challenge in many fields of Science and Technology. Many clustering methods have been successfully developed over the years. However, most of the existing clustering methods need multiple data scans to get converged. Therefore, these methods cannot be applied for cluster analysis in large datasets. Data summarization can be used as a pre-processing step to speed up classical clustering methods for large datasets. In this paper, we propose a data summarization scheme based on tolerance rough set theory termed *rough bubble*. *Rough bubble* utilizes leaders clustering method to collect sufficient statistics of the dataset, which can be used to cluster the dataset. We show that proposed summarization scheme outperforms recently introduced *data bubble* as a summarization scheme when agglomerative hierarchical clustering (single-link) method is applied to it. We also introduce a technique to reduce the number of distance computations required in leaders clustering method. Experiments are conducted with synthetic and real world datasets which show effectiveness of our methods for large datasets.

Keywords: Data summarization, hierarchical clustering method, tolerance rough set theory, large datasets, leaders clustering, rough bubble.

1 Introduction

Clustering problem appears in many different fields like data mining, pattern recognition, bio-informatics, *etc.* Clustering problem can be defined as follows. Let $\mathcal{D} = \{x_1, x_2, x_3, \ldots, x_n\}$ be the set of n patterns, where each x_i is an N-dimensional vector in the given feature space. The clustering activity is to find groups of patterns, called clusters of data in such a way that patterns in a cluster are more similar to each other than patterns in distinct clusters. The clustering methods are mainly divided into two categories *viz.*, partitional and hierarchical method [1].

Partitional clustering methods create a single clustering of $\mathcal{D}$ optimizing a criterion function defined locally or globally. Clustering methods like k-means [2], CLARA [3], DBSCAN [4] are typical examples of partitional methods. These partitional clustering methods scan the dataset many times.

J.F. Peters et al. (Eds.): Transactions on Rough Sets XIV, LNCS 6600, pp. 139–158, 2011.
© Springer-Verlag Berlin Heidelberg 2011

Another partitional clustering method called leaders clustering method [5,6] scans the dataset once. For a given threshold distance τ, it produces a clustering incrementally as follows. For each pattern x, leaders method starts searching a cluster representative called leader l such that $||x - l|| <= \tau$, then x is assigned to this cluster; otherwise x becomes a new cluster representative. The time and space complexity of leaders method are $O(mn)$ and $O(m)$, respectively, where m is the number of leaders.

T. Ho *et al.* [7] proposed a tolerance rough set model (TRSM) based document clustering. Tolerance relation and tolerance class are defined based on co-occurrence of index terms present in documents.

However, all these partitional clustering methods either scan the dataset many times or can find only compact and convex shaped clusters. Therefore, these methods are not suitable for detecting arbitrary shaped clusters in large datasets.

Hierarchical clustering methods create a sequence of clusterings π_1, π_2, π_3, ... π_i ... π_p. It can be generated in two ways.

1. π_{i+1} is a refinement [8] of π_i and is called Divisive (Top-down) approach. Divisive approach starts with single cluster consisting of all patterns in it and forms a sequence of clusterings splitting previous clustering.
2. π_i is a refinement of π_{i+1}. This is called Agglomerative (Bottom-up) approach. In the agglomerative approach, first clustering in the sequence consists of n singleton clusters and successive clusterings are obtained by merging a pair of closest clusters in the previous clustering.

Main advantage of the hierarchical clustering approach over partitional clustering is that hierarchical method can find hierarchy/taxonomy present in data. Clustering methods like single-link (SL) [9], complete-link (CL) [10], average-link (AL) [11], OPTICS [12] are typical examples of hierarchical methods. OPTICS is hierarchical version of DBSCAN method. It can find arbitrary shaped clusters. However, clustering results are influenced by the selection of parameters. The hierarchical clustering methods namely single-link, complete-link and average-link differ in the "distance measure" between a pair of clusters. In complete-link, distance between a pair of clusters is the distance between two farthest points each of them taken from different clusters. In average-link, distance between a pair of clusters is average pairwise distance. Complete-link and average-link are suitable for compact clusters only.

In single-link, distance between a pair of clusters is distance between two closest points (points are from two different clusters). The single-link method starts with n singleton clusters and progressively merges two closest clusters until number of cluster is one. Single-link can find arbitrary shaped clusters. The time and space requirements are $O(n^2)$.

S. K. De et al. [13] proposed rough-approximation based agglomerative clustering method for web access log databases. In this method, a pair of clusters is merged based on similarity upper-approximation of the clusters. The similarity upper-approximation is calculated based on tolerance relation of user transactions. In the same line, P. Kumar *et al.* [14] proposed a hierarchical agglomerative clustering method for sequential datasets. In this method, relative similarity is used to merge a pair of clusters. They showed that their method outperforms classical complete-link clustering method. S. Kawasaki et al. [15] proposed tolerance rough set model (TRSM) based hierarchical document clustering. In this method, tolerance relation is defined based on co-occurrence of index terms present in document sets.

All these clustering methods assume that entire dataset remains in main memory of the machine. This assumption may not be feasible for large datasets. Therefore, these clustering methods are not suitable for large datasets.

To speed up clustering activity in large datasets, data summarizations (data compression) became popular in recent years [16,17,18,19,20]. Data summarization is a scheme to obtain a representative set along with a summary of the dataset. Subsequently, a clustering method can be applied to this summary to find clustering structures of whole dataset.

One of the widely used compression schemes is to use the CF tree constructed by the BIRCH [16] clustering method. The core concept of BIRCH is *Clustering Feature (CF)*. The CF utilizes vector space (Euclidean space) properties to store the summary of k data points $\{\overrightarrow{X_i}\}_{i=1..k}$. A CF is defined as $CF = (k, \overrightarrow{LS}, ss)$, where $\overrightarrow{LS} = \sum_{i=1}^{k} \overrightarrow{X_i}, ss = \sum_{i=1}^{k} \overrightarrow{X_i}^2$. Bradley [19] proposed a data summarization scheme to speed up k-means clustering method. However, these schemes are suitable for finding convex-shaped clusters only.

There exist few data summarization schemes for hierarchical methods. Breunig et al. [17,18] proposed a data summarization scheme called *data bubble* to speed up the hierarchical clustering (OPTICS) method utilizing CF values.

It has two steps. In the first step, data bubble selects a small numbers of seed patterns randomly which requires one scanning of the dataset. In the second step, it assigns each pattern to its nearest seed point and updates statistics of the corresponding seed pattern in the form $CF = (k, \overrightarrow{LS}, ss)$. Each seed pattern with its CF value is called a data bubble. From a pair of CF values, the scheme calculates effective distance between a pair of data bubbles. Breunig et al. [18] showed that data bubble outperforms BRICH [16] as a summarization scheme when the OPTICS method is applied to it. This approach works fine if data bubbles are well separated in the given feature space. However, the approach may underestimate effective distance between a pair of bubbles in some cases. For example, let two data bubbles (O_1, O_2) are close (neighbor) to each other and one data bubble (say O_1) contains patterns from two different clusters, then cluster separation ("gap") becomes invisible to data bubble O_1. Therefore, effective distance between two data bubbles is underestimated and bubbles get merged. This leads to errors in final clustering results.

In order to handle this situation, Zhou and Sander [20] introduced "directional" notion to the data bubble. In this approach, relative position of each point $x \in O_1$ is located by two ways - whether x is in the direction of data bubble O_2 or in the reverse direction of O_2. The point $x \in O_1$ is in the direction of O_2, if $||x - r_{O_2}|| \leq ||r_{O_1} - r_{O_2}||$, where r_{O_1} and r_{O_2} are the representative patterns of O_1 and O_2, respectively. Otherwise, x is in the reverse direction (*i.e.* $||x - r_{O_2}|| > ||r_{O_1} - r_{O_2}||$). Each data bubble computes statistics of its patterns with respect to (in the direction of) all data bubbles. They showed that this approach outperforms other approaches of data bubble [17,18] when the OPTICS method is applied to it. They suggested to apply this approach for single-link clustering method. However, we argue that this approach is not suitable for single-link method due to the following artifacts.

1. Formation of data bubbles are dependent on the initial selection of patterns (seed patterns), which are randomly picked.
2. Lost of cluster separation ("gap") is handled using only distance information between patterns to their representatives.
3. This scheme may detect "gap" in a data bubble (say O_1). However, patterns in O_1 are not reassigned to proper data bubbles. This step is necessary to obtain proper cluster structures.

In this paper, we propose a summarization scheme termed *rough bubble* which is based on tolerance rough set model (TRSM). This scheme utilizes leaders clustering method to collect statistics of each *rough bubble*. We modify single-link clustering method to work with *rough bubble* for clustering large datasets. Modified clustering method is termed as *rough-single-link (R-SL)* method. Our proposed method is considerably faster than that of the single-link method and clustering produced by the proposed method is close to output produced by the single-link method. We show that *rough-single-link* method outperforms hierarchical (single-link) method which uses a recently proposed data bubble [20] as a summarization scheme in clustering results. It is found that proposed methods are suitable for large datasets.

The preliminary version of this paper was presented at a conference (RSFD-GrC 2009) [21]. Our contributions are summarized as follow. (i) A new summarization scheme for speeding up hierarchical clustering method is proposed, which is based on TRSM model and leaders method. (ii) A technique is proposed to reduce the number of distance computations in leaders method, (iii) Clustering results obtained by the proposed method are analyzed using clustering similarity measures (*i.e. Rand Index and Purity*).

The rest of the paper is organized as follows. Section 2 discusses some background of the proposed clustering method. Section 3 describes the proposed summarization scheme and the proposed clustering method. Experimental evaluations and conclusion are discussed in Section 4 and Section 5, respectively.

2   Background of Our Work

Rough set theory has been extensively used in many applications in recent years [22,23]. The fundamental idea of the rough set theory is based on an approximation space $\mathcal{A} = (U, R)$, where U is a nonempty set of objects and R is an equivalence relation (reflexive, symmetric, and transitive) called indiscernibility relation on U [22]. R creates a partition U/R of U, *i.e.*

$$U/R = \{X_1, \ldots, X_i, \ldots, X_p\}$$

where each X_i is an equivalence class of R. These equivalence classes and the empty set are considered as the Elementary sets in $\mathcal{A}$. Elementary sets form the basic granules of knowledge about any set of objects U. Any arbitrary set $X \subseteq U$ can be defined by two crisp sets called lower and upper approximation of X. Each crisp set is finite union of the Elementary sets in $\mathcal{A}$. More formally, one can define the lower and upper approximation as follow.

$$\underline{R}(X) = \bigcup_{X_i \subseteq X} X_i; \quad \overline{R}(X) = \bigcup_{X_i \cap X \neq \emptyset} X_i.$$

The set $BND(X) = \overline{R}(X) - \underline{R}(X)$ is called *boundary* of X in $\mathcal{A}$. Sets $\underline{Edg}(X) = X - \underline{R}(X)$ and $\overline{Edg}(X) = \overline{R}(X) - X$ are called *internal* and *external* edge of X in $\mathcal{A}$, respectively [22].

It is found that the transitive property does not hold in certain application domains (*e.g.* document clustering, information retrieval [7]). In that case, a tolerance relation (reflexive, symmetric) is used relaxing transitive property [7,24,25,26,27]. Other direction of advancement of classical rough set theory is found in [28]. R. Slowinski in [25] argued in favor of not imposing the symmetric property in tolerance relation. A similarity relation is defined instead. In tolerance relation based rough set model (TRSM), the basic granules of knowledge are the tolerance classes, which are overlapped. Therefore, the tolerance relation T does not create a partition of U. It creates a cover of the dataset. Any set $X \subseteq U$, can be characterized by the lower and upper approximations as follow.

$$\underline{T}(X) = \{x \in X : T(x) \subseteq X\}; \quad \overline{T}(X) = \{x \in U : T(x) \cap X \neq \emptyset\}$$

where $T(x)$ is a tolerance class. In accordance with the classical rough set theory, we can define the set $T_{BND} = \overline{T}(X) - \underline{T}(X)$ as *tolerant boundary*. The sets $T_{\underline{Edg}} = X - \underline{T}(X)$ and $T_{\overline{Edg}} = \overline{T}(X) - X$ are termed as *tolerant internal* and *tolerant external* edge of X, respectively.

Our proposed summarization scheme discussed in the next section uses above tolerance rough set model.

3   The Proposed Method

In general, clustering methods either scan dataset several times or assume dataset to be available in main memory of the machine. These assumptions are not feasible for large dataset because computational complexity (including I/O) increases significantly with the size of the dataset. Whereas, summarization or a suitable representative set of the dataset can be kept in the main memory for processing. This makes clustering scheme based on the summarization be a potential contender for clustering large dataset. In this section, a new summarization scheme called *rough bubble* is proposed. Subsequently, single-link method is applied to *rough bubble* for final clustering. This hybrid scheme speeds up the clustering method from $O(n^2)$ to $O(mn)$, where n and m are the size of the dataset and number of the *rough bubbles*, respectively. To further speed up the computation of *rough bubble*, we use triangle inequality for reduction in distance computation.

3.1   Speeding Up Leader Clustering Method

Triangle inequality property of the metric space is used to reduce the number of distance computations of the leaders method. We term this approach as *Speedy leader*. In recent years, triangle inequality property of the metric space has been used to reduce the distance computations in the clustering methods [29,30,31,32]. Elkan [29] used triangle inequality to reduce the distance computation in k-means clustering method. Nassar [30] used for speeding up summarization scheme (data bubble). Recently, Marzena et al. [31] proposed to speed up DBSCAN method using triangle inequality. Triangle inequality based reduction technique for leaders method has also been proposed recently for speeding up average-link method [32]. In this paper, another approach for reduction in distance computation using triangle inequality is proposed and discussed next.

In this approach, reduction in distance computations is achieved using the condition given in lemma 1.

Lemma 1. *[29,30] Let l_1, l_2 be the two leaders produced by the leaders clustering method and x be a pattern of dataset $\mathcal{D}$. If $d(l_1, l_2) \geq 2 * d(l_1, x)$, then $d(x, l_2) \geq d(x, l_1)$, where d is a distance function over a metric space $M = (\mathcal{D}, d)$.*   □

Lemma 1 states that given the distances between a pair of leaders (l_1, l_2) and distance between $(x \in \mathcal{D}, l_1)$, we can avoid the distance calculation between a pattern x and leader l_2 if distance between the leaders is greater than twice of the distance between x and leader l_1. Because x is closer to l_1 than leader l_2.

Corollary 1. *Let τ be the threshold of leaders clustering method and l_1, l_2 be two leaders produced by the leaders method. For any $x \in \mathcal{D}$, if $d(x, l_1) > \tau$ and $d(l_1, l_2) \geq 2 * d(x, l_1)$. Then, x can neither be follower of l_1 nor l_2.*   □

Corollary 1 states that given the distance $d(l_1, l_2)$ between a pair of leaders (l_1, l_2) and distance $d(x, l_1)$ between a pattern x and the leader l_1, pattern x cannot be the follower of l_2 if $d(l_1, l_2) \geq 2 * d(x, l_1)$. If l_1 is the nearest leader of x and distance between x and l_1 is greater than leaders threshold τ, then x will be the new leader.

The *Speedy leader* works as follows. The scheme requires a distance matrix for leaders. This distance matrix can be generated hand-in-hand during the generation of leaders. Therefore, we may avoid actual distance computation between x and l_2 only calculating distance $d(l_1, x)$ using Corollary 1.

Let τ be the leader's threshold. Let $\mathcal{L} = \{l_1, l_2, \ldots, l_k\}$ be the set of leaders generated at an instant and all be marked as "unprocessed" leaders. The technique starts with calculating the distance between a new pattern x and leader l_f (where l_f is the earliest generated leader among the set of "unprocessed" leaders). If $d(x, l_f) \leq \tau$, then x becomes the follower of leader l_f. If $d(x, l_f) > \tau$, we can avoid the distance computations from leaders $l_i \in L - \{l_f\}$ where $d(l_f, l_i) \geq 2 * d(x, l_f)$. Leaders l_i, l_f are marked as "processed" (pruned) leaders. If all leaders are pruned then x becomes a new leader and added to $\mathcal{L}$. If

---

Algorithm 1. Speedy leader($\mathcal{D}, \tau$)

1: $\mathcal{L} \leftarrow \{l_1\}$; { Let $l_1 \in \mathcal{D}$ be the first scanned pattern}
2: **for** each $x \in \mathcal{D} \setminus \{l_1\}$ **do**
3:     $S \leftarrow \mathcal{L}$; $MIN = \infty$;
4:     **while** (x does not become a follower and S is not empty) **do**
5:         Pick a leader l_i and delete it from S. { l_i is earliest generated leader in S}
6:         **if** $d(x, l_i) \leq \tau$ **then**
7:             x becomes a follower of l_i; break;
8:         **else if** $d(x, l_i) < MIN$ **then**
9:             $MIN = d(x, l_i)$;
10:            **for** each leader $l_k \in S(l_k \neq l_i)$ **do**
11:                **if** $d(l_i, l_k) \geq 2 * d(l_i, x)$ **then**
12:                    delete l_k from set S.
13:                **end if**
14:            **end for**
15:        **end if**
16:    **end while**
17:    **if** (x is not a follower of any existing leaders in $\mathcal{L}$) **then**
18:        x becomes new leader and added to $\mathcal{L}$.
19:    **end if**
20: **end for**
21: Output $\mathcal{L}^\star = \{(l, followers(l)) \mid l \in \mathcal{L}\}$.

---

all leaders are not marked as "processed", we repeat same procedure of calculating distance between x with next unprocessed leader $l_u \in \mathcal{L}$ if $d(x, l_u) < d(x, l_f)$. If no (unprocessed) $l_u \in \mathcal{L}$ is found such that $d(x, l_u) < d(x, l_f)$, then there cannot be a leader l_j such that $d(x, l_j) \leq \tau$; so x becomes a new leader and added to $\mathcal{L}$. The whole procedure of *Speedy leaders* is depicted in Algorithm 1. Summarization scheme proposed in next subsection uses *Speedy leader* to obtain a summary of the dataset.

3.2   Summarization Scheme

The proposed summarization scheme utilizes the leaders clustering method (as given in Algorithm 1) and tolerance rough set theory. Let the leaders threshold distance be τ and $\mathcal{L} = \{l_1, l_2, \ldots, l_m\}$ be the set of all leaders of the dataset $\mathcal{D}$. For a pattern x and leader l_1, l_2, we may observe a scenario such that $||l_1 - x|| <= \tau$, $||l_2 - x|| <= \tau$. Then x can be eligible to be the follower of both leaders. However, classical leaders clustering method assigns x to a leader which one is observed first (say l_1). It may happen that pattern x and leaders l_2 belong to a same cluster, whereas l_1 belongs to different cluster. So, leader l_1 contains patterns from two different clusters and it also contains cluster separation ("gap"). Clapping of pattern x with leader l_1 makes cluster separation invisible to l_1. This leads to error in obtaining final clustering structures in data.

If we allow x to be the member of l_2 also, we may recover cluster-structure in further analysis. Therefore, we virtually allow l_1, l_2 to share pattern x.

Let $\mathcal{L}^r = \{l_1^r, l_2^r, \ldots, l_m^r\}$ be the set of leaders where a leader may share patterns with others leaders. We call $l_i^r \in \mathcal{L}^r$ as a *rough leader* and the $\mathcal{L}^r$ as the set of rough leaders. Note that there is a one to one correspondence between the set $\mathcal{L}$ and $\mathcal{L}^r$. In the other words, let $\{l_i\}$ and $\{l_i^r\}$ be the sets of followers of ith leader $l_i \in \mathcal{L}$ and rough leader $l_i^r \in \mathcal{L}^r$, respectively. Then $\{l_i\} \subseteq \{l_i^r\}$, $l_i = l_i^r$ and $|\mathcal{L}| = |\mathcal{L}^r|$. Therefore, we may use l_i and l_i^r interchangeably throughout the paper.

Now we can apply the TRSM to the dataset $\mathcal{D}$. We consider dataset $\mathcal{D} = U$ for TRSM. Let $T \subseteq \mathcal{D} \times \mathcal{D}$ be a tolerance relation (reflexive, symmetric). One can define a tolerance class based on the proximity of the patterns in the dataset.

Definition 1 (Tolerance class). *If $x \in \mathcal{D}$, then the tolerance class of x is $T(x) = \{x_j \in \mathcal{D} \mid \ \ ||x - x_j|| <= \delta\}_{\delta \in \mathbb{R}^+}$.* □

Therefore, tolerance class of a pattern x is a set of patterns whose distance from x is less than equal to a given threshold δ. With different values of δ, we can get different tolerance relations of $\mathcal{D}$. Here, we consider the value of $\delta = \tau/2$. One can define lower approximation and tolerant internal edge for follower set of a leader $\{l_i^r\} \subseteq \mathcal{D}$ are as follow.

Definition 2 (Lower approximation). *Let T be a tolerance relation on $\mathcal{D}$ and l^r be a rough leader obtained by leader threshold distance τ. The lower approximation of $\{l^r\}$ is $\underline{T}(\{l^r\}) = \{x_j \in \mathcal{D} \mid \ \ ||l^r - x_j|| <= \delta\}$, where $\delta = \tau/2$.* □

It may be noted that $\{l^r\}$ is a subset of $\mathcal{D}$. Therefore, we can define lower approximation of $\{l^r\}$ as follows. This definition is based on observation. $\underline{T}(\{l^r\}) = \{x_j \in \{l^r\} \mid T(x_j) \subseteq \{l^r\}\}$, where $T(x_j)$ is a tolerance class of x_j in the approximation space $(\mathcal{D}, T)$.

Definition 3 (Tolerant internal edge). *Let T be a tolerance relation on $\mathcal{D}$ and ^{r} be a leader obtained using threshold distance τ. The tolerant internal edge of $\{l^r\}$ is defined as $T_{Edg}(\{l^r\}) = \{x_j \in \mathcal{D} \mid \ \ \tau/2 < ||l^r - x_j|| <= \tau\}$.* □

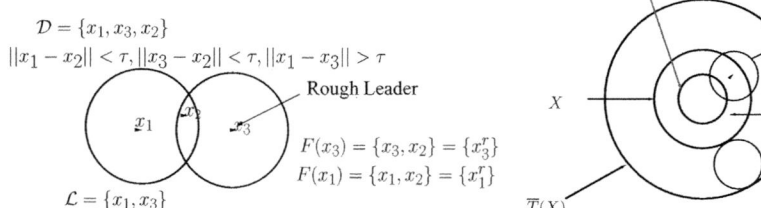

$\mathcal{D} = \{x_1, x_3, x_2\}$
$||x_1 - x_2|| < \tau, ||x_3 - x_2|| < \tau, ||x_1 - x_3|| > \tau$

Rough Leader

$F(x_3) = \{x_3, x_2\} = \{x_3^r\}$
$F(x_1) = \{x_1, x_2\} = \{x_1^r\}$

$\mathcal{L} = \{x_1, x_3\}$

$\underline{T}(X)$

$T(x)$

X

$T_{Edg} = X - \underline{T}(X)$

$\overline{T}(X)$

Fig. 1. Rough Leaders with follower set $F()$ **Fig. 2.**   Lower approximation and Tolerant internal edge of X

In the proposed summarization scheme, we need to know the information of followers of each leader in the direction of all other leaders. Let there be two leaders $l_1, l_2 \in \mathcal{L}$. The leader l_1 stores the information of its followers which are in the direction of l_2 (w. r. t l_2) and vice-versa. More formally :

Definition 4. *Let l_1 and l_2 be the two leaders. The set of followers of l_1^r in the direction of l_2^r and set of followers of l_2^r in the direction of l_1^r are $l_1^{(l_2)}$ and $l_2^{(l_1)}$, respectively, where*

$$l_1^{(l_2)} = \{x \in \{l_1^r\} : ||l_2 - x|| \leq ||l_1 - l_2||\},$$
$$l_2^{(l_1)} = \{x \in \{l_2^r\} : ||l_1 - x|| \leq ||l_1 - l_2||\}. \qquad \square$$

Based on discussion of TRSM, summary information of followers of each leader w.r.t. all other leaders is stored. For this purpose, dataset is scanned one more time in same order using leaders clustering method. Following information of leader l_1 is stored w.r.t. leader l_2.

$$(\; k^{(l_2)}, \;\; l_1, \;\; ld^{(l_2)}, \;\; sd^{(l_2)}, \;\; \underline{T}(\{l_1\}), \;\; |T_{\underline{Edge}}|, \;\; |Com_{l_1}^{(l_2)}|, \;\; ld_{Edge}^{(l_2)}, \;\; sd_{Edg}^{(l_2)} \;)$$

where $k^{(l_2)} = |l_1^{(l_2)}|$,

$$ld^{(l_2)} = \sum_{j=1}^{k^{(l_2)}} d_j, \quad sd^{(l_2)} = \sum_{j=1}^{k^{(l_2)}} d_j^2, \quad d_j = ||l_1 - x_j|| \text{ and } x_j \in l_1^{(l_2)}$$
$$\underline{T}^{(l_2)}(\{l_1^r\}) = \{x_j \in \underline{T}(\{l_1^r\}) : ||l_2^r - x_j|| \leq ||l_1 - l_2||\}.$$
$$T_{Edg}^{(l_2)}(\{l_1^r\}) = \{x_j \in T_{Edg}(\{l_1^r\}) : ||l_2^r - x_j|| \leq ||l_1 - l_2||\},$$
$$Com_{l_1}^{(l_2)} = \{x \in \mathcal{D} : \tau/2 < ||x - l_1||, ||x - l_2|| \leq \tau\}; \quad ld_{Edg}^{(l_2)} = \sum_{i=1} d_i;$$
$$sd_{Edg}^{(l_2)} = \sum_{i=1} d_i^2, d_i = ||l_1 - x_i|| \text{ and } x_i \in l_1^{(l_2)}, \;\; ||l_1 - x_i|| > \tau/2.$$

The set of shared patterns $Com_{l_1}^{(l_2)}$ between l_1^r and l_2^r is actually grouped with leader $l_1 \in \mathcal{L}$ as l_1 is observed prior to l_2. A leader l with statistics of its followers w.r.t. all other leaders is termed as *rough bubble*, $l(l^r)$ being the representative pattern. Let $\mathbb{S}$ be the set of *rough bubbles*, which is the summary of whole dataset. It is noted that followers of leader $l \in \mathcal{L}$ are also the members of a *rough bubble* $S \in \mathbb{S}$, whose representative is l^r. The whole process of summarization is depicted in Algorithm 2.

Complexity of the Rough bubble. The complexity of our summarization scheme is analyzed as follows.

1. The set $\mathcal{L}$ can be obtained in $O(mn)$ time, where $m = |\mathcal{L}|$. The space requirements for $\mathcal{L}$ and distance matrix are $O(m)$ and $O(m^2)$, respectively.
2. Computing statistics of each rough leaders (*rough bubble*) takes times of $O(n)$. For all leaders, scheme takes time of $O(mn)$. The space complexity is $O(m^2)$.

Overall time and space requirements of our summarization scheme are $O(mn)$ and $O(m^2)$, respectively.

Algorithm 2. *rough-bubble* $(\mathcal{D}, \tau)$

1: Apply leaders clustering method to $\mathcal{D}$ to obtain a set of leaders.
   Let $\mathcal{L} = \{l_1, l_2, \ldots, l_m\}$ be the set of leaders.
   (Distance Matrix for leaders can be generated along with the formation of $\mathcal{L}$.)
2: Apply leaders clustering method to collect statistics of each rough leader w.r.t. all other rough leaders incrementally. /* Dataset is scanned for second time in same order */
   Let $\mathbb{S} = \{S_1, S_2, \ldots, S_m\}$ be the summary of $\mathcal{D}$.
3: Output $\mathbb{S}$.

The summary information computed in this subsection for each *rough bubble* is used in next subsection for calculation of distances between a pair of *rough bubbles*.

3.3   Clustering Method

The summary information obtained for dataset in previous section is used here for clustering. We have applied single-link method to this summary information to obtain the clustering of the whole dataset. We term this method as rough-single-link (R-SL) method. Using the information of each *rough bubble*, we define a more specific distance measure between a pair of *rough bubbles*. The distance function *dist* is defined as follows. $dist : \mathbb{S} \times \mathbb{S} \to \mathbb{R}_{\geq 0}$, where $\mathbb{R}_{\geq 0}$ is set of non negative real numbers.

Let the average and standard deviation of distances from l_1 to its followers w.r.t. l_2 be $\mu_{l_1}^{(l_2)}$ and $\sigma_{l_1}^{(l_2)}$, respectively. Similarly, $\mu_{l_2}^{(l_1)}$ and $\sigma_{l_2}^{(l_1)}$ be the average and standard deviation of distances from l_2 to its followers w.r.t. l_1, respectively. The $\mu_{l_1}^{(l_2)}$ and $\sigma_{l_1}^{(l_2)}$ are calculated using information available in *rough bubble* S_1 as follow.

$$\mu_{l_1}^{(l_2)} = \frac{ld^{(l_2)}}{k^{(l_2)}}; \quad \sigma_{l_1}^{(l_2)} = \sqrt{\frac{sd^{(l_2)}}{k^{(l_2)}} - (\mu_{l_1}^{(l_2)})^2}$$

Similarly, we can calculate $\mu_{l_2}^{(l_1)}$ and $\sigma_{l_2}^{(l_1)}$ for the *rough bubble* S_2. The specific distance between a pair of *rough bubbles* S_1, S_2 (S_1 appears prior to S_2) are noted next.

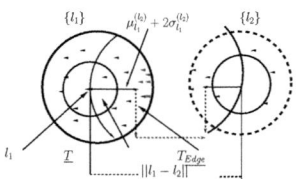

Fig. 3. $dist(S_1, S_2) = \max(||l_1 - l_2|| - (\mu_{l_1}^{(l_2)} + 2\sigma_{l_1}^{(l_2)}) - (\mu_{l_2}^{(l_1)} + 2\sigma_{l_2}^{(l_1)}), 0)$

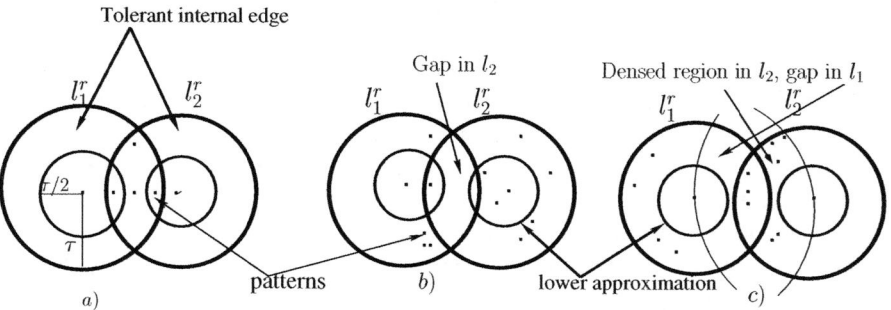

Tolerant internal edge

Gap in l_2

Densed region in l_2, gap in l_1

l_1^r l_2^r l_1^r l_2^r l_1^r l_2^r

$\tau/2$

τ

patterns   b)   lower approximation

a)   c)

Fig. 4. (a) Lower approx. of l_1^r and tolerant internal edge of l_2^r intersect each other.;   (b) Lower approx. of l_1^r intersects with tolerant internal edge of l_2^r, $\underline{T}^{(l_1)}(l_2^r) \cap T_{\underline{Edg}}^{(l_2)}(\{l_1^r\}) = \emptyset$; (c) Intersection of two tolerant internal edges.

1. $dist(S_1, S_2) = 0$, if $||l_1 - l_2|| < 2\tau$ and the following conditions are satisfied simultaneously,
   (a) Lower approximation of l_1^r intersects with tolerant internal edge of l_2^r and vice-versa.
   (b) There is a non empty set of patterns in tolerant internal edge of l_2^r. However, the set does not include points from lower approximation of l_1^r. *i.e.*
   ($T_{Edg}(\{l_2^r\}) \setminus \underline{T}(\{l_1^r\})$) $\neq \emptyset$.
2. If $||l_1 - l_2|| < 2\tau$, then one *rough bubble* (say S_1) may contain patterns from more than one clusters. As a result, S_1 may have gap (a region having no patterns) in the direction of $l_2 \in S_2$ or vice versa (Fig. 4). We handle these gaps reassigning correct position of patterns from one *rough bubble* to other. We have few scenarios as follow.
   (a) S_1 contains gap in the direction of S_2 if the following conditions are satisfied simultaneously, (i) Lower approximation of l_1^r does not intersect with tolerant internal edge of l_2^r. (ii) Lower approximation of l_2^r intersect with tolerant internal edge of l_1^r. (iii) There exsist only shared patterns $Com_{l_1}^{l_2}$ in the tolerant internal edge of l_1^r. *i.e.*   ($T_{Edg}(\{l_1^r\}) \setminus Com_{l_1}^{l_2}$) $= \emptyset$.
   In this case, the region of l_2^r is having more patterns towards l_1^r than the patterns of l_1^r w.r.t. l_2^r. Therefore, the shared patterns ($Com_{l_1}^{l_2}$) should be with S_2. The method removes the shared patterns from S_1 and patterns are added to S_2. The statistics of both *rough bubbles* (S_1, S_2) are updated accordingly.

$ld^{l_2} = ld^{l_2} - |Com_{l_1}^{(l_2)}| * (\frac{ld^{l_2}}{k^{l_2}})$;

$sd^{l_2} = sd^{l_2} - |Com_{l_1}^{(l_2)}| * (\frac{ld^{l_2}}{k^{l_2}})^2$;

$ld^{l_1} = ld^{l_1} + |Com_{l_1}^{(l_2)}| * (||l_1 - l_2|| - \frac{ld^{l_2}}{k^{l_2}})$;

$sd^{l_1} = sd^{l_1} + |Com_{l_2}^{(l_1)}| * (||l_1 - l_2|| - \frac{ld^{l_2}}{k^{l_2}})^2$.

(b) S_2 contains gap in the direction of S_1, if converse of the above conditions $((a)(i), (ii), (iii))$ hold simultaneously (Fig 4b). However, re-assignment is not required as shared pattern $(Com_{l_1}^{l_2})$ are already grouped with S_1.

(c) S_1 contains gap in the direction of S_2 if the following conditions hold (Fig. 4c) (i) Tolerant edges of l_1^r and l_2^r intersect each other. (ii) Let number of patterns of internal edge of l_1^r and l_2^r excluding the shared patterns $(Com_{l_1}^{l_2})$ be c_1 and c_2, respectively. If the ratio of c_1 and c_2 is not more than a given threshold h (0.5).

In this case re-assignment is necessary. The proposed method removes shared patterns from S_1 and adds to S_2. It updates the corresponding statistics (Same as $2(a)$).

According to the cases mentioned above, we restructure the *rough bubble* and recalculated $\mu_{l_1}^{(l_2)}$, $\sigma_{l_1}^{(l_2)}$, $\mu_{l_2}^{(l_1)}$ and $\sigma_{l_2}^{(l_1)}$. Distance between a pair of *rough bubbles* is defined in Definition 5.

Definition 5 (Distance between two *rough bubbles*). *The distance between a pair of rough bubbles S_1 and S_2 is defined as*

$$dist(S_1, S_2) = \max\{ \, [\, ||l_1 - l_2|| - (\mu_{l_1}^{(l_2)} + 2\sigma_{l_1}^{(l_2)}) - (\mu_{l_2}^{(l_1)} + 2\sigma_{l_2}^{(l_1)}) \,], \, 0 \, \}.$$

The $dist(.)$ function satisfies following properties.

1. $dist(S_1, S_2) \geq 0$
2. $dist(S_1, S_2) = 0$, if both of them are part of same cluster or $S_1 = S_2$.
3. $dist(S_1, S_2) = dist(S_2, S_1)$.

We apply an agglomerative hierarchical clustering method (single-link) using the distance measure defined in (Definition 5) to set of *rough bubbles* $\mathbb{S}$. This gives a hierarchy of clusterings of *rough bubbles*. Next, each *bubble* is replaced by its members. Finally, we get a hierarchy of clusterings of the dataset $\mathcal{D}$. The whole clustering method is noted in Algorithm 3.

Complexity Analysis. The complexity of R-SL method is analyzed as follows.

1. The summarization scheme takes time of $O(mn)$. The space complexity is $O(m^2)$.
2. Calculating distance-matrix for *rough-data sphere* takes $O(m^2)$ time.
3. Clustering the bubbles using single-link method takes $O(m^2)$.

The overall time and space complexity of R-SL are $O(mn)$ and $O(m^2)$, respectively. The proposed method scans the dataset twice. Experimentally, it is found that R-SL is significantly faster than the classical single-link clustering method.

4   Experimental Results

In this section, we study performance of our proposed computation reduction technique and rough-single-link (R-SL) method. We conducted experiments with

---

Algorithm 3. rough-single-link($\mathcal{D}, \tau$)

1: Apply *rough-bubble* $(\mathcal{D}, \tau)$ as given in Algorithm 2.
   Let $\mathbb{S} = \{S_1, S_2, \ldots, S_m\}$ be the output of *rough-bubble* $(\mathcal{D}, \tau)$.
2: **for** each leader l_i^r in $S_i \in \mathbb{S}$ **do**
3:   **for** each leader l_j^r in $S_j \in \mathbb{S} : l_i \neq l_j$ **do**
4:     **if** $(\|l_i^r - l_j^r\| \leq 2\tau)$ **then**
5:       **if** (lower approx. of $l_i^r(l_j^r)$ intersects with $l_j^r(l_i^r)$ and intersection of internal edges $\neq \emptyset$) **then**
6:         $dist(S_i, S_j) = 0$;   {Two data spheres are part of a cluster.}
7:       **else if** (lower approx. of l_i^r intersects with l_j^r and ($l_i^r \cap$ common edges) $\neq \emptyset$) **then**
8:         $dist(S_i, S_j) = 0$;   {No reassignment of patterns}
9:       **else if** (lower approx. of l_i^r not intersect with l_j^r but lower approx. of l_j^r does and ($l_j^r \cap$ common edges) $\neq \emptyset$) **then**
10:         Add $Com_{l_i}^{(l_j)}$ and lower approx. of l_i^r to S_j; Remove these from S_i.
11:         Update $ld^{l_j}, sd^{l_j}, ld^{l_i}, sd^{l_i}, k^{(l_i)}, k^{(l_j)}$; {There is a gap in S_1 w.r.t. l_2}
12:     **end if**
13:     **if** $(Com_{l_i}^{(l_j)} \neq \emptyset$ and $(\frac{|T_{Edg}^{(l_j)}(\{l_i^r\}) \setminus Com_{l_i}^{(l_j)}|}{|T_{Edg}^{(l_i)}(\{l_j^r\}) \setminus Com_{l_i}^{(l_j)}|}) <= h$) **then**
14:       Add $Com_{l_i}^{(l_j)}$ to S_j and remove from S_i;
15:       Updates $ld^{l_j}, sd^{l_j}, ld^{l_i}, sd^{l_i}, k^{(l_i)}, k^{(l_j)}$
16:     **end if**
17:     **end if**
18:     Calculate $\mu_{l_i}^{(l_j)}, \sigma_{l_i}^{(l_j)}, \mu_{l_j}^{(l_i)}, \sigma_{l_j}^{(l_i)}$
19:     $dist(S_i, S_j) = \max(\|l_i - l_j\| - (\mu_{l_i}^{(l_j)} + 2\sigma_{l_i}^{(l_j)}) - (\mu_{l_j}^{(l_i)} + 2\sigma_{l_j}^{(l_i)}), 0)$
20:   **end for**
21: **end for**
22: Apply single-link to $\mathbb{S}$. Let the result be $\pi_{\mathbb{S}} = \{\pi_1, \pi_2, \ldots, \pi_{|\mathbb{S}|}\}$, set of clusterings of $\mathbb{S}$.
23: Expand each $S_i \in \pi_i \in \pi_{\mathbb{S}}$. Therefore, $\pi_{\mathcal{D}} = \{\pi_{D_1}, \pi_{D_2}, \ldots, \pi_{D_{|\mathbb{S}|}}\}$.
24: Output $\pi_{\mathcal{D}}$

Table 1. Datasets Used

Data Set	# Patterns	# Features
Spiral(Synthetic)	3330	2
Pendigits	7494	16
Shuttle	40000	9
Circle(Synthetic)	28000	2
a8a	32561	123
GDS10	23709	28

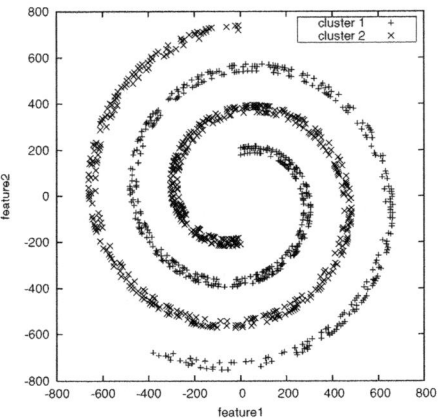

Fig. 5. Spiral Dataset

synthetic as well as real world datasets (http://archive.ics.uci.edu/ml/, (http://www.ncbi.nlm.nih.gov/geo/)) without class labels. Spiral and Circle are 2-dimensional synthetic datasets. The plot of Spiral dataset is given in (Fig 5). Brief description of the datasets are given in Table 1.

We used Rand-Index(RI) [33] and Purity [34] to show the effectiveness of our proposed clustering method. Rand Index (RI) is a similarity measure between a pair of clusterings of a dataset. Let $\mathcal{P}$ be the set of all clusterings of $\mathcal{D}$. RI is defined as follows.
$RI : \mathcal{P} \times \mathcal{P} \to [0, 1]$. For $\pi_1, \pi_2 \in \mathcal{P}$,

$$RI(\pi_1, \pi_2) = \frac{a + e}{a + b + c + e}$$

where, a is the number of pairs of patterns belonging to the same cluster in π_1 and to a same cluster in π_2, b is the number of pairs belonging to a same cluster in π_1 but to different clusters in π_2, c is number of pairs belonging to different clusters in π_1 but to a same cluster in π_2, and e is the number of pairs of patterns belonging to different clusters in π_1 and to different clusters in π_2.

Let π_1 and π_2 be two clusterings generated by classical single-link method and proposed rough-single-link method, respectively. Then, *Purity* is defined as follows.

$$Purity(\pi_1, \pi_2) = \frac{1}{|\mathcal{D}|} \sum_{i=1}^{|\pi_2|} \max_j | C_i^{(2)} \cup C_j^{(1)} |$$

where $C_i^{(2)} \in \pi_2$ and $C_j^{(1)} \in \pi_1$.

We show first performance of proposed *Speedy leader*, followed by proposed rough-single-link method in the following subsections.

4.1   Performance of Speedy Leaders

We implemented leaders clustering and *Speedy leaders* using C language and executed on Intel Core 2 Duo with $2GB$ RAM IBM PC. These two methods are tested with Circle and Shuttle datasets. The detailed results are shown in Table 2 and Fig 6. For the Circle dataset, with leader's threshold $\tau = 0.1$, our proposed *Speedy leader* does 60 millions less distance calculations to achieve same results as that of the classical leaders method. For the other values of τ $(0.2, 0.3, 0.4, 0.6)$, proposed *Speedy leader* performs significantly less computations compared to that of the classical leaders method (Table 2).

Table 2. Performance of Speedy leader for Circle dataset

Threshold (τ)	Method	# Computations (in Million)
0.1	Leaders	90.13
	Speedy leader	**28.87**
0.2	Leaders	20.03
	Speedy leader	**2.53**
0.3	Leaders	11.33
	Speedy leader	**1.09**
0.4	Leaders	5.97
	Speedy leader	**0.57**
0.5	Leaders	3.85
	Speedy leader	**0.38**
0.6	Leaders	2.81
	Speedy leader	**0.35**

To show the performance of the proposed computation reduction technique with variable dataset size, experiments were conducted with Shuttle dataset with leaders threshold $\tau = 0.001$. This is reported in Fig 6. It may be noted that with the increase of the data size, number of distance calculations in *Speedy leader* reduces significantly compared to that of the classical leaders.

4.2   Application of Single-Link Method to Data Bubble

As a summarization scheme data bubble works fine when OPTICS is applied to it. However, it is not clear whether it can produce consistent clusterings when single-link is applied to it. We tested data bubble with *Spiral* dataset. The clustering results of data bubble based single-link are compared with the classical single-link using the Rand Index [33]. We found that clustering results are inconsistent (Table 3). This is due to the following facts.

– Data bubbles select patterns randomly. Selected patterns do not cover all clusters.

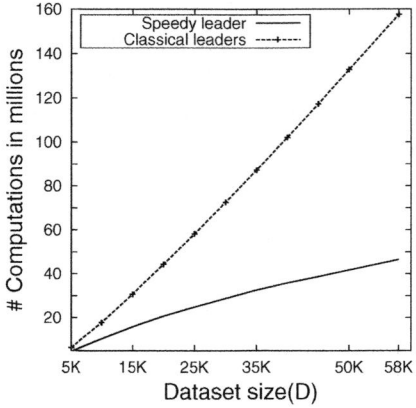

Fig. 6. Number of distance computations Vs dataset size for Shuttle data ($\tau = 0.001$)

Table 3. Results produced by data buble SL method

Dataset	Number of bubbles	Rand Index
Spiral	266	0.499-1.000
	208	0.503-0.767

– Data bubbles may detect gaps in a data bubble. However, reassignment of some patterns (re-structure the data bubble) are not suggested. Therefore, a cluster loses some patterns to other clusters.

Our proposed summarization scheme restructures *rough bubbles* if necessary. We compare performance of R-SL method in next subsection.

4.3 Performance of R-SL Clustering Method

To show the effectiveness of our proposed clustering method (R-SL), we implemented three clustering methods namely, classical SL, Data bubble based single-link [1] and our proposed *R-SL* method in Intel Xeon CPU ($3.6GHz$) with $8GB$ RAM Workstation .

Table 4 shows the detailed performance of the three clustering methods for different datasets. The Compression Ratio (CR) is defined as $CR = n/m$ where, m is the cardinality of the representative set of the data. We used Rand Index (RI) and Purity to compare the clusterings produced by the three methods. For this purpose, we considered a clustering in which number of clusters are same as the number of classes of the datasets for each method. RI and *Purity* (Table 4) are computed between the partitions produced by single-link (SL), proposed *R-SL* and single-link, data bubble based single-link (Data bubble-SL).

---

[1] There is no clustering method called Data bubble based single-link in literature. We simulate single-link clustering method using data bubble summarization scheme.

Table 4. Results for different datasets

Dataset	Compression Ratio($\frac{n}{m}$)	Method	Time (in sec.)	Rand Index (RI)	Purity
Spiral (Synthetic)	12.5	R-SL	0.1	1.000	1.000
		Data bubble-SL	0.1	0.885	0.887
		SL	40.1	–	–
Circle (Synthetic)	50.9	R-SL	1.5	1.000	1.000
		Data bubble-SL	0.9	0.995	1.000
		SL	36,415.3	–	–
a8a	12.4	R-SL	145.5	0.998	1.000
		Data bubble-SL	141.6	0.849	0.899
		SL	74,108	-	-

Table 5. Results for *Circle* dataset

Number of cluster	Method	Rand Index (RI)	Purity
2	R-SL	1.000	1.000
	Data bubble-SL	1.000	1.000
3	R-SL	0.749	0.811
	Data bubble-SL	0.745	0.788
4	R-SL	1.000	1.000
	Data bubble-SL	1.000	1.000
5	R-SL	0.973	0.993
	Data bubble-SL	0.968	0.978
6	R-SL	0.958	0.979
	Data bubble-SL	0.946	0.959
7	R-SL	0.928	0.948
	Data bubble-SL	0.913	0.901

In Spiral dataset, our clustering method produces same clusterings as produced by the SL method ($RI = 1.000, Purity = 1.000$) with compression ratio 12.5 (Table 4). With high compression ratio (50.9), our proposed clustering method produces same clusterings as produced by that of the single-link method in Circle dataset. However, R-SL method is at least two order of magnitude faster than that of the classical single-link method. It may be noted that our R-SL method outperforms Data bubble based SL method in clustering results for all three datasets namely, Spiral, Circle and a8a (Table 4).

The clustering similarity (RI, $Purity$) at different levels of hierarchy (dendogram) produced by R-SL are also superior than that of the Data bubble-SL method for Circle datasets (Table 5).

Experiments with real world large datasets. We conducted experiments with three real world datasets (Pendigits, Shuttle and GDS10). Pendigits and

Table 6. Results for real world datasets

Dataset	Compression Ratio($\frac{n}{m}$)	Method	Time (in sec.)	Rand Index (RI)	Purity
Pendigits	12.2	R-SL	1.6	0.985	0.998
		Data bubble-SL	1.5	0.742	0.812
		SL	517.9	-	-
Shuttle	18.7	R-SL	58.7	0.786	0.812
		Data bubble-SL	53.4	0.699	0.754
		SL	96,400	-	-
GDS10	10.64	R-SL	31.6	0.930	0.992
		Data bubble-SL	24.54	0.887	0.910
		SL	24,105	-	-
GDS10	19.23	R-SL	13.55	0.901	0.921
		Data bubble-SL	8.63	0.898	0.911
		SL	24,105	-	-

Shuttle are taken from UCI Machine Learning Repository (`http://archive.ics.uci.edu/ml/`). We took $40,000$ patterns out of $58,000$ of Shuttle dataset for our experiments due to memory shortage. To store the entire distance matrix for whole dataset, single-link method consumes $58000 \times 58000 \times 4 = 12.54GB$ memory.

From GEO database (`http://www.ncbi.nlm.nih.gov/geo/`) we selected a dataset named GDS10. We removed patterns having missing values from the dataset. Finally, we took $23,709$ out of $39,114$ patterns for our experiments. The experimental results are reported in Table 6 with different compression ratio (n/m).

In Pendigits dataset, R-SL method is more than 300 times faster that classical single-link method and clustering results are very close ($RI = 0.985, Purity = 0.998$) to that of the single-link method (Table 6).

In Shuttle dataset, R-SL method is considerably faster than classical single-link method (Table 6). We conducted experiment with GDS10 dataset and similar trends are observed.

5   Conclusions

Data summarization is an effective way of speeding up clustering methods for large dataset. In this paper, we proposed *rough bubble* summarization scheme, which is used to speed up single-link method. The summarization scheme exploits leaders method and tolerance rough set theory. A technique is proposed to reduce number of distance computations in leaders method. The proposed *rough bubble* scans the dataset twice only. The R-SL clustering method produces clustering results, which is close to single-link method and takes considerably less time compared to that of the single-link method. The R-SL outperforms data bubble based single-link method at clustering results. So, proposed R-SL method is a suitable candidate for clustering large datasets.

Acknowledgments. This work is supported by Council of Scientific & Industrial Research (CSIR), New Delhi, India. We are extremely thankful to **Prof. Andrzej Skowron, University of Warsaw, Poland** for his invaluable comments and suggestions in improving quality of the paper.

References

1. Jain, A.K., Murty, M.N., Flynn, P.J.: Data clustering: A review. ACM Comput. Surv. 31(3), 264–323 (1999)
2. MacQueen, J.B.: Some Methods for Classification and Analysis of MultiVariate Observations. In: Proceedings of the Fifth Berkeley Symposium on Mathematical Statistics and Probability, pp. 281–297 (1967)
3. Kaufman, L., Rousseeuw, P.J.: Finding Groups in Data:an Introduction to Cluster Analysis. John Wiley & Sons, USA (1990)
4. Ester, M., Kriegel, H.P., Sander, J., Xu, X.: A Density-Based Algorithm for Discovering Clusters in Large Spatial Databases with Noise. In: Proceedings of 2nd ACM SIGKDD, SIGKDD 1996, pp. 226–231 (1996)
5. Hartigan, J.A.: Clustering Algorithms. John Wiley & Sons, Inc., New York (1975)
6. Spath, H.: Cluster Analysis Algorithms for Data Reduction and Classification of Objects. Ellis Horwood, UK (1980)
7. Ho, T.B., Nguyen, N.B.: Nonhierarchical document clustering based on a tolerance rough set model. Int. J. Intell. Syst. 17, 199–212 (2002)
8. Tremblay, J.P., Manohar, R.: Discreate Mathematical Structures with Applications to Computer Science. Tata McGraw-Hill Publishing Company Limited, New Delhi (1997)
9. Sneath, A., Sokal, P.H.: Numerical Taxonomy. Freeman, London (1973)
10. King, B.: Step-Wise Clustering Procedures. Journal of the American Statistical Association 62(317), 86–101 (1967)
11. Murtagh, F.: Complexities of hierarchic clustering algorithms: state of the art. Computational Statistics Quarterly 1, 101–113 (1984)
12. Ankerst, M., Breunig, M.M., Kriegel, H.P., Sander, J.: Optics: Ordering points to identify the clustering structure. SIGMOD Rec. 28, 49–60 (1999)
13. De, S.K., Krishna, P.R.: Clustering web transactions using rough approximation. Fuzzy Sets and Systems 148(1), 131–138 (2004)
14. Kumar, P., Krishna, P.R., Bapi, R.S., De, S.K.: Rough clustering of sequential data. Data Knowl. Eng. 63, 183–199 (2007)
15. Kawasaki, S., Nguyen, N.B., Ho, T.B.: Hierarchical document clustering based on tolerance rough set model. In: Zighed, D.A., Komorowski, J., Żytkow, J.M. (eds.) PKDD 2000. LNCS (LNAI), vol. 1910, pp. 458–463. Springer, Heidelberg (2000)
16. Zhang, T., Ramakrishnan, R., Livny, M.: Birch: An efficient data clustering method for very large databases. In: Proceedings of the ACM SIGMOD Conference, SIGMOD 1996, pp. 103–114 (1996)
17. Breunig, M.M., Kriegel, H.P., Sander, J.: Fast hierarchical clustering based on compressed data and optics. In: Zighed, D.A., Komorowski, J., Żytkow, J.M. (eds.) PKDD 2000. LNCS (LNAI), vol. 1910, pp. 232–242. Springer, Heidelberg (2000)

18. Breunig, M.M., peter Kriegel, H., Kröger, P., Sander, J.: Data bubbles: Quality preserving performance boosting for hierarchical clustering. In: Proceedings of the ACM SIGMOD Conference, SIGMOD 2001, pp. 79–90 (2001)
19. Bradley, P.S., Fayyad, U.M., Reina, C.: Scaling clustering algorithms to large databases. In: Proceedings of the KDD, KDD 1998, pp. 9–15 (1998)
20. Zhou, J., Sander, J.: Data bubbles for non-vector data: speeding-up hierarchical clustering in arbitrary metric spaces. In: Proceedings of VLDB 2003, pp. 452–463 (2003)
21. Patra, B.K., Nandi, S.: A fast single link clustering method based on tolerance rough set model. In: Sakai, H., Chakraborty, M.K., Hassanien, A.E., Ślęzak, D., Zhu, W. (eds.) RSFDGrC 2009. LNCS, vol. 5908, pp. 414–422. Springer, Heidelberg (2009)
22. Pawlak, Z.: Rough sets. Int. J. of Computer and Information Sc. 11, 341–356 (1982)
23. Lin, T.Y., Cercone, N. (eds.): Rough Sets and Data Mining: Analysis of Imprecise Data. Kluwer Academic Publishers, Norwell (1996)
24. Skowron, A., Stepaniuk, J.: Tolerance approximation spaces. Fundamenta Informaticae 27, 245–253 (1996)
25. Slowinski, R., Vanderpooten, D.: A generalized definition of rough approximations based on similarity. IEEE Trans. on Knowl. and Data Eng. 12, 331–336 (2000)
26. Kryszkiewicz, M.: Rough set approach to incomplete information systems. Information Sciences 112(1-4), 39–49 (1998)
27. Ślezak, D., Wasilewski, P.: Granular sets — foundations and case study of tolerance spaces. In: An, A., Stefanowski, J., Ramanna, S., Butz, C.J., Pedrycz, W., Wang, G. (eds.) RSFDGrC 2007. LNCS (LNAI), vol. 4482, pp. 435–442. Springer, Heidelberg (2007)
28. Bedi, P., Chawla, S.: Use of fuzzy rough set attribute reduction in high scent web page recommendations. In: Sakai, H., Chakraborty, M.K., Hassanien, A.E., Ślęzak, D., Zhu, W. (eds.) RSFDGrC 2009. LNCS, vol. 5908, pp. 192–200. Springer, Heidelberg (2009)
29. Elkan, C.: Using the triangle inequality to accelerate k-means. In: Proceedings of the Twentieth International Conference on Machine Learning, ICML 2003, pp. 147–153 (2003)
30. Nassar, S., Sander, J., Cheng, C.: Incremental and effective data summarization for dynamic hierarchical clustering. In: Proceedings of SIGMOD Conference, SIGMOD 2004, pp. 467–478 (2004)
31. Kryszkiewicz, M., Lasek, P.: TI-DBSCAN: Clustering with DBSCAN by means of the triangle inequality. In: Szczuka, M., Kryszkiewicz, M., Ramanna, S., Jensen, R., Hu, Q. (eds.) RSCTC 2010. LNCS, vol. 6086, pp. 60–69. Springer, Heidelberg (2010)
32. Patra, B.K., Hubballi, N., Biswas, S., Nandi, S.: Distance based fast hierarchical clustering method for large datasets. In: Szczuka, M., Kryszkiewicz, M., Ramanna, S., Jensen, R., Hu, Q. (eds.) RSCTC 2010. LNCS, vol. 6086, pp. 50–59. Springer, Heidelberg (2010)
33. Rand, W.M.: Objective Criteria for Evaluation of Clustering Methods. J. of American Statistical Association 66, 846–850 (1971)
34. Zhao, Y., Karypis, G.: Criterion functions for document clustering: Experiments and analysis. Technical report, University of Minnesota (2002)

Projected Gustafson-Kessel Clustering Algorithm and Its Convergence

Charu Puri and Naveen Kumar

Department of Computer Science, University of Delhi, Delhi, India
cpuri.cs.du@gmail.com, nk.cs.du@gmail.com

Abstract. Fuzzy techniques have been used for handling vague boundaries of arbitrarily oriented clusters. However, traditional clustering algorithms tend to break down in high dimensional spaces due to inherent sparsity of data. We propose a modification in the objective function of Gustafson-Kessel clustering algorithm for projected clustering and prove the convergence of the resulting algorithm. We present the results of applying the proposed projected Gustafson-Kessel clustering algorithm to synthetic and UCI data sets, and also suggest a way of extending it to a rough set based algorithm.

Keywords: Convergence, Gustafson-Kessel algorithm, high dimensional data, rough sets, subspace clustering, validity measures.

1 Introduction

Clustering is a technique that aims to group data objects so that the objects within a group are similar while the objects belonging to different groups are dissimilar. Various similarity measures such as Euclidean, city-block, Mahalanobis distances, and cosine similarity [31] have been used for discovering the underlying structures in data. Applications of clustering include image processing, pattern recognition, analysis of microarray data in bioinformatics, spatial analysis, social network analysis, intrusion detection in networks, document classification, spam filtering, and market research [22][25][50]. Formally, the problem of clustering may be described as follows: Given a set of data objects $X = \{x_1, x_2, ..., x_n\}$, a clustering algorithm determines a suitable number k of homogeneous groups, and maps the data points to the labels in the set $C = \{1, 2, ..., k\}$, where each label identifies a homogeneous group of objects. A good clustering algorithm should have the following characteristics: it should be scalable, i.e., perform well on data sets having large number of objects and also large number of attributes, should be able to determine clusters of varying shape and size, should have least requirement of domain knowledge (e.g., number of clusters, thresholds, termination condition parameters), should work well in the presence of noise and outliers, and should be insensitive to order of objects [22].

Mainly, the clustering algorithms fall in the following categories: partition based, hierarchical and density based methods. Partition based methods such as k-means, k-medoid, CLARA, and CLARANS that attempt to minimize a

J.F. Peters et al. (Eds.): Transactions on Rough Sets XIV, LNCS 6600, pp. 159–182, 2011.

similarity measure so that similar objects are grouped together. Hierarchical methods can be broadly classified into two categories namely, agglomerative and divisive. Agglomerative approaches such as BIRCH begin by placing each object in a separate cluster and iteratively merge similar clusters to form large clusters, whereas divisive approaches such as DIANA begin by placing all data objects in a single cluster and iteratively divide large clusters into smaller ones based on the dissimilarity criterion. Density based methods such as DBSCAN and OPTICS work by defining a neighborhood of a point and connecting the dense neighborhoods in close proximity. Another density based method DENCLUE finds clusters by determining local maxima of the density function [50].

Determination of number of clusters in a data set has been a challenging task, often requiring domain expertise. Several clustering algorithms require the number of clusters as one of the input parameters. If the number of attributes is very small, visualizing the data by projecting it onto a two dimensional space, can give some estimate of the number of clusters. Xu and Wunsch [50] describe a number of measures such as Calinski and Harabasz index, Akaikes information criterion, Bayesian inference criterion, and the number of dominant principal components to estimate the number of clusters. However, none of these measures works well on data sets with varying characteristics. As such one often uses these measures in a collaborative manner to estimate the number of clusters. Constructive algorithms such as ART networks [50] generate new clusters only in the presence of sufficient dissimilarity. Self-splitting competitive learning algorithm begins with a single cluster of all objects and splits larger clusters into smaller ones based on a splitting criterion, and thus automatically determines the number of clusters. In this paper, we presume that the number of clusters is already known.

While dealing with high dimensional data, traditional clustering algorithms [22] fail in identifying hidden relationships of the underlying structure due to the reason that the nearest neighbor of a pattern may be nearly as distant as the farthest neighbor, since they compute the distance in full dimensional space [13]. In high dimensional data sets, clusters are often hidden in specific subspaces of the original feature space [28]. To cope with the problem of high dimensional feature spaces, feature reduction and feature selection techniques have been used in the literature [1]. Feature reduction techniques such as principal component analysis (PCA) suffer from usability problem as it becomes hard to interpret the results intuitively [28]. Feature selection techniques project the whole feature space to a lower dimensional subspace so that cluster structures become apparent [15][16][36][37]. However, these techniques do not deal effectively with clusters in varying subspaces. Hence, there is a need for more generalized techniques that can be used to obtain meaningful clusters in varying subspaces.

Pawlak [41] introduced rough set theory as a new framework for dealing with imperfect knowledge. Rough set theory provides a methodology for addressing the problem of relevant feature selection, by selecting a set of information rich features from a data set, without transforming the data. It is often possible to arrive at a minimal feature set (called reduct in rough set theory) that can be used for data analysis tasks such as classification and clustering [33] [39]. In [14], Bazan et al. discuss various techniques for rough set reduct generation and argue that the classical reducts being static may not be stable in randomly chosen samples of a given decision table. To deal with such situations they focus on reducts that are stable over different subsets of samples chosen from a given decision table. Such reducts are called dynamic reducts. They compute reducts using an order based gentic algorithm and subsequently extract dynamic reducts which are used to generate classification rules. Each rule set is associated with a measure called the rule strength which is used later to resolve conflicts when several rules are applicable. In [48] Slezak generalized the concept of reduct by introducing the notion of association reducts corresponding to both association rules and rough set reducts. He defined association reduct as a pair (A, B) of disjoint subsets of attributes such that all data supported patterns involving A approximately determine those involving B. He developed an information theory based algorithm to compute association reducts. As the algorithm needs to examine all association reducts, it has exponenetial time requirements. In order to aleviate this hardship, in [47], Slezak targeted significantly smaller ensembles of dependencies providing reasonably rich knowledge, and developed an order based genetic algorithm to achieve this. Shen and Jensen [46] proposed the concept of retainer as an approximation of a reduct. The authors suggest a heuristic to compute the retainer and demonstrate its usefulness for the classification task. For clustering textual database consisting of n documents, with a vocabulary of size V, Li et al. [32] developed an algorithm based on approximate reducts that works in time O(VN).

In this paper, we present a novel adaptation of Gustafson-Kessel (GK) algorithm [21] for high dimensional data by extending its objective function for projected clustering, that automatically detects the relevant cluster dimensions of high dimensional data set. As the assignment of weights to attributes is specific to each cluster, an efficient clustering scheme is generated. We have also proved the convergence of the proposed algorithm using Zangwill's convergence theorem [54]. The remainder of this paper is organised as follows: in section 2 on related work, we describe how classical clustering methods have been adapted to suit the high dimensional data, in section 3, we extend the Gustafson- Kessel algorithm for subspace clustering and also suggested a method of extending this algorithm to rough set based clustering, in section 4, we prove convergence of Projected Gustafson-Kessel (PGK) clustering algorithm, in section 5, we present the results based on UCI and synthetic data sets, and finally section 6 contains conclusions and suggestions for future work.

2   Related Work

Classical clustering approaches have been modified in literature [2][4][40], to detect relevant cluster dimensions in high dimensional data. Agrawal et al. [4] introduced the concept of subspace clustering, also called projected clustering to generate clusters in distinctive subspaces of multidimensional space so that intra cluster similarity is maximized and inter cluster similarity is minimized. Elke et al. proposed an algorithm to discover clusters in arbitrarily oriented subspaces [2]. The pioneering approaches to projected clustering can be classified into three categories: grid based approaches such as CLIQUE [4] and MAFIA [40], approaches based on density connectivity such as SUBCLU [28], and partitioning approaches like PROCLUS [3]. CLIQUE is a grid based algorithm that partitions the whole data space into non-overlapping rectangular units. It uses an apriori-like method to recursively navigate through the set of possible subspaces in a bottom up way. It works on the principle that if a k-dimensional unit is dense, then all associated (k-1)-dimensional units would also be dense. It merges the dense units to form clusters over large subspaces. SUBCLU is a greedy algorithm that computes all density connected sets. It detects arbitrarily shaped and positioned clusters in subspaces. PROCLUS is based on the concept of k-medoid clustering. It explores the locality of space near medoids to determine relevant dimensions. It uses a greedy hill climbing technique to iteratively search for medoids. SCHISM [45] extends CLIQUE using a variable threshold in order to cope up with varying number of dimensions and makes use of grid based discretization for pruning. FIRES [30] is a generic framework based on approximate subspace cluster computation. EDSC [7] is a multi step filter and refinement algorithm meant for density based subspace clustering. DUSC [6] introduces the concept of dimensionality bias and performs density based clustering using statistical techniques for pruning.

Classically, clustering has been based on the disjointness condition that no two data points belong to same cluster [40]. However, in real data sets, the clusters may be overlapping and the disjointness condition may be too hard to realize. A data point may belong to various clusters and a dimension may be relevant to various clusters with varying degree [51]. Hence, such situations require weakening of disjointness condition. Two major approaches one based on fuzzy set theory, introduced by Zadeh [53] to deal with uncertainties and another based on rough set theory introduced by Pawlak [41] have been proposed in literature to deal with such situations. Ruspini [43] developed the first fuzzy clustering algorithm based on least-squares optimization. Dunn [17] developed a fuzzy extension of this approach to clustering and proposed the fuzzy c-means algorithm. Bezdek [9] extended Dunn's formulation and proposed a generalization of conventional hard c-means clustering by fuzzy partitioning of data. However, fuzzy c-means algorithm cannot detect arbitrarily oriented clusters [1]. Gustafson and Kessel [21] proposed an algorithm, known as Projected Gustafson-Kessel clustering algorithm, based on an adaptive distance measure. In [8], Babuka et al. proposed improved covariance estimation for Gustafson-Kessel clustering algorithm.

In literature, rough sets have been widely used for classification and clustering [33][39][41]. As mentioned earlier, clusters in real data sets do not necessarily have crisp boundaries. Rough set theory deals with such situations by approximating a set [33][41]. Lingras et al. [33], extended the classical k-means algorithm to rough k-means algorithm. In rough k-means algorithm, a data point in the lower approximation surely belong to a cluster, however, membership of the objects in an upper approximation is uncertain. In the lower approximation, the core cluster is surrounded by a buffer or boundary set having objects with unclear membership status [33]. Each cluster is represented by a center and a pair of lower and upper approximations defined by means of weighted parameters. If lower and upper bounds are equal then clusters obtained are crisp in nature as the boundary region becomes empty. The process of updating the cluster centers and assignment of data points to clusters is repeated until the convergence criterion is satisfied. In [5] an extended version of rough k-means algorithm which does not require prior specification of the number of clusters has been presented. It is a two phase algorithm, in the first phase it identifies a set of leaders which act as prototypes and a set of supporting leaders which can subsequently act as leaders if they yield better partitioning. The evolutionary rough k-medoids algorithm [42] has characteristics of family of rough clustering algorithms and classical k-medoids algorithm [29]. Malyszko et al. [38] extended rough k-means clustering to rough entropy clustering. It proceeds iteratively as follows: a predefined number of weights pairs is selected, a new offspring clustering is determined for each weight pair and rough entropy is computed afresh, and finally the partition with the best rough entropy is selected.

3   Projected Gustafson-Kessel Clustering

In this section, we introduce the necessary notations, review the Gustafson-Kessel Clustering algorithm [21], and formulate a new objective function for adapting it to projected clustering. We also suggest a method of extending this algorithm to rough set based clustering. The Gustafson-Kessel algorithm [21] associates each cluster with the cluster centre and its covariance. The main feature of Gustafson-Kessel clustering is the local adaptation of distance matrix in order to identify ellipsoidal clusters.

Given a data set $X = \{x_1, x_2, ..., x_n\}$ in the d-dimensional space and the number of clusters k, the objective is to partition the data set X into clusters and determine the cluster prototype $Z = \{z_1, z_2, ..., z_k\}$. A fuzzy partition of the data set X can be represented by a $k \times n$ matrix $U = [\mu_{ij}]$, where μ_{ij} denotes the degree of membership with which j^{th} object belongs to the ith cluster, for $1 \leq i \leq k$, $1 \leq j \leq n$. The matrix U is called the fuzzy partition matrix. It satisfies the following constraints:

$$\sum_{i=1}^{k} \mu_{ij} = 1, \ 1 \leq j \leq n; \quad \mu_{ij} \in [0,1], \ 1 \leq j \leq n, \ 1 \leq i \leq k \tag{1}$$

The above constraint expresses the fact that the sum of memberships of an object over the set of clusters must be equal to 1. The fact that the number of clusters is at least two, is expressed by the following constraint:

$$0 < \sum_{j=1}^{n} \mu_{ij} < n, \quad 1 \leq i \leq k \tag{2}$$

The constraints expressed in eq. 1 lead to the following fuzzy partition space for (X, k):

$$M_f^{kn} = \{U = [\mu_{ij}]_{k \times n} \in \Re^{k \times n} | \mu_{ij} \in [0,1] \ \forall \, i,j; \sum_{i=1}^{k} \mu_{ij} = 1, \ \ 1 \le j \le n\} \ (3)$$

The objective function of GK algorithm is defined as follows [1]:

$$J_m = \sum_{j=1}^{n} \sum_{i=1}^{k} \mu_{ij}^{m} d_{ij}^{2}, \tag{4}$$

where

$$d_{ij}^{2} = (x_j - z_i) A_i (x_j - z_i)^{T} \tag{5}$$

J_m defines the within-group sum of squared-errors as the objective function i.e. it measures the total within-group scatter, or variance of X from k cluster centers $\{z_i, \ 1 \le i \le k\}$. J_m is used for finding the optimal pair (U, Z) where U denotes partitioning of the data, Z denotes the set of cluster centers of the partitions. A_i is a symmetric, positive definite matrix that induces for each cluster a norm of its own [21] [24]. J_m being linear in A_i, is prone to singularity problem. This leads to the difficulty in minimizing J_m directly. In order to obtain feasible solution, A_i is constrained in such a way that $det(A_i) = \rho_i > 0$, ρ_i being fixed for each i permitting different sizes of cluster. The exponent $m \in (1, \infty)$ is a fuzzification parameter, that controls the extent by which clusters may overlap. The objective function J_m is minimized using an alternating optimization (AO) technique. The AO optimization technique leads to the local optimum as it proceeds by fixing a set of parameters and optimizing the rest of parameters in an alternating manner. Iteratively updating in such a fashion yields the optimum value of J_m. However, such techniques do not ensure the global optimum and the algorithm may get stuck in the local optimum. In order to counter the possibility of getting stuck at local optimum, one often performs several runs of the algorithm.

The objective function J_m (eq. 4)is designed to find the clusters in the entire feature space and therefore cannot determine the respective natural subspaces of each cluster in high dimensional data set. We associate with i^{th} cluster, the weight vector ω_i, which represents the relevance of different attributes for the ith cluster. Thus in the matrix $W = [\omega_{ir}]_{k \times d}$ ω_{ir} denotes the contribution of rth dimension to i^{th} cluster. The sum of contributions from all dimensions adds to one for any cluster, as expressed by the following constraint.

$$\sum_{r=1}^{d} \omega_{ir} = 1, \ 1 \le i \le k; \ \ \omega_{ir} \in [0,1] \, , \ 1 \le i \le k, \ 1 \le r \le d \tag{6}$$

As we look for clusters in the projected subspaces, we assume that the number of dimensions is at least two. This is expressed in the form of following constraint:

$$0 < \sum_{i=1}^{k} \omega_{ir} < k, \quad 1 \le r \le d \tag{7}$$

Using eq. 6 we get the following partitioning space for the set of dimensions:

$$M_f^{kd} = \{W = [\omega_{ir}]_{k \times d} \in \Re^{k \times d} | \omega_{ir} \in [0,1] \; \forall \, i, r; \sum_{r=1}^{d} \omega_{ir} = 1, \; \forall \, i\} \tag{8}$$

Now, we formulate the new objective function as:

$$J_{\alpha, \beta} = \sum_{j=1}^{n} \sum_{i=1}^{k} \sum_{r=1}^{d} \mu_{ij}^{\alpha} \omega_{ir}^{\beta} d_{ijr}^2, \tag{9}$$

where,

$$d_{ijr}^2 = \sum_{s=1}^{d} (x_{jr} - z_{ir}) a_{rs} (x_{js} - z_{is}), \quad A_i = [a_{rs}]_{d \times d} \tag{10}$$

Parameters $\alpha \in (1, \infty)$, $\beta \in (1, \infty)$ are weight components. These parameters control the fuzzification of μ_{ij} and ω_{ir} respectively. The necessary condition for minimization of the objective function in eq. 9 yields the following update equations [see section 3]:

$$A_i = ((det(F_i)\rho_i))^{1/d} F_i^{-1} \tag{11}$$

$$\mu_{ij} = 1 \left/ \sum_{l=1}^{k} \left[\frac{\sum_{r=1}^{d} (\omega_{ir})^{\beta} d_{ijr}^2}{\sum_{r=1}^{d} (\omega_{lr})^{\beta} d_{ljr}^2} \right]^{1/(\alpha-1)} \right. \tag{12}$$

$$\omega_{ir} = 1 \left/ \sum_{l'=1}^{d} \left[\frac{\sum_{j=1}^{n} (\mu_{ij})^{\alpha} d_{ijr}^2}{\sum_{j=1}^{n} (\mu_{ij})^{\alpha} d_{ijl'}^2} \right]^{1/(\beta-1)} \right. \tag{13}$$

$$z_{ir} = \sum_{j=1}^{n} \omega_{ir}^{\beta} \mu_{ij}^{\alpha} x_{jr} \left/ \sum_{j=1}^{n} \omega_{ir}^{\beta} \mu_{ij}^{\alpha} \right. \tag{14}$$

3.1   Projected Gustafson-Kessel(PGK) Clustering Algorithm

Based on the update equations obtained above, we describe below the PGK algorithm.

Algorithm 1. Projected Gustafson-Kessel (PGK) Clustering Algorithm

Inputs:

n: size of data set X

k: number of clusters, $1 < k < n$

α: the weight exponent of matrix U, $\alpha > 1$

β: the weight exponent of matrix W, $\beta > 1$

ϵ: the termination tolerance, $\epsilon > 0$

A: the norm-inducing matrix A

Outputs:

U: membership matrix of objects in clusters

W: matrix indicating relevance of dimensions for clusters

Z: cluster centers

Initialize the partition matrices U, W randomly

t=0

while $\left\| U^t - U^{t-1} \right\| > \in$ **do**

> Step 1. Compute the cluster prototypes
>
> $$z_{ir}^t = \frac{\sum_{j=1}^n (\omega_{ir}^{t-1})^\beta (\mu_{ij}^{t-1})^\alpha x_{jr}}{\sum_{j=1}^n (\omega_{ir}^{t-1})^\beta \mu_{ij}^\alpha}$$
>
> Step 2. Compute the cluster covariance matrices
>
> $$F_i^t = \sum_{j=1}^n (\omega_{ir}^{t-1})^\beta (\mu_{ij}^{t-1})^\alpha (x_{jr} - z_{ir}^t)(x_{js} - z_{is}^t) \, 1 \le r \le d, 1 \le s \le d.$$
>
> Step 3. Compute the distances:
>
> $$d_{ijr}^2 = [(x_{j1} - z_{i1}^t) \dots (x_{jd} - z_{id}^t)] A_i [(x_{j1} - z_{i1}^t) \dots (x_{jd} - z_{id}^t)]^T$$
>
> Step 4. Update the partition matrices:
>
> $$\mu_{ij}^t = 1 \bigg/ \sum_{l=1}^k \left[\frac{\sum_{r=1}^d (\omega_{ir}^{t-1})^\beta d_{ijr}^2}{\sum_{r=1}^d (\omega_{lr}^{t-1})^\beta d_{ljr}^2} \right]^{1/\alpha - 1}$$
>
> $$\omega_{ir}^t = 1 \bigg/ \sum_{l'=1}^d \left[\frac{\sum_{j=1}^n (\mu_{ij}^{t-1})^\alpha d_{ijr}^2}{\sum_{j=1}^n (\mu_{ij}^{t-1})^\alpha d_{ijl'}^2} \right]^{1/\beta - 1}$$
>
> Step 5. t = t+1

PGK clustering algorithm proceeds by initializing U, W, and Z. Subsequently U, W, and Z are computed iteratively by fixing the other two out of these. Also, the covariance matrix is computed using U, W, and Z.

Computation cost of PGK clustering algorithm for each iteration in updating Z, U, W and objective function is O(nkd) and for F it is $O(nkd^2)$.

3.2   Rough Fuzzy Projected Clustering

Rough fuzzy c-means algorithm [39] is an extension of rough c-means algorithm, which incorporates fuzzy membership of data points in each cluster. In each iteration, cluster centers are updated and data points are assigned to either lower approximations or boundary regions of two or more clusters. This process is repeated until convergence criterion is met. In order to generalize our proposed approach of fuzzy subspace clustering to the class of rough c-means algorithms, the proposed rough projected Gustafson-Kessel clustering (RPGK) algorithm proceeds as follows:

The steps in the computation of cluster prototypes remain exactly same as in Rough fuzzy c-Means (RFCM) algorithm [39]. However, RPGK algorithm computes $\mu_{i,j}$ differently from RFCM algorithm using eq. 12 which also takes into account weights of dimensions. The weights of dimensions are computed using eq. 13 as in PGK algorithm. The data points are assigned to lower or upper approximation of a cluster in the following manner: in an iteration, let us assume that a point x_j takes highest membership $\mu_{i,j}$ and next to highest membership $\mu_{l,j}$ in clusters i and l respectively. If $\mu_{i,j} - \mu_{l,j}$ is less than a threshold value then the data point x_j does not belong to lower approximation of any cluster. However, it belongs to upper approximation of each of the clusters i and l. In case, $\mu_{i,j} - \mu_{l,j}$ is greater than the threshold value, x_j clearly belongs to the lower approximation of cluster i.

4   A Convergence Theorem for the PGK Clustering Algorithm

In [17] Dunn derived necessary conditions for iterative minimization of J_2. In [10], Bezdek extended J_2 to an infinite family of objective functions $\{J_m, 1 \leq m \leq \infty\}$. He used Zangwill's convergence theory [54] for proving global convergence of the iterate sequence, used to minimize J_m. This result has been used in extensions of J_m to the fuzzy c-varieties algorithms [11][12]. FCM proposed by Bezdek is based on iterative optimization of least square error function J_m. Convergence of an iterative algorithm implies that every iterate sequence stabilizes after a certain point, in the range of specified error. Global convergence property of FCM states that for any data set and initialization parameters, an iteration sequence of FCM algorithm either (i) converges to a local minimum or (ii) there exists a subsequence of the iteration sequence that converges to a stationary point. Even if FCM converges to its global minima it may not reflect the visually apparent substructure of data [23]. Recently, Gan et al. [19] have proposed the convergence of fuzzy subspace clustering algorithm using Zangwill's convergence theorem. They have proven the convergence properties of Fuzzy Subspace Clustering (FSC) algorithm and also carried out the experimental evaluation of rate of convergence. To establish the convergence properties of PGK algorithm by applying Zangwill's convergence theroem, we reproduce some basic definitions and notations from [19] and prove the convergence in a similar a manner.

Definition 1. *A point-to-set map τ from a set X into a set Y is defined as: $\tau : X \rightarrow P(Y)$, which associates a subset of Y with each point of X, where $P(Y)$ denotes the power set of Y.*

Definition 2. *A point-to-set map $\tau : X \rightarrow P(Y)$ is said to be open at a point $\bar{x} \in X$ if $\{x^{(p)}\} \subset X, x^{(p)} \rightarrow \bar{x}$ and $\bar{y} \in \tau(\bar{x})$ imply the existence of an integer p_0 and a sequence $\{y^{(p)}\} \subset Y$ such that $y^{(p)} \in \tau(x^{(p)})$ for $p \geq p_0$ and $y^{(p)} \rightarrow \bar{y}$.*

Definition 3. *A point-to-set map $\tau : X \rightarrow P(Y)$ is said to be closed at a point $\bar{x} \in X$ if $\{x^{(p)}\} \subset X, x^{(p)} \rightarrow \bar{x}, y^{(p)} \in \tau(x^{(p)})$, and $y^{(p)} \rightarrow \bar{y}$ imply $\bar{y} \in \tau(\bar{x})$.*

Definition 4. *A point-to-set map $\tau : X \rightarrow P(Y)$ is said to be continuous at a point $\bar{x} \in X$ if it is both open and closed at $\bar{x}$.*

Theorem 1. *Zangwill's Convergence Theorem*
Given a point $a^{(0)} \in Y$, let the point-to-set map $\tau : Y \rightarrow P(Y)$ determine an algorithm that generates the sequence $\{a^{(p)}\}$. Also, a solution set $S \subset Y$ is given. Assume that

(1) All points $a^{(p)}$ are in a compact subset of Y.
(2) There is a continuous function $J : Y \rightarrow \Re$ such that:
    (a) if $a \notin S$, then for any $y \in \tau(a), J(y) < J(a)$,
    (b) if $a \in S$, then either the algorithm terminates or for any $y \in \tau(a)$, $J(y) \leq J(a)$
(3) The map τ is closed at a if $a \notin S$.

Then either the algorithm terminates at a solution or the limit of any convergent subsequence is a solution.

Let $F : M_f^{kd} \times M_f^{kn} \rightarrow \Re^{kd}$ be a function defined as: $F(W,U){=}Z{=}(z_1, z_2, ..., z_k)^T$, where the vectors $z_i = (z_{i1}, z_{i2}, ..., z_{id}), 1 \leq i \leq k$, are computed by eq. 14 using W and U.

Let $G : \Re^{kd} \times M_f^{kn} \rightarrow M_f^{kd}$ be a function defined as: $G(Z,U){=}W{=}(\omega_1, \omega_2, .., \omega_k)^T$, where the vectors $\omega_i = (\omega_{i1}, \omega_{i2}, ..., \omega_{id}), 1 \leq i \leq k$, are computed by eq. 13. using Z and U.

Let $H : M_f^{kd} \times \Re^{kd} \rightarrow M_f^{kn}$ be a function defined as: $H(W,Z){=}U{=}(\mu_1, \mu_2, ..., \mu_k)^T$ where the vectors $\mu_i = (\mu_{i1}, \mu_{i2}, ..., \mu_{iN}), 1 \leq i \leq k$, are computed by eq. 12 using W and U.

Similar to the fuzzy c-means iteration, PGK algorithm can be expressed using a point-to-set map $T_{\alpha,\beta} : M_f^{kd} \times \Re^{kd} \times M_f^{kn} \rightarrow P(M_f^{kd} \times \Re^{kd} \times M_f^{kn})$ defined by the composition $T_{\alpha,\beta} = A_3 o A_2 o A_1$, where

$$A_1 : M_f^{kd} \times \Re^{kd} \times M_f^{kn} \rightarrow \Re^{kd} \times M_f^{kn}, \quad A_1(W,Z,U) = (F(W,U),U) \quad (15)$$

$$A_2 : \Re^{kd} \times M_f^{kn} \to M_f^{kd} \times \Re^{kd}, \quad A_2(Z,U) = (G(Z,U), Z) \tag{16}$$

$$A_3 : M_f^{kd} \times \Re^{kd} \to P(M_f^{kd} \times \Re^{kd} \times M_f^{kn}), \quad A_3(W,Z) = (W, Z, H(W,Z)) \tag{17}$$

Thus

$$T_{\alpha,\beta}(W,Z,U) = \{(\hat{W}, \hat{Z}, \hat{U}) | \hat{Z} = F(W,U), \hat{W} = G(\hat{Z}, U), \text{ and } \hat{U} = H(\hat{W}, \hat{Z})\} \tag{18}$$

Let,

$$
\begin{aligned}
(\hat{W}, \hat{Z}, \hat{U}) \in T_{\alpha,\beta}(W,Z,U) &= A_3 o A_2 o A_1(W,Z,U) \\
&= A_3 o A_2(F(W,U), U) \\
&= A_3(G(\hat{Z}, U), F(W,U))
\end{aligned}
$$

$$(\hat{W}, \hat{Z}, \hat{U}) \in T_{\alpha,\beta}(W,Z,U) = (G(\hat{Z}, U), F(W,U), H(\hat{W}, \hat{Z})) \tag{19}$$

Definition 5. *A sequence $\{(W^{(p)}, Z^{(p)}, U^{(p)})\}$ is said to be PGK clustering algorithm iteration sequence if $(W^{(1)}, Z^{(1)}, U^{(1)}) \in M_f^{kd} \times \Re^{kd} \times M_f^{kn}$ and $(W^{(p)}, Z^{(p)}, U^{(p)}) \in T_{\alpha,\beta}(W^{(p-1)}, Z^{(p-1)}, U^{(p-1)})$ for $p = 2, 3, 4,....$*

Theorem 2. *Let $\eta : M_f^{kd} \to \Re$, $\eta(U) = J_{\alpha,\beta}(U, W, Z)$, where $W \in M_f^{kd}$ and $Z \in \Re^{kd}$ are fixed. Then $U \in M_f^{kn}$ is the strict local minima of η if and only if U is computed by equation:*

$$\mu_{ij} = 1 \left/ \sum_{l=1}^{k} \left[\frac{\sum_{r=1}^{d} (\omega_{ir})^{\beta} d_{ijr}^2}{\sum_{r=1}^{d} (\omega_{lr})^{\beta} d_{ljr}^2} \right]^{1/(\alpha-1)} \right. \tag{20}$$

Proof. We have to minimize J with respect to U, W, subject to constraints 1, 6 and $det(A_i) = \rho_i > 0$ where $\alpha \in (1, \infty)$, and $\beta \in (1, \infty)$. In order to ensure non-negativity of μ_{ij} and ω_{ir}, we set $\mu_{ij} = S_{ij}^2$ and $\omega_{ir} = P_{ir}^2$. The constraints 1 and 6 have been adjoined to J with a set of Lagrange multipliers $\{\lambda_j, \ 1 \le j \le n\}$ and $\{\phi_i, \ 1 \le i \le k\}$ to formulate the new objective function as follows:

$$
\begin{aligned}
J_{\alpha,\beta} = \sum_{j=1}^{n} \sum_{i=1}^{k} \sum_{r=1}^{d} S_{ij}^{2\alpha} P_{ir}^{2\beta} d_{ijr}^2 + \sum_{j=1}^{n} \lambda_j \left(\sum_{i=1}^{k} S_{ij}^2 - 1 \right) \\
+ \sum_{i=1}^{k} \phi_i \left(\sum_{r=1}^{d} P_{ir}^2 - 1 \right) + \sum_{i=1}^{k} \psi_i (det(A_i) - \rho_i)
\end{aligned}
\tag{21}
$$

In order to obtain the first order necessary condition for optimality, we set the gradient of J w.r.t A_i equal to zero. We use the follwing identities [21] and obtain:

$$\nabla x_j^T A_i x_j = x_j x_j^T \text{ and } \nabla det(A_i) = det(A_i) A_i^{-1}$$

$$\frac{\partial J}{\partial A_i} = \sum_{j=1}^{n} \mu_{ij}^\alpha \omega_{ir}^\beta (x_{jr} - z_{ir})(x_{js} - z_{is}) + \psi_i det(A_i) A_i^{-1} = 0 \qquad (22)$$

Writing first part of summation in eq. 22, as

$$F_i = \sum_{j=1}^{n} \mu_{ij}^\alpha \omega_{ir}^\beta (x_{jr} - z_{ir})(x_{js} - z_{is}), \quad 1 \le r \le d, \ 1 \le s \le d \qquad (23)$$

Eq. 22 can be rewritten as:

$$F_i = \psi_i \rho i A_i^{-1} \qquad (24)$$

Multiplying by A_i on both sides we get:

$$F_i A_i = \psi_i \rho_i I$$

Taking determinant on both sides we get:

$$det(F_i A_i) = det(\psi_i \rho_i I)$$

or equivalently

$$det(F_i A_i) = (\psi_i \rho_i)^d \qquad (25)$$

Using eq. 24 and eq. 25 we get:

$$A_i = (det(F_i) \rho_i)^{1/d} F_i^{-1} \qquad (26)$$

In order to obtain the first order necessary condition for optimality, we set the gradient of J w.r.t S_{ij} equal to zero.

$$\frac{\partial J}{\partial S_{ij}} = 2\alpha \sum_{r=1}^{d} S_{ij}^{2\alpha-1} P_{ir}^{2\beta} d_{ijr}^2 + 2\lambda_j S_{ij} = 0 \qquad (27)$$

$$\frac{\partial J}{\partial S_{ij}} = 2S_{ij} \left[\alpha \sum_{r=1}^{d} S_{ij}^{2\alpha-2} P_{ir}^{2\beta} d_{ijr}^2 + \lambda_j \right] = 0$$

Assuming that $S_{ij} \ne 0, \ 1 \le j \le n, \ 1 \le i \le k$, we get:

$$\alpha \sum_{r=1}^{d} S_{ij}^{2\alpha-2} P_{ir}^{2\beta} d_{ijr}^2 + \lambda_j = 0$$

$$\text{or} \quad \lambda_j = -\alpha \sum_{r=1}^{d} S_{ij}^{2\alpha-2} P_{ir}^{2\beta} d_{ijr}^2$$

$$\text{or} \quad S_{ij}^{2\alpha-2} = \frac{-\lambda_j}{\alpha \sum_{r=1}^{d} P_{ir}^{2\beta} d_{ijr}^2}$$

$$\text{or} \quad S_{ij}^2 = \left[\frac{-\lambda_j}{\alpha \sum_{r=1}^{d} P_{ir}^{2\beta} d_{ijr}^2} \right]^{\frac{1}{(\alpha-1)}}$$

$$\mu_{ij} = \left[\frac{-\lambda_j}{\alpha \sum_{r=1}^{d} P_{ir}^{2\beta} d_{ijr}^2} \right]^{\frac{1}{(\alpha-1)}} \tag{28}$$

Using constraint eq. 1 in eq. 28, we get:

$$\sum_{i=1}^{k} \mu_{ij} = \sum_{i=1}^{k} \left[\frac{-\lambda_j}{\alpha \sum_{r=1}^{d} P_{ir}^{2\beta} d_{ijr}^2} \right]^{\frac{1}{(\alpha-1)}} = 1$$

Substituting the value of λ_j in eq. 28, we obtain:

$$\mu_{ij} = 1 \left/ \sum_{l=1}^{k} \left[\frac{\sum_{r=1}^{d} \omega_{ir}^{\beta} d_{ijr}^2}{\sum_{r=1}^{d} \omega_{lr}^{\beta} d_{ljr}^2} \right]^{1/(\alpha-1)} \right. \tag{29}$$

Now, to prove the sufficiency condition we compute the second order partial derivative

$$\frac{\partial^2 J}{\partial S_{ij} S_{i'j'}} = \begin{cases} 2\alpha(2\alpha-1)\sum_{r=1}^{d} S_{ij}^{2\alpha-2} P_{ir}^{2\beta} d_{ijr}^2 + 2\lambda_j & \text{if } i = i' \ j = j' , \\ 0 & \text{otherwise.} \end{cases}$$

$$= 2\alpha(2\alpha-1)\sum_{r=1}^{d} \mu_{ij}^{(\alpha-1)} P_{ir}^{2\beta} d_{ijr}^2 + 2\lambda_j \tag{30}$$

$$= 2\alpha(2\alpha-1)\mu_{ij}^{(\alpha-1)} d_{ij}^{2'} + 2\lambda_j \tag{31}$$

where $d_{ij}^{2'} = \sum_{r=1}^{d} P_{ir}^{2\beta} d_{ijr}^2$.

Substituting the value of μ_{ij} and λ_j in 31, we get:

$$2\alpha(2\alpha - 1)d_{ij}^{2}{}' \left[1 \Big/ \sum_{l=1}^{k} \left[\frac{d_{ij}^{2}{}'}{d_{lj}^{2}{}'} \right]^{1/(\alpha-1)} \right]^{(\beta-1)} - 2 \left[1 \Big/ \sum_{l=1}^{k} \left[\frac{1}{\alpha d_{lj}^{2}{}'} \right]^{1/(\alpha-1)} \right]^{(\alpha-1)}$$

$$= (2\alpha(2\alpha - 1) - 2\alpha) \times \left[1 \Big/ \sum_{l=1}^{k} \left[\frac{1}{d_{lj}^{2}{}'} \right]^{1/(\alpha-1)} \right]^{(\alpha-1)} \tag{32}$$

$$= 4\alpha(\alpha - 1) \left[\sum_{l=1}^{k} \left[d_{lj}^{2}{}' \right]^{1/(1-\alpha)} \right]^{(1-\alpha)} \tag{33}$$

$Letting, a_j = \left[\sum_{l=1}^{k} (d_{lj}^{2}{}')^{1/(1-\alpha)} \right]^{(1-\alpha)}, 1 \le j \le n,$

$$\frac{\partial^2 J}{\partial S_{ij} S_{i'j'}} = \gamma_j \quad where, \quad \gamma_j = 4\alpha(\alpha - 1)a_j \ 1 \le j \le n. \tag{34}$$

Hence there are n distinct eigen values each of multiplicity k, of Hessian matrix of U which is a diagonal matrix. With the assumptions $\alpha > 1, \beta > 1$ and $d_{ij}^{2} > 0 \ \forall \, l, j$ it implies $\gamma_j > 0 \ \forall j$. Thus, Hessian matrix of U is positive definite and hence, the sufficiency condition is proved.

Theorem 3. *Let* $\zeta : M_f^{kd} \to \Re$, $\zeta(W) = J_{\alpha,\beta}(U, W, Z)$, *where* $U \in M_f^{kn}$ *and* $Z \in \Re^{kd}$ *is fixed. Then* $W \in M_f^{kd}$ *is the strict local minima of* ζ *if and only if* W *is computed by equation:*

$$\omega_{ir} = 1 \Big/ \sum_{l'=1}^{d} \left[\frac{\sum_{j=1}^{n} (\mu_{ij})^{\alpha} d_{ijr}^{2}}{\sum_{j=1}^{n} (\mu_{ij})^{\alpha} d_{ijl'}^{2}} \right]^{1/(\beta-1)} \tag{35}$$

Proof. In order to obtain the first order necessary condition for optimality, we set the gradient of J w.r.t P_{jr} equal to zero.

$$\frac{\partial J}{\partial P_{ir}} = 2\beta \sum_{j=1}^{n} S_{ij}^{2\alpha} P_{ir}^{2\beta-1} d_{ijr}^{2} + 2\phi_i P_{ir} = 0 \tag{36}$$

We assume that $P_{ir} \ne 0$, $1 \le i \le k$, $1 \le r \le d$. Computing in a manner as in theorem 2 we obtain:

$$P_{ir}^{2} = \left[\frac{-\phi_i}{\beta \sum_{j=1}^{n} S_{ij}^{2\alpha} d_{ijr}^{2}} \right]^{\frac{1}{(\beta-1)}}$$

Since, $\omega_{ir} = P_{ir}^2$ we get:

$$\omega_{ir} = \left[\frac{-\phi_i}{\beta \sum_{j=1}^{n} S_{ij}^{2\alpha} d_{ijr}^2} \right]^{\frac{1}{(\beta-1)}} \tag{37}$$

Using constraint eq. 6 we get:

$$\sum_{r=1}^{d} \omega_{ir} = \sum_{r=1}^{d} \left[\frac{-\phi_i}{\beta \sum_{j=1}^{n} S_{ij}^{2\alpha} d_{ijr}^2} \right]^{\frac{1}{(\beta-1)}} = 1$$

Substituting the value of ϕ_i in eq. 37, we obtain:

$$\omega_{ir} = 1 \left/ \sum_{l'=1}^{d} \left[\frac{\sum_{j=1}^{n} \mu_{ij}^{\alpha} d_{ijr}^2}{\sum_{j=1}^{n} \mu_{ij}^{\alpha} d_{ijl'}^2} \right]^{1/(\beta-1)} \right. \tag{38}$$

Now, to prove the sufficiency condition we compute the second order partial derivative

$$\frac{\partial^2 J}{\partial P_{ir} P_{i'r'}} = \begin{cases} 2\beta(2\beta-1) \sum_{j=1}^{n} P_{ir}^{2\beta-2} S_{ij}^{2\alpha} d_{ijr}^2 + 2\phi_i & \text{if } i = i', \ r = r' \\ 0 & \text{otherwise.} \end{cases}$$

$$= 2\beta(2\beta-1) \sum_{j=1}^{n} \omega_{ir}^{(\beta-1)} S_{ij}^{2\alpha} d_{ijr}^2 + 2\phi_i \tag{39}$$

$$= 2\beta(2\beta-1) \omega_{ir}^{(\beta-1)} \hat{d}_{ir}^2 + 2\phi_i \tag{40}$$

where $\hat{d}_{ir}^2 = \sum_{j=1}^{n} S_{ij}^{2\alpha} d_{ijr}^2$.
Substituting the value of ω_{ir} and ϕ_i in 40, we get:

$$= 2\beta(2\beta-1)\hat{d}_{ir}^2 \left[1 \left/ \sum_{l'=1}^{d} \left[\frac{\hat{d}_{ir}^2}{\hat{d}_{il'}^2} \right]^{1/(\beta-1)} \right. \right]^{(\beta-1)} - 2 \left[1 \left/ \sum_{l'=1}^{d} \left[\frac{1}{\beta \hat{d}_{il'}^2} \right]^{1/(\beta-1)} \right. \right]^{(\beta-1)}$$

$$= (2\beta(2\beta-1) - 2\beta) \times \left[1 \left/ \sum_{l'=1}^{d} \left[\frac{1}{\hat{d}_{il'}^2} \right]^{1/(\beta-1)} \right. \right]^{(\beta-1)}$$

$$= 4\beta(\beta-1) \left[\sum_{l'=1}^{d} (\hat{d}_{il'}^2)^{1/(1-\beta)} \right]^{(1-\beta)}$$

$$Letting, b_i = \left[\sum_{l'=1}^{d} (d_{il'}^{\hat{2}})^{1/(1-\beta)} \right]^{(1-\beta)} 1 \leq i \leq k,$$

$$where, \quad \frac{\partial^2 J}{\partial P_{ir} P_{i'r'}} = \kappa_i \quad \kappa_i = 4\beta(\beta-1)b_i 1 \leq i \leq k. \tag{41}$$

Hence there are k distinct eigen values each of multiplicity r, of Hessian matrix of W which is a diagonal matrix. With the assumption $\alpha > 1, \beta > 1$ and $d_{il'}^{\hat{2}} > 0 \; \forall i, l'$ it implies $\kappa_i > 0 \; \forall i$. Thus, Hessian matrix of W is positive definite and hence, the sufficiency condition is proved.

Theorem 4. *Let* $\xi : \Re^{kd} \to \Re, \xi(z) = J_{\alpha,\beta}(U, W, Z)$, *where* $U \in M_f^{kn}$ *and* $W \in M_f^{kd}$ *is fixed. Then* Z *is the strict local minima of* ξ *if and only if* $z_i 1 \leq i \leq k$ *is computed by equation:*

$$z_{ir} = \sum_{j=1}^{n} \omega_{ir}^{\beta} \mu_{ij}^{\alpha} x_{jr} \Bigg/ \sum_{j=1}^{n} \omega_{ir}^{\beta} \mu_{ij}^{\alpha} \; 1 \leq i \leq k, 1 \leq r \leq d \tag{42}$$

Proof. Proof follows imediately from proposition 2 of [10].

Theorem 5. *Let* $\alpha > 1$ *and* $\beta > 1$ *be fixed and* $X = \{x_1, x_2, ..., x_n\}$ *contain at least* $k(< n)$ *distinct points. Let the solution set* S *of the optimization problem*

$$\min_{(W,Z,U) \in M_f^{kd} \times \Re^{kd} \times M_f^{kn}} J_{\alpha,\beta}(W, Z, U))$$

$S = \{(\bar{W}, \bar{Z}, \bar{U}) \in M_f^{kd} \times \Re^{kd} \times M_f^{kn} | J_{\alpha,\beta}(\bar{W}, \bar{Z}, \bar{U}) < J_{\alpha,\beta}(\bar{W}, \bar{Z}, U) \; \forall U \neq \bar{U}$ *and* $J_{\alpha,\beta}(\bar{W}, \bar{Z}, \bar{U}) < J_{\alpha,\beta}(W, \bar{Z}, \bar{U}) \; \forall W \neq \bar{W}$ *and* $J_{\alpha,\beta}(\bar{W}, \bar{Z}, \bar{U}) < J_{\alpha,\beta}(\bar{W}, Z, \bar{U}) \; \forall Z \neq \bar{Z}\}$
Let $(\bar{W}, \bar{Z}, \bar{U}) \in M_f^{kd} \times \Re^{kd} \times M_f^{kn}$. *Then* $J_{\alpha,\beta}(\hat{W}, \hat{Z}, \hat{U}) \leq J_{\alpha,\beta}(\bar{W}, \bar{Z}, \bar{U})$ *for every* $(\hat{W}, \hat{Z}, \hat{U}) \in T(\bar{W}, \bar{Z}, \bar{U})$ *and the inequality is strict if* $(\bar{W}, \bar{Z}, \bar{U}) \notin S$.

Proof. Proof follows as in Theorem 11 of [19].

Lemma 1. *Let* $M : X \to Y$ *be a function and* $\tau : Y \to P(Y)$ *point-to-set map. If* M *is continuous at* $a^{(0)}$ *and* τ *is closed at* $M(a^{(0)})$, *then the point-to-set map* $\tau o M : X \to P(V)$ *is closed at* $a^{(0)}$.

Theorem 6. *Let* $\alpha > 1$ *and* $\beta > 1$ *be fixed and* $X = \{x_1, x_2, ..., x_n\}$ *contain at least* $k(< n)$ *distinct points. Then the point-to set map* $T_{\alpha,\beta} : M_f^{kd} \times \Re^{kd} \times M_f^{kn} \to P(M_f^{kd} \times \Re^{kd} \times M_f^{kn})$ *is closed at every point in* $M_f^{kd} \times \Re^{kd} \times M_f^{kn}$.

To prove the above theorem we need to prove the following:

a) The function $A_1 : M_f^{kd} \times \Re^{kd} \times M_f^{kn} \to \Re^{kd} \times M_f^{kn}$ as defined in eq. 22 is continuous at every point in $M_f^{kd} \times \Re^{kd} \times M_f^{kn}$.

b) The function $A_2 : \Re^{kd} \times M_f^{kn} \to M_f^{kd} \times \Re^{kd}$ as defined in eq. 16 is continuous at every point in $\Re^{kd} \times M_f^{kn}$.

c) To prove $A_3 : M_f^{kd} \times \Re^{kd} \to P(M_f^{kd} \times \Re^{kd} \times M_f^{kn})$ as defined in eq. 17 is closed on $M_f^{kd} \times \Re^{kd} \times M_f^{kn}$.

Proof. Proof follows imediately from lemma 19, 20, 21 respectively in [19].

Theorem 7. *Let $[conv(X)]^k$ be the k-fold cartesian product of the convex hull of X, and let $(W^{(0)}, Z^{(0)}, U^{(0)})$ be the starting point of iteration $T_{\alpha,\beta}$ where $W^{(0)} \in M_f^{kd}, Z^{(0)} \in \Re^{kd}$ and $U^{(0)} \in M_f^{kn}$. Then $T_{\alpha,\beta}^p(W^{(0)}, Z^{(0)}, U^{(0)}) \in M_f^{kd} \times [conv(X)]^k \times M_f^{kn}$ for p=1, 2, 3,..., where $T_{\alpha,\beta}^p = T_{\alpha,\beta}oT_{\alpha,\beta}oT_{\alpha,\beta}o...oT_{\alpha,\beta}$ (p times) and $M_f^{kd} \times [conv(X)]^k \times M_f^{kn}$ is compact in $M_f^{kd} \times \Re^{kd} \times M_f^{kn}$*

Proof. Proof follows imediately from theorem 22 of [19].

Theorem 8. *Let $X = \{x_1, x_2, ..., x_n\}$ contain atleast $k(< n)$ distinct points, and let $J_{\alpha,\beta}(W, Z, U)$ defined as in eq. 9. Let $(W^{(0)}, Z^{(0)}, U^{(0)})$ be the starting point of iteration $T_{\alpha,\beta}$, where $W^{(0)} \in M_f^{kd}, Z^{(0)} \in \Re^{kd}$ and $U^{(0)} \in M_f^{kn}$. Then the iteration sequence $\{(W^{(p)}, Z^{(p)}, U^{(p)})\}$, p=1, 2, 3,...either terminates at a point (W^*, Z^*, U^*) in a solution set S; or there is a subsequence converging to a point in S.*

Proof. Proof of this theorem follows by theorems 2 to 7 and Zangwill's Convergence theorem.

5 Experimental Results

For evaluating the accuracy and efficiency of the PGK clustering algorithm, we have compared its performance with PROCLUS and GK algorithms using real and synthetic data sets. Forest Fire, Breast Cancer, Parkinson and Alzheimr data sets from the UCI data repository [55] were used for experimentation. These data sets had no missing values. The parameters of each algorithm were fine tuned and multiple runs of the experiments were conducted to minimize the effect of initialization.

Table 1. Data Sets

Data Sets	Instances	Attributes	Classes
Forest Fire	517	13	3
Alzheimr	45	8	3
Parkinson	197	23	2
Breast Cancer	569	32	2

Table 2. Comparative Accuracy of PGK, PROCLUS and GK Algorithm

Data Sets	PGK	PROCLUS	GK
Forest Fire	0.8588	0.8696	0.8337
Alzheimr	0.7556	0.6667	0.6000
Parkinson	0.7538	0.7641	0.7538
Breast Cancer	0.8875	0.7907	0.8401

To measure the cluster purity, we use clustering accuracy measure defined in [18] as follows: $r = \sum_{i=1}^{c} s_i/n$ where s_i denotes the number of correctly classified instances in cluster i and n denotes the total number of instances in the data set. In order to compare the cluster purity results of PROCLUS, GK algorithm, and PGK algorithm, we defuzzified the fuzzy assignments. As in [51] we use the following maximum membership rule for defuzzification and resolve the conflicts arbitrarily: A data point x_j belongs to cluster c_l if and only if:

$$l = arg \ max_{1 \leq i \leq k} \ \mu_{ij}$$

Table 2 shows, while the PGK algorithm achieves highest accuracy for Alzehmir and Breast Cancer data sets, the performance of the PGK algorithm is comparable to PROCLUS and GK algorithm for the other two data sets. Tables 3 and 4 show the projected dimensions associated with each cluster for the aforementioned data sets. It may be noted that the PGK algorithm discovers clusters in lower dimensional subspaces.

Table 3. Dimensions found by PGK and PROCLUS Algorithm

Cluster no.	Forest Fire		Alzheimr	
	PGK	PROCLUS	PGK	PROCLUS
1	12	3, 4, 5, 12	1, 7	4, 5, 6, 7
2	12, 13	1, 2, 3, 4, 11, 12	1, 6, 7	4, 5, 6, 7
3	12, 13	1, 2, 3, 4, 5, 8, 11, 12	1	4,5, 6, 7

Table 4. Dimensions found by PGK and PROCLUS Algorithm

Cluster no.	Breast Cancer		Parkinson	
	PGK	PROCLUS	PGK	PROCLUS
1	1, 19	5, 6, 8, 9, 10, 15, 16 17, 18, 19, 20, 25, 29, 30	1, 3, 5, 19, 20	4, 5, 6, 7, 15, 17, 18, 19, 20, 21, 22
2	15, 16, 19	5, 6, 8, 9, 10, 15, 16 17, 18, 19, 20, 25, 29, 30	3, 5	4, 5, 6, 7, 8, 12

Cluster validity measures have been used to find whether a clustering algorithm generates the best partitioning of a data set with respect to various parameters[1][24]. We have used the validity measures Partition Coefficient(PC), Coefficient of Entropy(CE) and Sum of Squared Errors (SSE) in our experiments. These are described below:

- Partition Coefficient: It measures the amount of overlapping between the clusters.
$PC = (1/n) \sum_{i=1}^{k} \sum_{j=1}^{n} \mu_{ij}^2$
As value of partition coefficient increases from $1/k$ to 1, the assignments become most fuzzy to crisp.

- Coefficient of Entropy: It measures the fuzziness of the cluster partitions.
$CE = (-1/n) \sum_{i=1}^{k} \sum_{j=1}^{n} \mu_{ij} \log(\mu_{ij})$
When μ_{ij} values are close to $1/k$ the measure CE takes a high value which means that the result of clustering is unsatisfactory. Similarly if all μ_{ij} values are close to zero or one CE takes a low value indicating a good result of clustering [44].

- Sum of Squared Errors: It is used as a measure of compactness. Lower values of SSE measure indicate compact clusters. As we are looking for clusters in subspaces of the entire set of dimensions, we modify the notion of the sum of squared errors by computing the distance using only the dimensions relevant to a cluster and call it Sum of Projected Squared Errors (SPSE).

$$SPSE = \sum_{i=1}^{k} \sum_{x_j \in c_i} d(x_j, z_i)$$

Table 5 shows the values of PC and CE obtained for different UCI data sets on applying PGK and GK algorithm. It shows that for Breast Cancer and Parkinson data sets PGK achieves higher values for PC and lower values for CE, as compared to GK. Table 6 shows the SSE values on applying the different algorithms to various data. We observed that the PGK algorithm achieves a lower SPSE value and hence, generates more compact clusters in comparison to other algorithms.

We have also evaluated the sensitivity of clusters which is defined as the ratio between the number of true positives and sum of the number of true positives and the number of false negatives. Table 7 shows that PGK has higher sensitivity than GK, although PROCLUS has higher sensitivity for Alzehmir and Forest Fire data sets.

We studied the scalability of PGK algorithm, GK algorithm, and PROCLUS clustering algorithms on increasing the number of data instances, clusters, and dimensions. As in [3], in the experiment on scalability, we have used synthetic data sets so that various parameters can be controlled. In our experiments also relevant dimensions follow normal distribution and irrelevant dimensions follow

Table 5. PC and CE comparison

Data Sets	PC		CE	
	PGK	GK	PGK	GK
Forest Fire	0.4376	0.5899	0.9600	0.5054
Alzheimr	0.5935	0.7965	0.6848	0.3063
Parkinson	0.5934	0.5741	0.5944	0.6103
Breast Cancer	0.7724	0.5205	0.3610	0.6706

Table 6. Sum of Projected Squared Errors

	PGK	PROCLUS	GK
Forest Fire	3.4452×10^3	8.2966×10^3	32.1127×10^3
Alzheimr	109.0263	193.9787	308.0943
Parkinson	2.8385	6.5523×10^3	17.2162×10^3
Breast Cancer	131.4887	1.1060×10^3	9.5432×10^3

Table 7. Comparative Sensitivity of PGK, PROCLUS and GK Algorithm

Data Sets	PGK	PROCLUS	GK
Forest Fire	0.9528	0.9750	0.7333
Alzheimr	0.8602	1	0.5591
Parkinson	0.0010	0.0417	0.0031
Breast Cancer	1	0.4528	0.7204

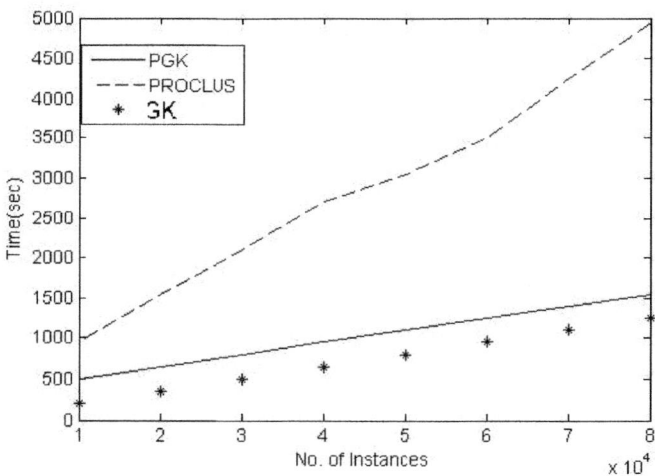

Fig. 1. Scalability as a function of number of instances

uniform distribution. We observe that in all the experiments, the PGK algorithm scales significantly better than the PROCLUS however, it is somewhat slower by a linear factor as compared to the GK algorithm. For measuring the scalability of algorithm with increasing number of instances, various data sets having data instances in the range 10,000-80,000 were generated. Each data set had 5 clusters and 15 dimensions, out of which 10 followed normal distribution. The results are presented in Figure 1. Scalability with the increasing number of attributes is measured using data sets containing 50,000 instances and 5 clusters. We varied the number of dimensions in the range 10-50. In each case, 5 dimensions followed normal distribution and others followed uniform distribution. The results of experiments are presented in Figure 2. We considered data sets containing 50,000

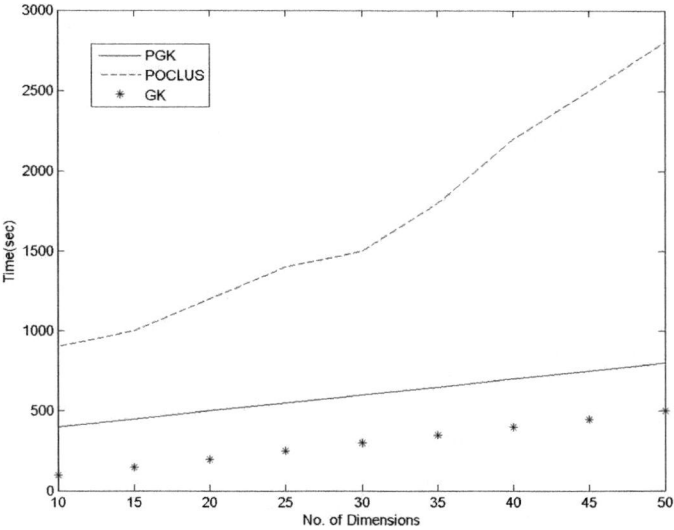

Fig. 2. Scalability as a function of number of dimensions

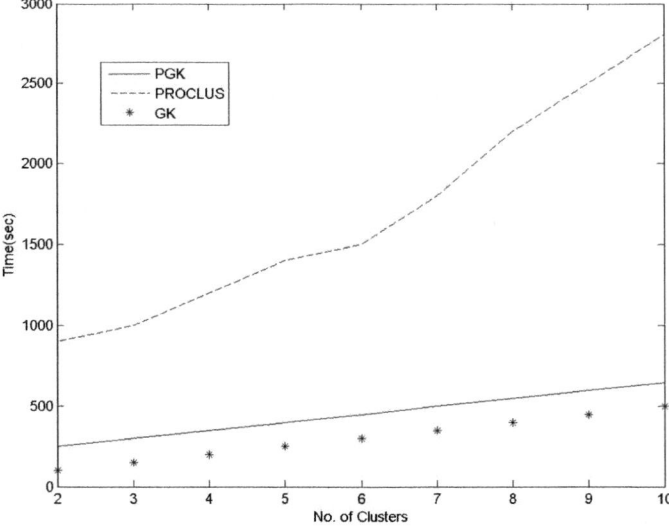

Fig. 3. Scalability as a function of number of clusters

instances with 15 dimensions from which 10 followed normal distribution and 5 followed uniform distribution. The number of clusters was varied from 2-10. The results are shown in Figure 3.

6   Conclusion and Future Work

Our main contribution in this paper is the adaptation of GK algorithm for projected clustering such that each of the dimensions has possibly different weights associated with different clusters. PGK algorithm minimizes the objective function to determine prototype parameters, the weight matrix and membership matrix for each cluster. It determines the best feature weights corresponding to each cluster. We proved the convergence of the PGK algorithm and also suggested a way of extending projected Gustafson-Kessel clustering algorithm to a rough set based projected clustering algorithm. PGK discovers cluster in projected subspaces with high accuracy for each of the UCI data sets used. It may be interesting to explore the applicability of PGK algorithm for categorical data. By taking outliers into account, the proposed algorithm can be generalized.

Acknowledgment. This work was supported by University of Delhi, research grant no. Dean(R)/ R&D/2007/Ph-III/382 and UGC Research grant no. Ref. no. Sch/JRF/AA/16/ 2005-2006/7372. We express our thanks to Prof. Ajay Kumar and Mukund, Department of Mathematics, University of Delhi, for valuable mathematical discussion. We would also like to thank the anonymous reviewers whose critical comments helped us to improve the paper significantly.

References

1. Abonyi, J., Feil, B.: Cluster Analysis for Data Mining and System Identification. Birkhäuser, Basel
2. Achtert, E., Böhm, C., David, J., Kröger, P., Zimek, A.: Robust Clustering in Arbitrarily Oriented Subspaces. In: SDM, pp. 763–774 (2008)
3. Aggarwal, C., Wolf, J., Yu, P., Procopiuc, C., Park, J.: Fast algorithms for projected clustering. In: ACM SIGMOD, pp. 61–72 (1999)
4. Agrawal, R., Gehrke, J., Gunopolos, D., Raghavan, P.: Automatic Subspace Clustering of High Dimensional Data for Data Mining Applications. In: ACM SIGMOD (1998)
5. Asharaf, S., Murty, M.N.: A Rough Fuzzy Approach to Web Usage Categorization. In: Fuzzy Sets and Systems, pp. 119–129 (2004)
6. Assent, I., Krieger, R., Müller, E., Seidl, T.: DUSC: Dimensionality Unbiased Subspace Clustering. In: ICDM (2007)
7. Assent, I., Krieger, R., Müller, E., Seidl, T.: EDSC: Efficient density-based subspace clustering. In: Proceeding of the 17th ACM Conference on Information and Knowledge Management (2008)
8. Babuka, R., van der Veen, P.J., Kaymak, U.: Improved covariance estimation for Gustafson-Kessel clustering. In: FUZZ-IEEE 2002, vol. 2, pp. 1081–1085 (2002)
9. Bezdek, J.C.: Pattern recognition with Fuzzy Objective Function Algorithm. Plenum Press, New York (1981)
10. Bezdek, J.C.: An convergence theorem for the fuzzy ISODATA clustering algorithms. IEEE Trans. Pattern Anal. Mach. Intell. 2(1), 1–8 (1980)
11. Bezdek, J.C., Hathaway, R., Sabin, M., Tucker, W.: Convergence theory for fuzzy c-means: counterexamples and repairs. In: Bezdek, J., Pal, S. (eds.) Fuzzy Models for Pattern Recognition: Methods that Search for Approximate Structures in Data, pp. 138–142. IEEE Press, New York (1992)

12. Bezdek, J.C., Coray, C., Gunderson, R., Watson, J.: Detection and characterization of cluster substructure. SIAM J. Appt. Math. 40(2), 339–372 (1981)
13. Beyer, K., Goldstein, J., Ramakrishnan, R., Shaft, U.: When is "nearest neighbor" meaningful? In: Beeri, C., Bruneman, P. (eds.) ICDT 1999. LNCS, vol. 1540, pp. 217–235. Springer, Heidelberg (1998)
14. Bazan, J.G., Nguyen, H.S., Nguyen, S.H., Synak, P., Wroblewsk, J.: Rough Set Algorithms in Classification Problem. In: Polkowski, L., Tsumoto, S., Lin, T.Y. (eds.) New Developments in Knowledge Discovery in Information Systems, pp. 49–88. Physica Verlag, Heidelberg (2000)
15. Dash, M., Choi, K., Scheuermann, P., Liu, H.: Feature Selection for Clustering - A Filter Solution. In: ICDM, pp. 115–122 (2002)
16. Dhillon, I., Kogan, J., Nicholas, M.: A Comprehensive Survey of Text Mining, pp. 73–100. Springer, Heidelberg (2003)
17. Dunn, J.: A Fuzzy Relative of the ISODATA Process and Its Use in Detecting Compact Well-Separated Clusters. J. Cybernetics 3, 32–57 (1974)
18. Gan, G., Wu, J., Yang, Z.: PARTCAT: A Subspace Clustering Algorithm for High Dimensional Categorical Data. In: IJCNN, pp. 4406–4412 (2006)
19. Gan, G., Wu, J.: A convergence theorem for the fuzzy subspace clustering (FSC) algorithm. Pattern Recognition 41, 1939–1947 (2008)
20. Gan, G., Wu, J., Yang, Z.: A fuzzy subspace algorithm for clustering high dimensional data. In: Li, X., Zaïane, O.R., Li, Z.-h. (eds.) ADMA 2006. LNCS (LNAI), vol. 4093, pp. 271–278. Springer, Heidelberg (2006)
21. Gustafson, D.E., Kessel, W.: Fuzzy clustering with a Fuzzy Covariance Matrix. In: Proc. IEEE-CDC, vol. 2, pp. 761–766 (1979)
22. Han, J., Kamber, M.: Data Mining: Concepts and Techniques, 2nd edn. Morgan Kaufmann, San Francisco (2006) (1st ed., 2000)
23. Hathaway, J.R., Bezdek, C.J.: Local convergence of the fuzzy c-Means algorithms. In: Pattern Recognition, pp. 477–480 (1986)
24. Hoppner, F., Klawonn, F., Kruse, R., Runkler, T.: Fuzzy Cluster Analysis: Methods for Classification. In: Data Analysis, and Image Recognition. John Wiley & Sons, Chichester
25. Jain, A.K., Murty, M.N., Flynn, P.J.: Data Clustering: A Review. ACM Comput. Surv. 31(3), 264–323 (1999)
26. Jensen, R., Shen, Q.: A Rough Set – Aided system for Sorting WWW Book-marks. In: Zhong, N., et al. (eds.) Web Intelligence: Research and Development, pp. 95–105 (2001)
27. John, G.H., Kohavi, R., Pfleger., K.: Irrelevant Features and the Subset Selection Problem. In: ICML, pp. 121–129 (1994)
28. Kailing, K., Kriegel, H.-P., Kröger, P.: Density-Connected Subspace Clustering for High Dimensional Data, pp. 246–257. SIAM, Philadelphia (2004)
29. Kaufman, J., Rousseeuw, P.: Finding Groups in Data: An Introduction to Cluster Analysis. John Wiley, New York (1990)
30. Kelling, K., Peter, H., Kröger, P.: A Generic Framework for Efficient Subspace Clustering of High-Dimensional Data. In: ICDM, pp. 205–257 (2005)
31. Tan, P.-N., Steinbach, M., Kumar, V.: Introduction to Data Mining. Addison-Wesley, Reading (2005)
32. Li, Y., Shiu, S.C.K., Pal, S.K.: Combining feature reduction and case selection in building CBR classifiers. In: Pal, S.K., Aha, D.W., Gupta, K.M. (eds.) Case-Based Reasoning in Knowledge Discovery and Data Mining. Wiley, New York (2005)

33. Lingras, P., West, C.: Interval set clustering of Web users with rough k-means. Technical Report No. 2002-002, Dept. of Mathematics and Computer Science, St. Mary's University, Halifax, Canada (2002)
34. Lingras, P.: Applications of Rough Set Based K-Means, Kohonen, GA Clustering. T. Rough Sets, VII, pp. 120–139 (2007)
35. Liu, H., Setiono, R.: A probabilistic approach to feature selection – a filter solution. In: Proceedings of the 9th International Conference on Industrial and Engineering Applications of AI and ES, pp. 284–292 (1996)
36. Luo, Y., Xiong, S.: Clustering Ensemble for Unsupervised Feature Selection. In: FSKD, pp. 445–448 (2009)
37. Luss, R., d'Aspremont, A.: Clustering and Feature Selection using Sparse Principal Component Analysis. Optimization & Engineering 11(1), 145–157 (2010)
38. Małyszko, D., Stepaniuk, J.: Rough entropy based k-means clustering. In: Sakai, H., Chakraborty, M.K., Hassanien, A.E., Ślęzak, D., Zhu, W. (eds.) RSFDGrC 2009. LNCS, vol. 5908, pp. 406–413. Springer, Heidelberg (2009)
39. Mitra, S., Banka, H., Pedrycz, W.: Rough-fuzzy collaborative clustering. IEEE Transactions on Systems, Man, and Cybernetics, Part B: Cybernetics 36, 795–805 (2006)
40. Nagesh, H., Goil, S., Choudhary, A.: MAFIA: Efficient and Scalable Subspace Clustering for Very Large Data Sets, Technical Report, Northwestern Univ. (1999)
41. Pawlak, Z.: Rough Sets, Theoretical Aspects of Reasoning about Data. Kluwer Academic, Dordrecht (1991)
42. Peters, G., Lampart, M., Weber, R.: Evolutionary Rough k-Medoid Clustering. T. Rough Sets 8, 289–306 (2008)
43. Ruspini, E.H.: A New Approach to Clustering. Information and Control, 22–32 (1969)
44. Rutkowski, L.: Computational Intelligence Methods and Techniques. Springer, Heidelberg (2008)
45. Sequeira, K., Zaki, M.: SCHISM: A new approach for interesting subspace mining. In: Proc. IEEE ICDM, Hong Kong (2004)
46. Shen, Q.: Rough Feature Selection for Intelligent Classifiers. T. Rough Sets 7, 244–255 (2007)
47. Slezak, D.: Rough Sets and Functional Dependencies in Data: Foundations of Association Reducts. T. *Computational Science* 5 (2009) 182-205.
48. Slezak, D.: Association Reducts: A Framework for Mining Multi-attribute Dependencies. In: Hacid, M.-S., Murray, N.V., Raś, Z.W., Tsumoto, S. (eds.) ISMIS 2005. LNCS (LNAI), vol. 3488, pp. 354–363. Springer, Heidelberg (2005)
49. Spath, H.: Cluster Analysis Algorithms for Data Reduction and Classification. Ellis Horwood, Chichester (1980)
50. Xu, R.: Survey of Clustering Algorithms. IEEE Transactions on Neural Networks 16(3) (2005)
51. Wiswedel, B., Berthold, M.R.: Fuzzy clustering in parallel universes. In: NAFIPS, pp. 567–572 (2005)
52. Xie, X.L., Beni, G.: A validity measure for fuzzy clustering. Pattern Analysis and Machine Intelligence 13, 841–847 (1991)
53. Zadeh, L.A.: Fuzzy logic, neural networks, and soft computing. Communications of the ACM 37, 77–84 (1994)
54. Zangwill, W.: Non-Linear programming: A unified Approach. Prentice-Hall, Englewood Cliffs (1969)
55. http://www.ics.uci.edu/~mlearn/MLRepository.html

Generalized Rough Sets and Implication Lattices

Pulak Samanta[1] and Mihir Kumar Chakraborty[2]

[1] Department of Mathematics, Katwa College
Katwa, Burdwan, West Bengal, India
pulak_samanta06@yahoo.co.in
[2] Honorary Visiting Professor, Center for Soft Computing Research,
Indian Statistical Institute, Kolkata, India &
UGC Visiting Professor, Centre for Cognitive Science
Jadavpur University, Kolkata, India
mihirc4@gmail.com

Abstract. This paper consists of an extensive survey of various general-ized approaches to the lower and upper approximations of a set, the two approximations being first defined by Pawlak while introducing rough set theory. Particularly, relational, covering based and operator based ap-proaches are considered. Categorization of various approaches in terms of implication lattices is shown. Significance of this categorization in rough logics is briefly mentioned.

Keywords: rough sets, partition, covering, implication lattice, modal logic.

1 Introduction

Pawlak's rough set theory begins with an approximation space $< U, R >$ where U is a non empty set and R is an equivalence relation on U. So, the set U is partitioned. Given any subset A of U, the lower and upper approximations $\underline{A}_R$ and $\overline{A}^R$ are then defined by $\underline{A}_R = \{x|\ [x]_R \subseteq A\}$ and $\overline{A}^R = \{x|\ [x]_R \cap A \neq \phi\}$ where $[x]_R$ is the equivalence class of x with respect to R. One can immediately observe that the following properties of lower and upper approximations hold.

(1a) $\underline{U} = U$ (Co-normality)   (1b) $\overline{U} = U$ (Co-normality)
(2a) $\underline{\phi} = \phi$ (Normality)   (2b) $\overline{\phi} = \phi$ (Normality)
(3a) $\underline{A} \subseteq A$ (Contraction)   (3b) $A \subseteq \overline{A}$ (Extension)
(4a) $\underline{A \cap B} = \underline{A} \cap \underline{B}$ (Multiplication)   (4b) $\overline{A \cup B} = \overline{A} \cup \overline{B}$ (Addition)
(5a) $\underline{(\underline{A})} = \underline{A}$ (Idempotency)   (5b) $\overline{(\overline{A})} = \overline{A}$ (Idempotency)
(6) $\underline{(\sim A)} =\sim (\overline{A})$, $\overline{(\sim A)} =\sim (\underline{A})$ (Duality)
(7a) $A \subseteq B \Rightarrow \underline{A} \subseteq \underline{B}$ (Monotone)   (7b) $A \subseteq B \Rightarrow \overline{A} \subseteq \overline{B}$ (Monotone)
(8a) $A \subseteq \overline{(\underline{A})}$ (8b) $\overline{(\underline{A})} \subseteq A$

However, we shall have to expand this list of properties for addressing more generalized approaches. For example the property $4(a)$ will be split into two

J.F. Peters et al. (Eds.): Transactions on Rough Sets XIV, LNCS 6600, pp. 183–201, 2011.
© Springer-Verlag Berlin Heidelberg 2011

components viz. $\underline{A \cap B} \subseteq \underline{A} \cap \underline{B}$ and $\underline{A} \cap \underline{B} \subseteq \underline{A \cap B}$ and similarly the other equalities. It will be observed that in some cases only one component holds and in some cases both the components do not hold.

Almost from inception of the theory, various generalizations took place from various stand points of which some are purely theoretical and some emerging for specific applications. We can mark at least three clearly distinct approaches towards generalization.

Relational Approach

In this approach instead of taking an equivalence relation on the universe U an arbitrary binary relation R is taken and gradually conditions like reflexivity, symmetry and transitivity are imposed on it. The construction is accomplished in two stages. Firstly for each element $x \in U$, the set $R_x = \{y | xRy\}$ is defined. Then for each subset A of U sets $\underline{A}_R$ and $\overline{A}^R$ are defined by

$$\underline{A}_R = \{x \in U | R_x \subseteq A\},$$
$$\overline{A}^R = \{x \in U | R_x \cap A \neq \phi\}.$$

It may be noted that at this stage not necessarily $\underline{A}_R \subseteq A$ or $A \subseteq \overline{A}^R$. Yet many of the properties listed above appear. For example $1(a)$, $2(b)$, $3(a)$, $4(b)$, $7(a)$, $7(b)$ and 6 hold. That the other listed properties do not hold in general may be verified by simple counterexamples. Addition of conditions to R adds more properties from the above list gradually. We give the results in a tabular form in the Table no. 2(Section 2).

In Pawlak's original work [20] the equivalence relation (the indiscernibility) emerged out of some information system. An information system is a list of attributes and values applicable to the objects of an universe. The indiscernibility relation is defined by relating two objects having the same value with respect to each attribute. The objects are clustered in terms of the attribute-values and the universe is partitioned into disjoint classes of mutually equivalent elements. From the angle of application however, it is found to be important to look into situations when the relation is not an equivalence but only tolerance (reflexive and symmetric) or simply reflexive. J.W.G.Busse and W.J.G.Busse [7] and others [12,18,25,30,32,36] have done significant research in this direction. From the theoretical point of view also a generalization from equivalence is quite important.

Mathematically however, the converse question is more interesting which may be formulated as follows.

Let U be the universe and $\_$ and $\bar{}$ be mappings from $\mathcal{P}(U)$ the power set of U to $\mathcal{P}(U)$ satisfying a subset of properties given in the expanded list. Does there exist a relation R on U having certain specified conditions s.t. the $\underline{A}_R = \underline{A}$ and $\overline{B}^R = \overline{B}$?

A noticeable amount of research has been and is being done in this direction during the last decades [8,12,32,34,36].

Covering based Approach

In this approach the partition of U is replaced by a covering which is a collection $C = \{C_i\}$ of subsets of U such that $\cup C_i = U$, $C_i \neq \phi$. The lower and upper approximations of a subset A of U with respect to a covering are not defined in a unique way but all of them reduce to Pawlakian definition when the covering is taken as a partition. In our paper [29] we made a survey of various pairs of lower-upper approximation but realized that some more had been left out. In this paper we, however, update the survey but are sure neither of the exhaustiveness with respect to the extant literature nor of impossibility of further additions.

In both the above approaches the differences in fixing lower / upper approximations of a subset emerge due to various ways of granule formation. A granule G_x usually related to an object x of the Universe is a subset of the universe. In the extreme case x may not be a member of G_x. In the relation based approach R_x is the granule at x and if R is not reflexive $x \notin R_x$ in general. But this kind of granule formation is not of much significance in the theory of rough sets.

We shall discuss more on granulation in a latter subsection of the introduction.

Operator based Approach

This is also called axiomatic approach [32] in which two operators $\_$ and $^-$ are defined on $\mathcal{P}(U)$, the powerset of U satisfying certain axioms which are some properties like those of the list given at the beginning or its expanded version. That means lower and upper approximations $\underline{A}$ and $\overline{A}$ of a set A are taken straightway satisfying certain axioms. This approach is important mathematically and logically since in whatever way $\underline{A}$ and $\overline{A}$ might have been defined what matters here are the properties that they possess. The entire group of algebras / topologies / logics that have emerged from rough set approach care only for these properties. These in turn has pushed towards another kind of generalization of rough sets viz. the abstract algebraic and topological approaches [1,2,4,9,19,22,23,24,26].

Granulation

The starting point of rough set theory is granule formation. As mentioned before a granule G_x in the universe U is a subset of it usually associated with a point x of U. In case of Pawlak rough set the granule at x is the indiscernibility class $[x]_R$ which is the equivalence class with respect to the equivalence relation R to which x belongs.

This notion is directly generalized to the case of any arbitrary relation R by taking the set $R_x = \{y | xRy\}$ as the granule at x as mentioned earlier. Obviously, the granules are no more disjoint in general as in the case of $[x]_R$. Even the set $\{R_x | x \in U\}$ may or may not form a covering of U. However, if R_x is to contain x which is a natural demand to a granule, R has to be reflexive and then the set $\{R_x\}$ forms a cover. From this observation we step into the next immediate generalization by taking a covering $\{C_i\}$ straightway and then look for granules relative to $x \in U$.

This is done in various ways. Let $\mathcal{C} = \{C_i\}$ be a covering of U and let us consider the following sets.

$N_x^{\mathcal{C}} = \cup\{C_i \in \mathcal{C} : x \in C_i\} = Friends(x)$ [17,25,28]

$P_x^{\mathcal{C}} = \{y \in U : \forall C_i(x \in C_i \Leftrightarrow y \in C_i)\}$ (Partition generated by a covering) [11,21,25,28,38]

$N(x) = \cap\{C_i : C_i \in \mathcal{C}, x \in C_i\} = Neighbour(x)$ [13,35,38]

$e.f(x) = U - Friends(x)$ [17]

All these are granules at x. Interpretations and significances of these subsets of the universe from the angle of real-world scenario is an important research that will not be done here. That will divert us too far from the present objective. However one may look into the original papers where these granules were first introduced. It should be mentioned once again that x itself is not included in $e.f(x)$ and hence its significance is quite intriguing. A more abstract study of granules within the framework of category theory is done in [3]. Granule is a fundamental notion in rough set theory. For some interesting papers on granules the reader is referred to [14,15,16,19].

Beside the above sets, one more notion has been used viz.

$Md(x) = \{C_i : x \in C_i \in \mathcal{C} \wedge (\forall S \in \mathcal{C} \wedge x \in S \subseteq C_i \Rightarrow C_i = S)\}$ [5,27].

These are sets of subsets of U around x. Thus these are similar to the notion of neighborhood system in topology.

It should be mentioned that the references by the side of each of the above sets indicate the papers from which the definitions are taken. They do not necessarily include the first work where it has appeared. In fact, it is often difficult to locate the first occurrence of each notion in rough set literature.

A passage from partition to covering was natural from the point of view of applications also. The equivalence relation R in U originates from an attribute-value system $(U, \{A_i\}, V)$ where $\{A_i\}$ is a set of attributes and V is a set of values, each attribute A_i giving a unique value from V to each object in the universe U. Thus a partition emerges, elements having the same attribute-values being clustered together form an equivalence class. Elements belonging to the same class are indiscernible with respect to the given set of attributes. Now, in real cases, indiscernibility relation is not in general transitive. Particularly when very small changes take place, x_1 may be indistinguishable with x_2, x_2 with x_3, x_3 with x_4 and so on. But for some k, x_1 usually turns out to be distinguishable from x_k. In attribute value systems a non-transitive indiscernibility arises if there are some gaps in the information viz. for some objects the value of some attribute may not be known. In other words, this means that the attributes are partial functions over the universe of objects. Two objects x and y considered indiscernible if and only if for each attribute A_i which is defined at these objects, the values $A_i(x)$ and $A_i(y)$ coincide. Thus the relation turns out to be a tolerance relation and one can take the tolerance classes as the C_i's. However, value-gaps are not the only reason for generation of non transitive indiscernibility. But what is to be noted is that given a covering $\{C_i\}$ if one constructs the corresponding tolerance relation R by xRy iff $x, y \in C_i$ for some i, then defining lower and upper approximations with the help of $\{R_x | x \in U\}$ is just one of several ways. In the present paper this particular pair of operators is given by C^- (c.f. next section).

Organization of the paper

In the following section we have presented various types of lower and upper approximations available in extant literature on rough set theory and their properties in tabular form. In Section 3 for each type of lower-upper approximations, determined by their properties, a unique partially ordered set which turns out to be a lattice in some cases has been defined. These partially ordered sets or lattices are presented diagrammatically. Thus, pairs of approximation operators are categorized in terms of the set of properties they posses. The concluding section viz. section 4 gives some logical interpretation for this categorization and presents some open issues.

This work is an extension of our earlier work [29]. Besides a thorough rewriting, we have added the following new items in it;

(i) correcting some mistakes in [29]
(ii) addition of relational approach
(iii) updating covering based approaches
(iv) explicit categorization of all the lower-upper approximation pairs
(v) more explanation about the significance of this kind of categorization by the partially ordered sets and lattices called implication lattices.

2   Various Types of Lower and Upper Approximations

2.1   Covering-Based Operators

Let X be a subset of U. We give below a list of constructions of the lower-upper approximation of X given in various manners with respect to a covering.

We have used $\underline{P}_i, \overline{P}^i$ $i = 1, 2, 3, 4$ to acknowledge Pomykala, since to our knowledge he first studied the lower and upper approximations with respect to covering. However, the pair $< \underline{P}_4, \overline{P}^4 >$ was due to Pawlak but looking apparently different. $\underline{C}_i, \overline{C}^i$, $i = 1, 2, 3, 4, 5$ are other covering based approximations which are essentially duals. $\underline{C}$ and $\overline{C}$ with extra symbols are also covering based lower and upper approximation operators, the symbols being taken from the respective papers straightway. This group of pairs barring $\underline{C}_{Gr}$ and $\overline{C}^{Gr}$ are non-duals.

$\underline{P_1}(A) = \{x : N_x^{\mathcal{C}} \subseteq A\}$
$\overline{P}^1(A) = \cup\{C_i : C_i \cap A \neq \phi\}$ [25,28,33]

$\underline{P_2}(A) = \cup\{N_x^{\mathcal{C}} : N_x^{\mathcal{C}} \subseteq A\}$
$\overline{P}^2(A) = \{z : \forall y(z \in N_y^{\mathcal{C}} \Rightarrow N_y^{\mathcal{C}} \cap A \neq \phi)\}$ [25,28]

$\underline{P_3}(A) = \cup\{C_i : C_i \subseteq A\}$
$\overline{P}^3(A) = \{y : \forall C_i(y \in Ci \Rightarrow C_i \cap A \neq \phi)\}$ [13,25,28,31,33,38]

$\underline{P_4}(A) = \cup\{P_x^{\mathcal{C}} : P_x^{\mathcal{C}} \subseteq A\}$
$\overline{P}^4(A) = \cup\{P_x^{\mathcal{C}} : P_x^{\mathcal{C}} \cap A \neq \phi\}$ [4,11,13,21,25,27,28,31,32,35,38]

$\underline{C_1}(A) = \cup\{C_i : C_i \in \mathcal{C}, C_i \subseteq A\}$
$\overline{C}^1(A) = \sim \underline{C_1}(\sim A) = \cap\{\sim C_i : C_i \in \mathcal{C}, C_i \cap A = \phi\}$ [27]

$\underline{C_2}(A) = \{x \in U : N(x) \subseteq A\}$
$\overline{C}^2(A) = \{x \in U : N(x) \cap A \neq \phi\}$ [13,27]

$\underline{C_3}(A) = \{x \in U : \exists u(u \in N(x) \wedge N(u) \subseteq A)\}$
$\overline{C}^3(A) = \{x \in U : \forall u(u \in N(x) \rightarrow N(u) \cap A \neq \phi)\}$ [27]

$\underline{C_4}(A) = \{x \in U : \forall u(x \in N(u) \rightarrow N(u) \subseteq A)\}$
$\overline{C}^4(A) = \cup\{N(x) : N(x) \cap A \neq \phi\}$ [27]

$\underline{C_5}(A) = \{x \in U : \forall u(x \in N(u) \rightarrow u \in A)\}$
$\overline{C}^5(A) = \cup\{N(x) : x \in A\}$ [27]

With the same lower approximation there are a few different upper approximations. In the following we have borrowed the symbols from corresponding authors.

$\underline{C_*}(A) = \underline{C_-}(A) = \underline{C_\#}(A) = \underline{C_@}(A) = \underline{C_+}(A) = \underline{C_\%}(A)$
$= \cup\{C_i \in \mathcal{C} : C_i \subseteq A\} \equiv \underline{P_3}(A)$ [17]

$\overline{C}^*(A) = \underline{C_*}(A) \cup \{Md(x) : x \in A \setminus A_*\}$ [5,17,38,39]

$\overline{C}^-(A) = \cup\{C_i : C_i \cap A \neq \phi\}$ [17,37]

$\overline{C}^\#(A) = \cup\{Md(x) : x \in A\}$ [17,38]

$\overline{C}^@(A) = \underline{C_@}(A) \cup \{C_i : C_i \cap (A \setminus \underline{C_@}(A)) \neq \phi\}$ [17]

$\overline{C}^+(A) = \underline{C_+}(A) \cup \{Neighbour(x) : x \in A \setminus \underline{C_+}(A)\}$ [17,35]

$$\overline{C}^{\%}(A) = \underline{C}_{\%}(A) \cup \{\sim \cup \{Friends(y) : x \in A \setminus \underline{C}_{\%}(A), y \in e.f(x)\}\} \quad [17]$$

Yet another type of lower and upper approximation is defined with the help of covering.

Let, $Gr_*(A) = \cup\{C_i \in \mathcal{C} : C_i \subseteq A\} \equiv \underline{P}_3(A)$.

This is taken as lower approximation of A and is denoted by $\underline{C}_{Gr}(A)$.

$Gr^*(A) = \cup\{C_i \in \mathcal{C} : C_i \cap A \neq \phi\} \equiv \overline{P}^1(A)$.

The upper approximation is defined by $\overline{C}^{Gr}(A) = Gr^*(A) \setminus NEG_{Gr}(A)$, where, $NEG_{Gr}(A) = \underline{C}_{Gr}(\sim A)$, $\sim A$ being the complement of A [31].

The following table shows the properties satisfied by the respective lower-upper approximation pairs. Names of the columns indicate the type of lower and upper approximations considered. For Example the column C_* means the entries in this column are with respect to $\underline{C}_*(A)$ and $\overline{C}^*(A)$ as the lower and upper approximations of A. It may be noted that the properties in the list given in the introduction is now expanded because of obvious reasons.

In Table 1, Y means 'yes, the property holds' and N means 'no, the property does not hold'.

Properties that we have verified ourselves are marked $*$. Other results are taken straightway from the respective papers.

<u>Remark 1</u>. From C_* to $C_{\%}$ the lower approximations as stated earlier are the same as that of P_3 viz. $\underline{P}_3(A) = \cup\{C_i : C_i \subseteq A\}$. $\overline{P}^3(A)$ is its dual. Naturally all other upper approximations viz. $\overline{C}^*(A)$ to $\overline{C}^{\%}(A)$ are non duals. Now an attempt may be made to retain these upper approximations and take their respective duals as corresponding lower approximations. One such pair is suggested by Cattaneo [8]. He takes $\overline{C}^-(A)$ as the upper approximation and its dual the lower one. His motivation is completely mathematical making a connection with pre-topological (Čech) behavior.

<u>Remark 2</u>. In the same paper [8], Cattaneo has taken another interesting mathematical (Tarski Topological) approach to define lower and upper approximations in terms pseudo-open and pseudo-closed sets. Let $\mathcal{C} = \{C_i\}$ be a covering of the universe U. A subset $A \subseteq X$ is called pseudo-open if A is the union of some sets in $\mathcal{C}$ and is called pseudo-closed if A is the intersection of the complements of some sets in $\mathcal{C}$. Then $\underline{A} = \{P \subseteq A : P \text{ is pseudo-open}\}$ and $\overline{A} = \cap\{Q \subseteq A : Q \text{ is pseudo-closed}\}$. We can check that the lower and upper approximations are $\underline{P}_3(A)$ and $\overline{P}^3(A)$ respectively. Thus, this pair of operators gets an elegant topological interpretation.

<u>Remark 3</u>. We have met with some other covering based operators [34] which virtually coincide with some of those presented here.

Table 1. Properties of covering based approximations

	P_1	P_2	P_3	P_4	C_1	C_2	C_3	C_4	C_5	C_{Gr}	C_*	C_-	$C_\#$	$C_@$	C_+	$C_\%$
Dual	Y	Y	Y	Y	Y	Y	Y	Y	Y	Y	N	N	N	N	N	N
$\underline{\phi}=\phi=\overline{\phi}$	Y	Y	Y	Y	Y	Y	Y	Y	Y	Y	Y	Y	Y	Y*	Y	Y
$\underline{U}=U=\overline{U}$	Y	Y	Y	Y	Y	Y	Y	Y	Y	Y	Y	Y	Y	Y*	Y	Y
$\underline{A\cap B}\subseteq\underline{A}\cap\underline{B}$	Y	Y	Y	Y	Y	Y	Y	Y	Y	Y	Y	Y	Y	Y*	Y	Y
$\underline{A}\cap\underline{B}\subseteq\underline{A\cap B}$	Y	N*	N*	Y	N*	Y	N*	Y	Y	N	N*	N*	N*	N*	N*	N
$\overline{A\cup B}\subseteq\overline{A}\cup\overline{B}$	Y	N*	N*	Y	N*	Y	N*	Y	Y	N	N*	Y	N*	Y*	Y	N
$\overline{A}\cup\overline{B}\subseteq\overline{A\cup B}$	Y	Y	Y	Y	Y	Y	Y	N	Y	Y	Y	Y	Y	Y*	Y	N
$A\subseteq B\Rightarrow\underline{A}\subseteq\underline{B}$	Y	Y	Y	Y	Y	Y	Y	Y	Y	Y	Y	Y	Y	Y*	Y	Y
$A\subseteq B\Rightarrow\overline{A}\subseteq\overline{B}$	Y	Y	Y	Y	Y	Y	Y	Y	Y	Y	N	Y	Y	N*	Y	Y
$\underline{A}\subseteq A$	Y	Y	Y	Y	Y	Y	Y	N	Y	Y	Y	Y	Y	Y	Y*	Y
$A\subseteq\overline{A}$	Y	Y	Y	Y	Y	Y	Y	N	Y	Y	Y	Y	Y	Y	Y*	N
$\underline{A}\subseteq\overline{A}$	Y	Y	Y	Y	Y	Y	Y	N*	Y	Y	Y	Y	Y	Y	Y*	Y
$A\subseteq\overline{(\underline{A})}$	Y*	N*	Y*	Y	N*	N	N*	Y	N	N*	Y*	Y*	Y*	Y*	N*	N*
$\overline{(\underline{A})}\subseteq A$	Y*	N*	Y*	Y	N*	N*	N*	Y	N*	N*	N*	N*	Y*	Y*	Y*	N*
$\underline{A}\subseteq\underline{(\underline{A})}$	N	Y	Y	Y	Y	Y	Y	N*	N	Y	Y	Y	Y	Y*	Y	Y
$\overline{(\overline{A})}\subseteq\overline{A}$	N	Y	Y	Y	Y	Y	Y	N*	N	Y	Y	Y	N	N	Y*	Y
$\overline{A}\subseteq\overline{(\overline{A})}$	N*	N*	N*	Y	N*	N*	N*	N	N*	N*	Y*	Y	Y*	Y*	N*	N*
$\underline{(\underline{A})}\subseteq\underline{A}$	N*	N*	N*	Y	N*	N*	N*	N*	N*	N*	N*	N*	Y*	Y*	Y*	N*

2.2   Relation-Based Operators

We are now presenting the table corresponding to above mentioned properties but further expanded, with respect to relation based definition of lower-upper approximations. The 2nd and 3rd rows are now split into four rows reasons for which will be clear from the table. Here we have used R for any relation, r, s, t denote reflexivity, symmetry and transitivity. R with suffix(es) means that the relation possesses the corresponding property or properties. There are other important conditions that may be ascribed to R e.g. seriality or Archemedianness, but in this study we are not considering them.

No originality is claimed here. (Almost) all the results are nicely furnished in Yao's and Zhu's papers [32,33,36]. Besides, anybody familiar with elementary modal logic will recognize R as the accessibility relation and lower and upper approximation operators as the semantic counterparts of necessity and possibility operators.

Table 2. Properties of relation based approximations

	R	R_r	R_s	R_t	R_{rs}	R_{rt}	R_{st}	R_{rst}
Dual	Y	Y	Y	Y	Y	Y	Y	Y
$\underline{\phi} = \phi$	N	N	N	N	Y	Y	N	Y
$\phi = \overline{\phi}$	Y	Y	Y	Y	Y	Y	Y	Y
$\underline{U} = U$	Y	Y	Y	Y	Y	Y	Y	Y
$U = \overline{U}$	N	N	N	N	Y	Y	N	Y
$\underline{A \cap B} \subseteq \underline{A} \cap \underline{B}$	Y	Y	Y	Y	Y	Y	Y	Y
$\underline{A} \cap \underline{B} \subseteq \underline{A \cap B}$	Y	Y	Y	Y	Y	Y	Y	Y
$\overline{A \cup B} \subseteq \overline{A} \cup \overline{B}$	Y	Y	Y	Y	Y	Y	Y	Y
$\overline{A} \cup \overline{B} \subseteq \overline{A \cup B}$	Y	Y	Y	Y	Y	Y	Y	Y
$A \subseteq B \Rightarrow \underline{A} \subseteq \underline{B}$	Y	Y	Y	Y	Y	Y	Y	Y
$A \subseteq B \Rightarrow \overline{A} \subseteq \overline{B}$	Y	Y	Y	Y	Y	Y	Y	Y
$\underline{A} \subseteq A$	N	Y	N	N	Y	Y	N	Y
$A \subseteq \overline{A}$	N	Y	N	N	Y	Y	N	Y
$\underline{A} \subseteq \overline{A}$	N	Y	N	N	Y	Y	N	Y
$A \subseteq \overline{(\underline{A})}$	N	N	Y	N	Y	N	Y	Y
$\overline{(\underline{A})} \subseteq A$	N	N	Y	N	Y	N	Y	Y
$\underline{A} \subseteq \underline{(\underline{A})}$	N	N	N	Y	N	Y	Y	Y
$\overline{(\overline{A})} \subseteq \overline{A}$	N	N	N	Y	N	Y	Y	Y
$\overline{A} \subseteq \underline{(\overline{A})}$	N	N	N	N	N	N	N	Y
$\overline{(\underline{A})} \subseteq \underline{A}$	N	N	N	N	N	N	N	Y

Operator Based Approach

The next step of natural generalization is to go straightway to the abstract lower and upper approximation operators of the power set of the universe U and impose conditions on them from the list. Mathematically interesting problem would be to find relation based or covering based equivalents of them. It should, however, be marked that any relation based definition of lower-upper approximation will have certain minimal properties (vide 1st column of Table 2). So, if the axiom taken excludes any of them for instance the property $\underline{A} \cap \underline{B} \subseteq \underline{A \cap B}$ (as is the case P_2), no relation-based representation would be possible.

Let us call (loosely) $P_1 - P_4$, $C_1 - C_5$, $C_{Gr} - C_\%$ and $R - R_{rst}$, 'systems' of lower-upper approximation operators. So, with respect to the expanded list of properties in the two tables, the above systems may be categorized just by considering their behavior, that is, not considering how are they formed. Pomykala's system P_4 and Pawlak's R_{rst} belong to the same category for obvious reasons. The other clusters are $\{P_1, C_4, R_{rs}\}$, $\{P_2, C_1, C_{Gr}\}$, $\{C_2, C_5, R_{rt}\}$ and all the remaining systems form singleton sets.

Categorization with a subset of properties that we shall mention in the next section however, play a significant role in the systems of rough logics [1]. Given two wffs α and β in a logic, $\alpha \to \beta$ is a classical logic theorem if and only if $A_\alpha \subseteq B_\beta$ in every universe of interpretation U, A_α and B_β being subsets of U which are interpretations of α and β respectively. Thus inclusion of subsets of the domain S of interpretation are related with implication in logics. Also implication is strongly connected with the Modus Ponens rule viz.

$$\frac{\Gamma \vdash \alpha \ and \ \Gamma \vdash \alpha \to \beta}{\Gamma \vdash \beta}$$

where Γ is a set of wffs forming the premise and $\Gamma \vdash \alpha$ means that α follows from Γ.

In this paper we shall not discuss logic but categorize various systems presented so far in terms of various inclusion-properties. However, in the concluding section we shall indicate briefly how the modus ponens rule may be generalized to rough M.P. rules using the lower-upper approximations and what is the significance of the partially ordered sets and lattices in the formation of rough logics with the rough M.P. rules.

3   Partial Ordering of Inclusion Relations and Implication Lattices

Given two subsets A, B of the universe U there are nine possible inclusions $P \subseteq Q$, $P \in \{\underline{A}, A, \overline{A}\}$, $Q \in \{\underline{B}, B, \overline{B}\}$. In case the lower and upper approximations arise out of a partition on U, that is the systems P_4 or R_{rst}, we have the following equivalences, $\{\underline{A} \subseteq \overline{B}\}$, $\{\underline{A} \subseteq B, A \subseteq B\}$, $\{A \subseteq \overline{B}, \overline{A} \subseteq \overline{B}\}$, $\{A \subseteq B\}$ and $\{\underline{A} \subseteq \underline{B}, \overline{A} \subseteq \underline{B}, \overline{A} \subseteq B\}$ in the sense that inclusions belonging to the same group are equivalent that is, each implies the other. This implication relation is then extended for the equivalence classes of relations. It may be observed that with respect to this latter implication, the set of above equivalence classes in most cases form a lattice which has been called an implication lattice. The equivalence classes of inclusion relations are disjoint but one class may imply the other. This implication is defined by an arrow ($\to$) in the diagrams. Implication lattices were first introduced in [10]. In this more general context we shall see that implication relation between clusters of inclusions does not form a lattice in general - it forms a partial order relation. We shall categorize below the systems in terms of the partial ordering among inclusion relations and observe that some of them are lattices.

We observe that some of the properties of lower and upper approximation operators as taken in the tables are universals viz. $\underline{\phi} = \overline{\phi}$, $\underline{U} = U$, $\underline{A \cap B} \subseteq \underline{A} \cap \underline{B}$ and $A \subseteq B \Rightarrow \underline{A} \subseteq \underline{B}$. If duality holds, all the properties do no longer remain independent. The system P_4 or equivalently R_{rst} posses all the properties. For a categorization of inclusion relation in these systems, properties that are responsible are the following :

(i) $A \subseteq B \Rightarrow \underline{A} \subseteq \underline{B}$

(ii) $A \subseteq B \Rightarrow \overline{A} \subseteq \overline{B}$

(iii) $\underline{A} \subseteq A$

(iv) $A \subseteq \overline{A}$

(v) $\underline{A} \subseteq \overline{A}$

(vi) $A \subseteq \overline{(\underline{A})}$

(vii) $\overline{(\underline{A})} \subseteq A$

(viii) $\underline{A} \subseteq \underline{(\underline{A})}$

(ix) $\overline{(\overline{A})} \subseteq \overline{A}$

(x) $\overline{A} \subseteq \underline{(\overline{A})}$

Other properties do not play any role in it. Also properties $\underline{A \cap B} \subseteq \underline{A} \cap \underline{B}$ and $\underline{A} \cap \underline{B} \subseteq \underline{A \cap B}$ imply $A \subseteq B \Rightarrow \underline{A} \subseteq \underline{B}$ and $\overline{A \cup B} \subseteq \overline{A} \cup \overline{B}$ and $\overline{A} \cup \overline{B} \subseteq \overline{A \cup B}$ imply $A \subseteq B \Rightarrow \overline{A} \subseteq \overline{B}$. So, the above set of properties may be considered as the weakest set required for categorization of the inclusion relations in case of the most elegant systems viz. P_4 and R_{rst}. This is the reason why we have confined ourselves within these ten properties. For systems other than P_4 and R_{rst} not all of the above properties hold. Again among the properties that hold, not all are effective in clustering the inclusion relations. In what follows, we draw pictures of partially ordered sets and implication lattices of all the systems given in the two tables. The arrow indicates that the node at the tail implies that at the head. The properties used and equivalence classes formed with more than one elements are shown by the side of the diagrams. The other equivalence classes are singletons. In the diagrams a representative of each class is depicted without using brackets.

Diagram 1 (Systems C_3, R)

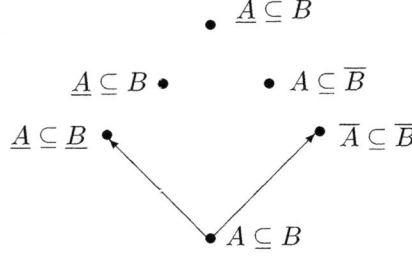

$\bullet\ \underline{A} \subseteq \overline{B}$

$\underline{A} \subseteq B\ \bullet$ $\qquad \bullet\ A \subseteq \overline{B}$

$\underline{A} \subseteq \underline{B}\ \bullet$ $\qquad\qquad \bullet\ \overline{A} \subseteq \overline{B}$

Properties used :

(i), (ii).

$\bullet\ A \subseteq B$

Equivalence classes :

All singletons

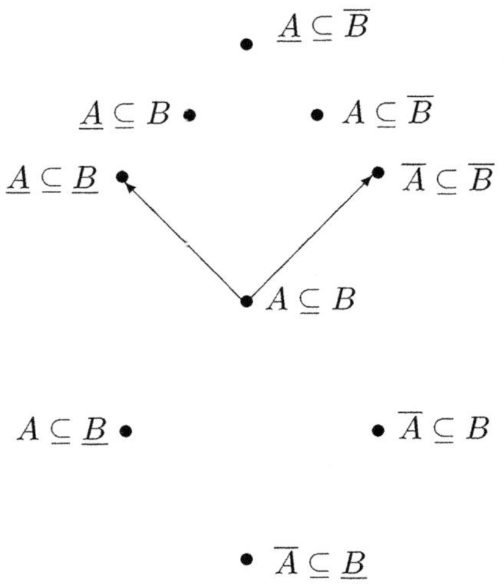

$A \subseteq \underline{B}\ \bullet$ $\qquad\qquad \bullet\ \overline{A} \subseteq B$

$\bullet\ \overline{A} \subseteq \underline{B}$

Diagram 2 (System R_t)

$\bullet\ \underline{A} \subseteq \overline{B}$

$\underline{A} \subseteq B\ \bullet$ $\qquad\qquad \bullet\ \overline{A} \subseteq \overline{B}$

Properties used :

(i), (ii), (viii) and (ix).

$\underline{A} \subseteq \underline{B}\ \bullet$ $\qquad \bullet\ A \subseteq B \qquad A \subseteq \overline{B}\ \bullet$

Equivalence classes :

All singletons

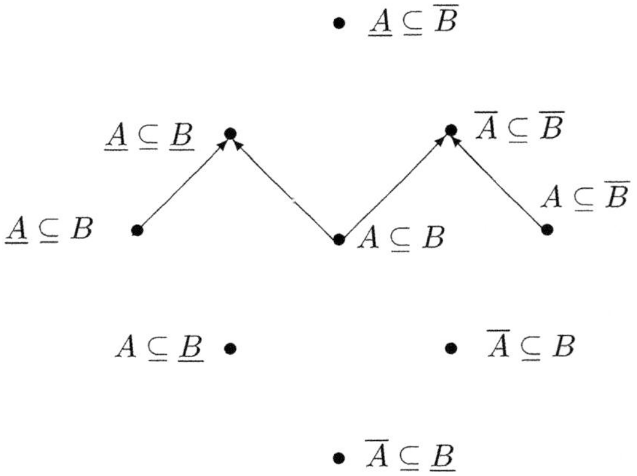

$A \subseteq \underline{B}\ \bullet$ $\qquad\qquad \bullet\ \overline{A} \subseteq B$

$\bullet\ \overline{A} \subseteq \underline{B}$

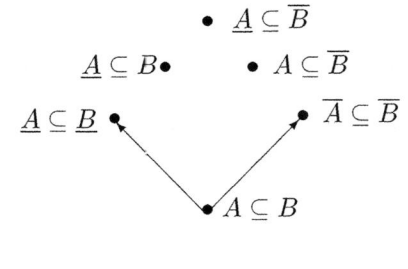

Diagram 3 (System R_s)

$\bullet\ \underline{A} \subseteq \overline{B}$

$\underline{A} \subseteq B\ \bullet$ $\qquad \bullet\ A \subseteq \overline{B}$

$\underline{A} \subseteq \underline{B}\ \bullet$ $\qquad\qquad \bullet\ \overline{A} \subseteq \overline{B}$

Properties used :

(i), (ii), (vi) and (vii).

$\bullet\ A \subseteq \underline{B}$

Equivalence classes :

$\{\overline{A} \subseteq B, A \subseteq \underline{B}\}$

and singletons

$\bullet\ \overline{A} \subseteq B$

$\bullet\ \overline{A} \subseteq \underline{B}$

Diagram 4 (System R_{st})

Properties used :

(i), (ii), (vi) to (ix).

Equivalence classes :
$\{\overline{A} \subseteq B, A \subseteq \underline{B}\}$
and singletons

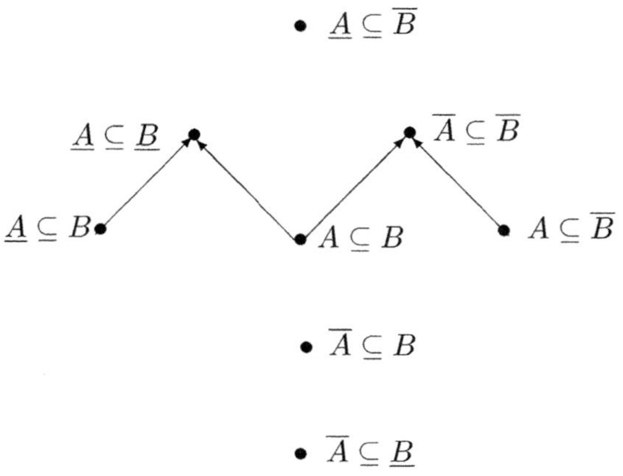

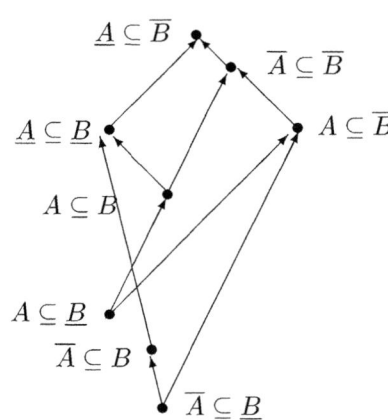

Diagram 5 (System $C_{\%}$)

Properties used :

(i), (ii), (iii), (v), (viii) and (ix).

Equivalence classes :
$\{\underline{A} \subseteq \underline{B}, \underline{A} \subseteq B\}$
and singletons

Diagram 6 (System $C_{@}$)

Properties used :

(i), (iii), (iv), (vi) to (viii).

Equivalence classes :
$\{\underline{A} \subseteq \underline{B}, \underline{A} \subseteq B\}$
and singletons

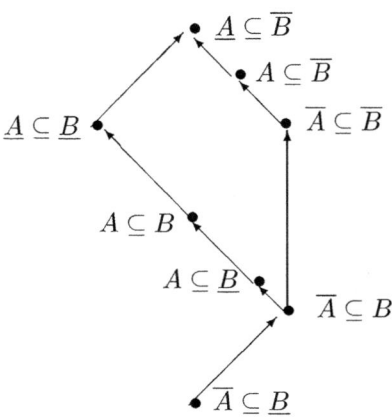

Diagram 7 (System R_r)

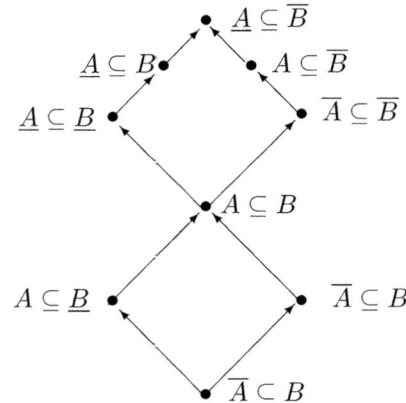

Properties used :

(i) to (iv).

Equivalence classes :
All singletons

Diagram 8 (System C_*)

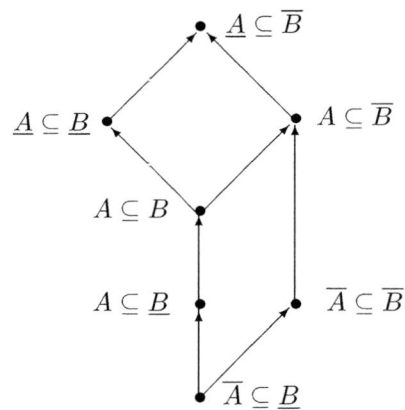

Properties used :

(i), (iii), (iv) and (viii).

Equivalence classes :
$\{\underline{A} \subseteq \underline{B}, \underline{A} \subseteq B\}$, $\{\overline{A} \subseteq \underline{B}, \overline{A} \subseteq B\}$
and singletons

Diagram 9 (Systems P_1 and C_4 and R_{rs})

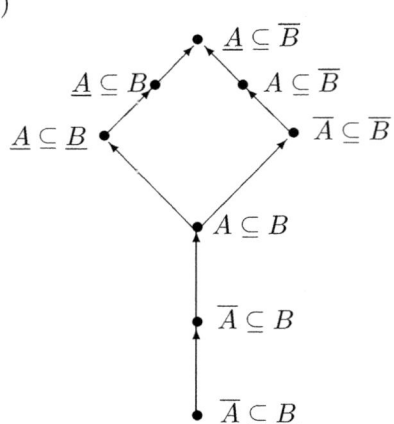

Properties used :

(i) to (iv), (vi) and (vii).

Equivalence classes :
$\{\overline{A} \subseteq B, A \subseteq \underline{B}\}$
and singletons

Diagram 10 (Systems $P_2, C_1, C_2, C_5, C_{Gr}, C_+$ and R_{rt})

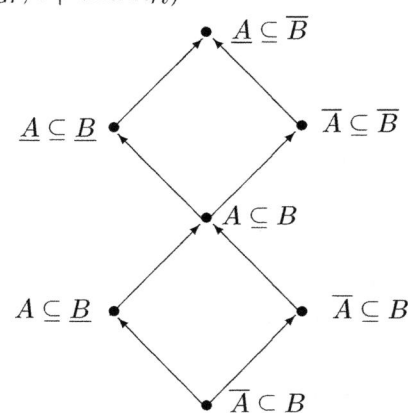

Properties used :

(i) to (iv), (viii) and (ix)

Equivalence classes :

$\{\underline{A} \subseteq \underline{B}, \underline{A} \subseteq B\}$,

$\{A \subseteq \overline{B}, \overline{A} \subseteq \overline{B}\}$

and singletons

Diagram 11 (System C_-)

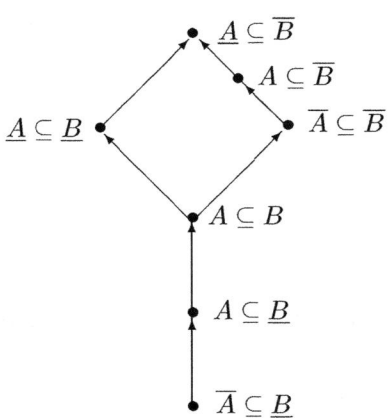

Properties used :

(i) to (iv) and (viii).

Equivalence classes :

$\{\overline{A} \subseteq \underline{B}, \overline{A} \subseteq B\}$, $\{\underline{A} \subseteq \underline{B}, \underline{A} \subseteq B\}$

and singletons

Diagram 12 (System P_3)

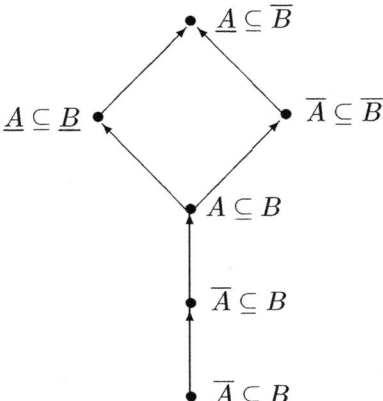

Properties used :

(i) to (iv) and (vi) to (ix).

Equivalence classes :

$\{\overline{A} \subseteq B, A \subseteq \underline{B}\}$, $\{\underline{A} \subseteq \underline{B}, \underline{A} \subseteq B\}$,

$\{A \subseteq \overline{B}, \overline{A} \subseteq \overline{B}\}$ and singletons

Diagram 13 (System $C_\#$)

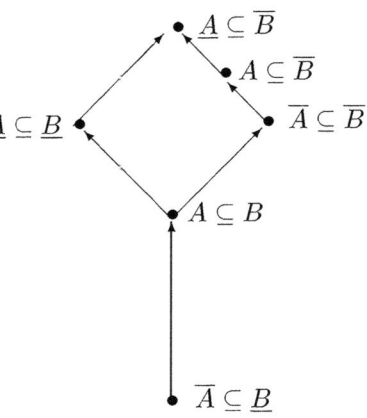

Properties used :

(i) to (iv), (vi) to (viii) and (x).

Equivalence classes :
$\{\underline{A} \subseteq \underline{B}, \underline{A} \subseteq B\}$,
$\{A \subseteq \underline{B}, \overline{A} \subseteq \underline{B}, \overline{A} \subseteq B\}$
and singletons

Diagram 14 (Systems P_4 and R_{rst})

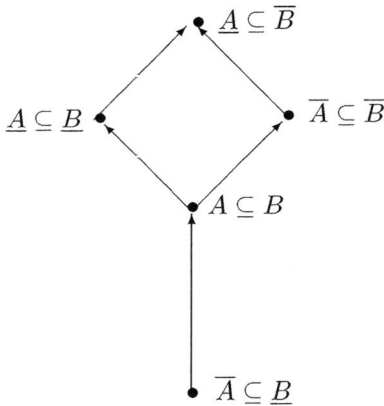

Properties used :

(i) to (iv) and (vi) to (x).

Equivalence classes :
$\{\underline{A} \subseteq \underline{B}, \underline{A} \subseteq B\}$, $\{A \subseteq \overline{B}, \overline{A} \subseteq \overline{B}\}$
and $\{A \subseteq \underline{B}, \overline{A} \subseteq \underline{B}, \overline{A} \subseteq B\}$
and singletons

4   Concluding Remarks

We wish to say a few words about this study.

Firstly almost all the existing lower / upper approximations in the rough set literature are at hand now. The genesis of them that is the mathematical procedure of their construction either from granules or from relations are available. They have been clustered according to their behavior pattern.

Given two Sets A and B, of the nine possible inclusions between the pairs from $\{\underline{A}, A, \overline{A}\}$ and $\{\underline{B}, B, \overline{B}\}$ how many are independent in a system is depicted by the nodes of the corresponding diagrams. Which inclusion entails which one is shown.

Any of the inclusion gives rise to a rough Modus Ponens rule [6] and a corresponding rough logic [6]. The underlying modal logical systems of various rough logics are also immediately visible from the tables.

We take a case. Let us consider the diagram 9 and the system P_1 having this diagram as its implication lattice. The base modal logic - let us call it P_1 again - should satisfy the property given by the column below P_1 in Table 1. (Obviously, a subset of it is to be taken as axiom and rules are to be framed in such a way that the other properties appear). A theorem of this modal system shall be denoted by $\vdash_{P_1} \alpha$. Now with respect to the node $\overline{A} \subseteq B$, the corresponding rough modus ponens rule may be taken as

$$\frac{\Gamma \mid\!\sim \alpha \quad \Gamma \mid\!\sim \beta \to \gamma \quad \vdash_{P_1} \Diamond\alpha \to \beta}{\Gamma \mid\!\sim \gamma}$$

and the rough logic may be developed as in [6]. Since, from the diagram it is clear that $\overline{A} \subseteq B$ implies $\overline{A} \subseteq \overline{B}$, the rough logic with modal system P_1 as base and rough modus ponens rule corresponding to this latter node namely

$$\frac{\Gamma \mid\!\sim \alpha \quad \Gamma \mid\!\sim \beta \to \gamma \quad \vdash_{P_1} \Diamond\alpha \to \Diamond\beta}{\Gamma \mid\!\sim \gamma}$$

will be a sublogic of the former one. In [6] one can see all the rough logics corresponding to diagram 14. It may also be noted that it will be necessary to define and investigate modal logic systems in which necessity and possibility are not duals because of the fact that the systems in table 1 include non-dual systems too. This observation may open up a new branch of investigation in modal logics.

Are the 14 diagrams the only possibilities? At present we cannot say 'yes'. One may draw a diagram arbitrarily and look for properties from the list given that would give the proposed diagram. Then a suitable granulation and definitions of lower-upper approximation may be sought for so as to make the properties hold. The present categorization technique opens up this direction of research.

Acknowledgement

We are thankful to the referees for their variable comments. The first author acknowledges the financial support from the University Grants Commission, Government of India.

References

1. Banerjee, M., Chakraborty, M.K.: Rough Sets Through Algebraic Logic. Fundam. Inform. 28(3-4), 211–221 (1996)
2. Benerjee, M., Chakraborty, M.K.: Algebras from Rough Sets. In: Pal, S.K., Polkowski, L., Skowron, A. (eds.) Rough-Neural Computing - Techniques for Computing with Words, Springer, Heidelberg (2004)
3. Banerjee, M., Yao, Y.: A Categorial Basis for Granular Computing. In: An, A., Stefanowski, J., Ramanna, S., Butz, C.J., Pedrycz, W., Wang, G. (eds.) RSFDGrC 2007. LNCS (LNAI), vol. 4482, pp. 427–434. Springer, Heidelberg (2007)

4. Bonikowski, Z.: A Certain Copnception of the Calculus of Rough Sets. Notre Dame J. Formal Logic 33, 412–421 (1992)
5. Bonikowski, Z., Bryniarski, E., Wybraniec-Skardowska, U.: Extensions and intentions in the rough set theory. Journal of Information Sciences 107, 149–167 (1998)
6. Bunder, M.W., Banerjee, M., Chakraborty, M.K.: Some Rough Consequence Logics and their Interrelations. In: Peters, J.F., Skowron, A. (eds.) Transactions on Rough Sets VIII. LNCS, vol. 5084, pp. 1–20. Springer, Heidelberg (2008)
7. Grzymała-Busse, J.W., Grzymala-Busse, W.J.: An Experimental Comparison of Three Rough Set Approaches to Missing Attribute Values. In: Peters, J.F., Skowron, A., Düntsch, I., Grzymała-Busse, J.W., Orłowska, E., Polkowski, L. (eds.) Transactions on Rough Sets VI. LNCS, vol. 4374, pp. 31–50. Springer, Heidelberg (2007)
8. Cattaneo, G.: Foundational and Mathematical Investigation of Roughness Theory (preprint)
9. Cattaneo, G., Ciucci, D.: Lattices with interior and closure operators and abstract approximation spaces. In: Peters, J.F., Skowron, A., Wolski, M., Chakraborty, M.K., Wu, W.-Z. (eds.) Transactions on Rough Sets X. LNCS, vol. 5656, pp. 67–116. Springer, Heidelberg (2009)
10. Chakraborty, M.K., Banerjee, M.: Rough dialogue and implication lattices. Fundamenta Informaticae 75(1-4), 123–139 (2007)
11. Chakraborty, M.K., Samanta, P.: Consistency-Degree Between Knowledges. In: Kryszkiewicz, M., Peters, J.F., Rybiński, H., Skowron, A. (eds.) RSEISP 2007. LNCS (LNAI), vol. 4585, pp. 133–141. Springer, Heidelberg (2007)
12. Jarinen, J.: Approximations and Rough Sets Based on Tolerances. In: Ziarko, W.P., Yao, Y. (eds.) RSCTC 2000. LNCS (LNAI), vol. 2005, pp. 182–189. Springer, Heidelberg (2001)
13. Li, T.-J.: Rough Approximation Operators in Covering Approximation Spaces. In: Greco, S., Hata, Y., Hirano, S., Inuiguchi, M., Miyamoto, S., Nguyen, H.S., Słowiński, R. (eds.) RSCTC 2006. LNCS (LNAI), vol. 4259, pp. 174–182. Springer, Heidelberg (2006)
14. Lin, T.Y.: Granular Computing on Binary Relations I: Data Mining and Neighbourhood Systems. In: Skowron, A., Polkowski, L. (eds.) Rough Sets in Knowledge Discovery, pp. 107–121. Physica-Verlag, Berlin (1998)
15. Lin, T.Y.: Granular Computing on Binary Relations II: Rough Set Representations on Belief Functions. In: Skowron, A., Polkowski, L. (eds.) Rough Sets in Knowledge Discovery, pp. 122–140. Physica-Verlag, Berlin (1998)
16. Lin, T.Y.: What is Granular Computing. Journal of Latex Class Files 1(11) (November 2002)
17. Liu, J., Liao, Z.: The sixth type of covering-based rough sets. In: IEEE International Conference on Granular Computing, GrC 2008, pp. 438–441 (2008)
18. Orlowska, E. (ed.): Incomplete Information: Rough Set Analysis. Physcia-Verlag, Heidelberg (1997)
19. Pal, S.K., Polkowski, L., Skowron, A. (eds.): Rough-Neural Computing. Springer, Heidelberg (2004)
20. Pawlak, Z.: Rough Sets. International Journal of Computer and Information Sciences 11(5) (1982)
21. Pawlak, Z.: ROUGH SETS - Theoritical Aspects of Reasoning About Data. Kluwer Academic Publishers, Dordrecht (1991)
22. Pagliani, P.: Pre-topologies and Dynamic Spaces. In: Wang, G., Liu, Q., Yao, Y., Skowron, A. (eds.) RSFDGrC 2003. LNCS (LNAI), vol. 2639, pp. 146–155.

Springer, Heidelberg (2003); Extended Version in Fundamenta Informatica 59(2-3), 221–239 (2004)

23. Pagliani, P., Chakraborty, M.K.: A geometry of Approximation - Rough Set theory: Logic, Algebra and Topology of Conceptual Patterns. Springer Science + Business Media B.V (2008)

24. Pomykala, J.A.: Approximation operations in approximation space. Bulletin of the Polish Academy of Sciences: Mathematics 35, 653–662 (1987)

25. Pomykala, J.A.: Approximation, Similarity and Rough Constructions. ILLC Prepublication Series for Computation and Complexity Theory CT-93-07, University of Amsterdam (1993)

26. Pomykala, J., Pomykala, J.A.: The Stone Algebra of Rough Sets. Bull. Polish Acad. Sci. Math. 36(7-8), 495–508 (1988)

27. Qin, K., Gao, Y., Pei, Z.: On Covering Rough Sets. In: Yao, J., Lingras, P., Wu, W.-Z., Szczuka, M.S., Cercone, N.J., Ślęzak, D. (eds.) RSKT 2007. LNCS (LNAI), vol. 4481, pp. 34–41. Springer, Heidelberg (2007)

28. Samanta, P., Chakraborty, M.K.: On Extension of Dependency and Consistency Degrees of Two Knowledges Represented by Covering. In: Peters, J.F., Skowron, A., Rybiński, H. (eds.) Transactions on Rough Sets IX. LNCS, vol. 5390, pp. 351–364. Springer, Heidelberg (2008)

29. Samanta, P., Chakraborty, M.K.: Covering Based Approaches to Rough Sets and Implication Lattices. In: Sakai, H., Chakraborty, M.K., Hassanien, A.E., Ślęzak, D., Zhu, W. (eds.) RSFDGrC 2009. LNCS (LNAI), vol. 5908, pp. 127–134. Springer, Heidelberg (2009)

30. Skowron, A., Stepaniuk, J.: Tolerance approximation spaces. Fundamenta Informaticae 27, 245–253 (1996)

31. Slezak, D., Wasilewski, P.: Granular Sets – Foundations and Case Study of Tolerance Spaces. In: An, A., Stefanowski, J., Ramanna, S., Butz, C.J., Pedrycz, W., Wang, G. (eds.) RSFDGrC 2007. LNCS (LNAI), vol. 4482, pp. 435–442. Springer, Heidelberg (2007)

32. Yao, Y.Y.: Constructive and algebraic methods of the theory of rough sets. Journal of Information Sciences 109, 21–47 (1998)

33. Yao, Y.Y.: On generalizing rough set theory. In: Wang, G., Liu, Q., Yao, Y., Skowron, A. (eds.) RSFDGrC 2003. LNCS (LNAI), vol. 2639, pp. 44–51. Springer, Heidelberg (2003)

34. Yun, Z., Ge, X: Some notes on covering-based rough sets (preprint)

35. Zhu, W.: Topological approaches to covering rough sets. ScienceDirect, Information Sciences 177, 1499–1508 (2007)

36. Zhu, W.: Relationship between generalized rough sets based on binary relation and covering. Information Sciences 179, 210–225 (2009)

37. Zhu, W.: Properties of the Second Type of Covering-Based Rough Sets. In: Proceedings of the 2006 IEEE/WIC/ACM International Conference on Web Intelligence and Intelligent Agent Technology (WI-IAT 2006 Workshops)(WI-IATW 2006) (2006)

38. Zhu, W., Wang, F.-Y.: Relationship among Three Types of Covering Rough Sets. In: Proc. IEEE Int'l Conf. Grannuler Computing (GrC 2006), May 2006, pp. 43–48 (2006)

39. Zhu, W., Wang, F.-Y.: Properties of the First Type of Covering-Based Rough Sets. In: Sixth IEEE International Conference on Data Mining - Workshops, ICDMW 2006 (2006)

Classification with Dynamic Reducts and Belief Functions

Salsabil Trabelsi[1], Zied Elouedi[1], and Pawan Lingras[2]

[1] Larodec, Institut Superier de Gestion de Tunis, Tunisia
[2] Saint Mary's University Halifax, Canada

Abstract. In this paper, we propose two approaches of classification namely, *Dynamic Belief Rough Set Classifier* (D-BRSC) and *Dynamic Belief Rough Set Classifier based on Generalization Distribution Table* (D-BRSC-GDT). Both the classifiers are induced from uncertain data to generate classification rules. The uncertainty appears only in decision attribute values and is handled by the Transferable Belief Model (TBM), one interpretation of the belief function theory. D-BRSC only uses the basic concepts of Rough Sets (RS). However, D-BRSC-GDT is based on GDT-RS which is a hybridization of Generalization Distribution Table (GDT) and Rough Sets (RS). The feature selection step relative to the construction of the two classifiers uses the approach of dynamic reduct which extracts more relevant and stable features. The reduction of uncertain and noisy decision table using dynamic approach generates more significant decision rules for the classification of unseen objects. To prove that, we carry experimentations on real databases according to three evaluation criteria including the classification accuracy. We also compare the results of D-BRSC and D-BRSC-GDT with those obtained from Static Belief Rough Set Classifier (S-BRSC) and Static Belief Rough Set Classifier based on Generalization Distribution Table (S-BRSC-GDT). To further evaluate our rough sets based classification systems, we compare our results with those obtained from the Belief Decision Tree (BDT).

Keywords: rough sets, generalization distribution table, belief function theory, uncertainty, dynamic reduct, classification.

1   Introduction

The idea of Rough Sets (RS) was introduced by Pawlak [13,14] to deal with imprecise and vague concepts. It has been successfully applied to many real-life problems. It constitutes a sound basis for decision support system and data mining. It is a mathematical approach to data analysis and classification problems. Learning from incomplete or uncertain data sets is usually more difficult than from complete data sets. Several extensions of induction systems based on rough sets have been developed to handle the problem of incomplete data sets which may be characterized by missing condition attribute values and not

J.F. Peters et al. (Eds.): Transactions on Rough Sets XIV, LNCS 6600, pp. 202–233, 2011.

with partially uncertain decision attributes [8,9,10]. Previously, we have developed two approaches to classification based on rough sets called *Belief Rough Set Classifier* (BRSC) [27] and *Belief Rough Set Classifier based on Generalization Distribution Table* (BRSC-GDT) [28] which are able to learn decision rules from uncertain data. The uncertainty appears only in decision attributes. This type of uncertainty exists in many real-world applications such as marketing, finance, management and medicine. For example, some decision attribute values in a client's database, used by the bank to plan a loan policy, are partially uncertain. Such an uncertainty can be represented by belief functions as defined in the Transferable Belief Model (TBM), which is one interpretation of the belief function theory [18]. TBM is a useful tool for representing and managing partially or totally uncertain knowledge because of its relative flexibility [15]. The belief function theory is widely applied in machine learning and also in real life problems related to decision making and classification.

Feature selection is one of the important steps used to build BRSC and BRSC-GDT. This step is based on computing reduct, a minimal set of attributes that preserves the discernibiliy of objects from different decision classes, from uncertain data. In many cases, finding reducts from uncertain and noisy data leads to results which are unstable and sensitive to the sampling process. Dynamic reducts [2] can provide better performance in very large datasets. In fact, rules induced with dynamic reducts are more appropriate to classify new objects, since these reducts are more stable and appear more frequently in sub-decision systems created by random samples of a given decision system.

This paper takes advantage of dynamic reducts in two approaches to classification denoted as Dynamic Belief Rough Set Classifier (D-BRSC) and Dynamic Belief Rough Set Classifier based on Generalization Distribution Table (D-BRSC-GDT), which are able to learn uncertain decision rules from uncertain data. As in our previous work, the uncertainty appears only in decision attributes and is handled by the TBM. The first classification technique is based only on the basic concepts of rough sets [26]. However, the second more sophisticated technique is based on the hybrid system named GDT-RS which is a combination of Generalization Distribution Table (GDT) and Rough Sets (RS). The feature selection step used to build the new proposed classifiers is based on dynamic reducts. The dynamic reducts are redefined [25] in order to extract more stable reducts from data with an uncertain decision attribute. To prove that the uncertain decision rules induced with dynamic reducts are more appropriate to classify unseen objects than static reducts, we carry experimentations on real databases to evaluate our proposed classifiers based on three evaluation criteria: the classification accuracy, the time requirement and the size. To further validate the experimental results, we compare them with those obtained from the Static Belief Rough Set Classifier (S-BRSC) [27] and Static Belief Rough Set Classifier based on Generalization Distribution Table (S-BRSC-GDT) [28]. Finally, the quality of our classification systems based on rough sets and generalized distribution table is compared with a similar classifier called Belief Decision Tree (BDT).

This paper is organized as follows: Section 2 provides an overview of the Rough Sets (RS), Generalization Distribution Table (GDT) and the hybrid system GDT-RS. Section 3 introduces the belief function theory as understood in the TBM. In Section 4, we propose two approaches to classification namely, Dynamic Belief Rough Set Classifier (D-BSRC) and Dynamic Belief Rough Set Classifier based on Generalization Distribution Table (D-BRSC-GDT). Finally, we report the experimental results using real databases for D-BRSC, D-BRSC-GDT, S-BRSC, S-BRSC-GDT and BDT.

2   Rough Sets (RS), Generalization Distribution Table (GDT) and the Hybrid System GDT-RS

The Rough Sets (RS), proposed by Pawlak [13,14], is a mathematical tool for dealing with imperfect knowledge. It has been successfully applied to data classification problems. This theory is not competitive but complementary to other methods and can be often used jointly with other approaches [17]. Here, we introduce the basics of Rough Sets (RS), the Generalization Distribution Table (GDT) and the hybridization system GDT-RS that will be useful in this paper.

2.1   Rough Sets (RS)

Let us recall some basic notions regarding rough sets [14]. A decision table (DT) is defined as A= (U, C, {d}), where $U = \{o_1, o_2, ..., o_n\}$ is a nonempty finite set of n objects called *the universe*, $C = \{c_1, c_2, ..., c_k\}$ is a finite set of k *condition* attributes and $d \notin C$ is a distinguished attribute called *decision*. The value set of d is called $\Theta = \{d_1, d_2, ..., d_s\}$. In this paper, the notation $c_i(o_j)$ is used to represent the value of a condition attribute $c_i \in C$ for an object $o_j \in U$. Similarly, $d(o_j)$ represents the value of the decision attribute d for an object o_j. We further extend these notations for a set of attributes $B \subseteq C$, by defining $B(o_j)$ to be value tuple of attributes in B for an object o_j.

Indiscernibility relation. A decision system expresses all the knowledge about the model. This table may be unnecessarily large. The same or indiscernible objects may be represented several times. The objects o_i and o_j are indiscernible on a subset of attributes $B \subseteq C$, if they have the same values for each attribute in the subset B of C. Rough sets adopt the concept of indiscernibility relation [13,14] to partition the object set U into disjoint subsets, denoted by U/B or IND_B, and the partition that includes o_j is denoted $[o_j]_B$.

$$[o_j]_B = \{o_i | B(o_i) = B(o_j)\} \tag{1}$$

$$IND_B = U/B = \{[o_j]_B | o_j \in U\} \tag{2}$$

The set of equivalence classes based on the decision attribute is denoted by $U/\{d\}$

$$IND_{\{d\}} = U/\{d\} = \{[o_j]_{\{d\}} | o_j \in U\} \tag{3}$$

Set approximation. The concept of indiscernibility relation is a natural tool for reducing data. Since only one element of the equivalence class is needed to represent the entire class. Subsets that are most often of interest have the same value of the outcome attribute. It may happen that a target concept cannot be defined in a crisp manner. In other words, it is not possible to induce a crisp description of such objects from table. In such cases, the notion of rough sets is useful.

Let $B \subseteq C$ and $X \subseteq U$. We can approximate X using $B-lower$ and $B-upper$ *approximations* of X, denoted respectively as $\underline{B}(X)$ and $\bar{B}(X)$ where

$$\underline{B}(X) = \{o_j | [o_j]_B \subseteq X\} \ and \ \bar{B}(X) = \{o_j | [o_j]_B \cap X \neq \emptyset\} \tag{4}$$

Objects in $\underline{B}(X)$ can be classified with certainty as members of X on the basis of knowledge in B, while objects in $\bar{B}(X)$ can only be classified as possible members of X on the basis of knowledge in B.

The set $BN_B(X)$ is called the $B-boundary \ region$ of X, and thus consists of those objects that we cannot decisively classify into X on the basis of knowledge in B:

$$BN_B(X) = \bar{B}(X) - \underline{B}(X) \tag{5}$$

A set is said to be rough if the boundary region is non-empty.

Dependency of attributes and positive region. Another important issue in data analysis is discovering dependencies between attributes [14]. Intuitively, the decision attribute d depends totally on a set of condition attributes C, denoted $C \Rightarrow \{d\}$, if all values of attribute d are uniquely determined by values of attributes from C. Formally, a functional dependency can be defined in the following way. The attribute d depends on the set of attributes C in a degree k ($0 \leq k \leq 1$), denoted $C \Rightarrow_k \{d\}$, if

$$k = \gamma_C(A, \{d\}) = \frac{|Pos_C(A, \{d\})|}{|U|} \tag{6}$$

Where

$$Pos_C(A, \{d\}) = \bigcup_{X \in U/\{d\}} \underline{C}(X) \tag{7}$$

$Pos_C(A, \{d\})$, called a positive region of the partition $U/\{d\}$ with respect to C, is the set of all elements of U that can be uniquely classified to blocks of the partition $U/\{d\}$, by means of C.

Reduct and core. With indiscernibility relation, we have investigated one natural dimension of reducing data which is to identify equivalence classes. The other dimension in reduction is to keep only those attributes that preserve the indisernibility and consequently set approximation. There are usually several such subsets of attributes and those which are minimal are called reducts [14]. In order to express the above idea more precisely, we need some auxiliary notions.

Let $c \in C$, the attribute c is dispensable in C with respect to d, if $Pos_C(A, \{d\}) = Pos_{C-c}(A, \{d\})$. Otherwise attribute c is indispensable in C with respect to d. If $\forall c \in C$ is indispensable in C with respect to d, then C is called independent.

1. **Reduct:** A subset $B \subseteq C$ is a reduct of C with respect to d, if B is independent and:

$$Pos_B(A, \{d\}) = Pos_C(A, \{d\}) \tag{8}$$

   A reduct of knowledge is its essential part, which is sufficient to define all basic concepts occurring in the considered knowledge.

2. **Core:** The set of all the condition attributes indispensable in the decision system A with respect to d is denoted by $Core_C(A, \{d\})$.

$$Core_C(A, \{d\}) = \bigcap Red_C(A, \{d\}) \tag{9}$$

   Where $Red_C(A, \{d\})$ is the set of all relative reducts of the decision system A.

   Since the relative core is the intersection of all relative reducts, it is included in every reduct. Thus, in a sense, the core is the most important subset of attributes which cannot be removed when reducing the data.

Value reduct and value core. To further simplify a decision table, we can eliminate some values of an attribute from the table. We also need a concept of value reduct and value core.

   We say that the value of attribute $c \in C$, is dispensable for o_j with respect to d, if $[o_j]_C \subseteq [o_j]_{\{d\}}$ implies $[o_j]_{C-\{c\}} \subseteq [o_j]_{\{d\}}$; otherwise the value of attribute c is indispensable for o_j with respect to d.

   If for every attribute $c \in C$ the value of c is indispensable for o_j, then C is called *independent* for o_j with respect to d.

   A subset $B \subseteq C$ is a value reduct of C for o_j, iff B is *independent* for o_j and $[o_j]_C \subseteq [o_j]_{\{d\}}$ implies $[o_j]_B \subseteq [o_j]_{\{d\}}$.

   The set of all indispensable attributes values from C for o_j with respect to d is called value core and is denoted by $Core_C^j(A, \{d\})$.

   We also have the following property:

$$Core_C^j(A, \{d\}) = \bigcap Red_C^j(A, \{d\}) \tag{10}$$

Where $Red_C^j(A, \{d\})$ is the family of all the reduct values of C for o_j with respect to d.

Dynamic reduct. If $A = (U, C \cup \{d\})$ is a decision table, then any system $A' = (U', C \cup \{d\})$ such that $U' \subseteq U$ is called a subtable of A. Let F be a family of subtables of A [2].

$$DR(A, F) = Red(A, \{d\}) \cap \bigcap_{A' \in F} Red(A', \{d\}) \tag{11}$$

Any element of $DR(A, F)$ is called an F-dynamic reduct of A. From the definition of dynamic reducts, it follows that a relative reduct of A is dynamic if it is also a reduct of all the subtables from a given family F. This notation can sometimes be too restrictive, so we apply a more general notion of dynamic reduct.

They are called (F, ε)-dynamic reducts, where $0 \leq \varepsilon \leq 1$. The set $DR_\varepsilon(A, F)$ of all (F, ε)-dynamic reducts is defined by:

$$DR_\varepsilon(A, F) = \left\{ R \in Red(A, \{d\}) : \frac{|\{A' \in F : R \in Red(A', \{d\})\}|}{|F|} \geq 1 - \varepsilon \right\} \tag{12}$$

If $R \in Red(B, \{d\})$, then the number $\frac{|\{A' \in F : R \in Red(A', \{d\})\}|}{|F|}$ is called the stability coefficient of R (relative to F).

2.2   Generalization Distribution Table (GDT)

The Generalization Distribution Table (GDT) proposed by Zhong et al. in [30] consists of three components:

1. Possible instances (PI), represented at the top row of GDT, are defined by all possible combinations of attribute values from a database.
2. Possible generalizations of instances (PG), represented by the left column of a GDT, are all possible cases of generalization for all possible instances. A wild card '*' denotes the generalization for instances.
3. Probabilistic relationships between possible instances (PI) and possible generalizations (PG), represented by entries G_{ij} of a given GDT, are defined by means of a probabilistic distribution describing the strength of the relationship between any possible instance and any possible generalization. The prior distribution is assumed to be uniform if background knowledge is not available. Thus, it is defined by:

$$G_{ij} = p(PI_j|PG_i) = \begin{cases} \frac{1}{N_{PG_i}} & \text{if } PG_i \text{ is a generalization of } PI_j \\ 0 & \text{otherwise.} \end{cases} \tag{13}$$

where PI_j is the j-th possible instance, PG_i is the i-th possible generalization and N_{PG_i} is the number of the possible instances satisfying the i-th possible generalization. The latter is defined as follows:

$$N_{PG_i} = \prod_{k \in \{l | PG_i[l] = *\}} n_k \tag{14}$$

where $PG_i[l]$ is the value of the l-th attribute in the possible generalization PG_i and n_k is the number of values of the k-th attribute. From the probability theory, we have $\sum_j G_{ij} = 1$ for any i.

2.3   Generalization Distribution Table and Rough Set System (GDT-RS)

The GDT-RS, a soft hybrid induction system, is based on a hybridization of the *Generalization Distribution Table* (GDT) and the *Rough Set* methodology (RS). It is proposed for discovering classification rules from databases with noisy data [3,29]. The GDT-RS system can generate a set of rules with the minimal description length, having large strength and covering all instances.

From the decision table (DT), we can generate decision rules expressed in the following form:

$$P \rightarrow Q \text{ with } S,$$

- P is a conjunction of descriptors over C,
- Q denotes a concept that the rule describes,
- S is a 'measure of the strength' of the rule.

According to the GDT-RS, the strength S of a given rule reflects incompleteness and noise and is equal to [3,29]:

$$S(P \rightarrow Q) = s(P) * (1 - r(P \rightarrow Q)) \qquad (15)$$

where $s(P)$ is the strength of the generalization P and r is the noise rate function. On the assumption that the prior distribution is uniform, the strength of the generalization $P = PG$ is given by:

$$s(P) = \sum_{l} p(PI_l|P) = \frac{1}{N_P}|[P]_{DT}| \qquad (16)$$

where $[P]_{DT}$ is the set of all the objects in DT satisfying the generalization P and N_P is the number of possible instances satisfying the generalization P which is computed using eqn. (14). The strength of the generalization P represents its ability of prediction for unseen instances.

On the other hand, the noise rate is given by:

$$r(P \rightarrow Q) = 1 - \frac{|[P]_{DT} \cap [Q]_{DT}|}{|[P]_{DT}|} \qquad (17)$$

It shows the quality of classification measured by the number of the instances satisfying the generalization P which cannot be classified into class Q.

3   Belief Function Theory

In this Section, we briefly review the main concepts underlying the belief function theory as interpreted in the TBM [18,19]. The latter is a useful model to represent quantified belief functions.

3.1   Definitions

Let Θ be a finite set of elementary events to a given problem, called the frame of discernment. All the subsets of Θ belong to the power set of Θ, denoted by 2^{Θ}.

The impact of a piece of evidence on the different subsets of the frame of discernment Θ is represented by a basic belief assignment (bba).

The bba is a function $m : 2^{\Theta} \rightarrow [0,1]$ such that:

$$\sum_{E \subseteq \Theta} m(E) = 1 \qquad (18)$$

The value $m(E)$, called a basic belief mass (bbm), represents the portion of belief committed exactly to the event E.

Associated with m is a belief function, denoted bel, which assigns the sum of masses of belief committed to every subset of $E \in 2^{\Theta}$ by m [15].

The belief function bel is defined for $E \subseteq \Theta$, $E \neq \emptyset$ as:

$$bel(E) = \sum_{\emptyset \neq F \subseteq E} m(F) \qquad (19)$$

The plausibility function pl quantifies the maximum amount of belief that could be given to a subset E of the frame of discernment. It is equal to the sum of the bbm's relative to subsets F compatible with E.

The plausibility function pl is defined as follows:

$$pl(E) = \sum_{E \cap F \neq \emptyset} m(F), \forall E \subseteq \Theta \qquad (20)$$

3.2   Combination

Handling information induced from different experts (information sources) requires an evidence gathering process in order to get the fused information. In the TBM, the basic belief assignments induced from distinct pieces of evidence are combined by either the conjunctive rule or the disjunctive rule of combination.

1. The conjunctive rule: When we know that both sources of information are fully reliable, then the bba representing the combined evidence satisfies [20]:

$$(m_1 \cap m_2)(E) = \sum_{F,G \subseteq \Theta : F \cap G = E} m_1(F) m_2(G) \qquad (21)$$

2. The disjunctive rule: When we only know that at least one of the sources of information is reliable but we do not know which is reliable, then the bba representing the combined evidence satisfies [20]:

$$(m_1 \cup m_2)(E) = \sum_{F,G \subseteq \Theta : F \cup G = E} m_1(F) m_2(G) \qquad (22)$$

3.3   Discounting

In the TBM, discounting takes into consideration the reliability of the information source that generates the bba m. For $\alpha \in [0,1]$, let $(1-\alpha)$ be the degree of reliability, we assign to the source of information. If the source is not fully reliable, the bba it generates is 'discounted' into a new less informative bba denoted m^{α}:

$$m^{\alpha}(E) = (1 - \alpha)m(E), \qquad for \ E \subset \Theta \qquad (23)$$

$$m^{\alpha}(\Theta) = \alpha + (1 - \alpha)m(\Theta) \qquad (24)$$

3.4   Decision Making

In the TBM, holding beliefs and making decisions are distinct processes. Hence, it proposes two levels:

- *The credal level* where beliefs are represented by belief functions.
- *The pignistic level* where beliefs are used to make decisions and represented by probability functions called pignistic probabilities denoted $BetP$ and defined as follows [19]:

$$BetP(\{a\}) = \sum_{F \subseteq \Theta} \frac{|\{a\} \cap F|}{|F|} \frac{m(F)}{(1 - m(\emptyset))} \text{, for all } a \in \Theta \qquad (25)$$

4   Dynamic Belief Rough Set Classifier (D-BRSC) and Dynamic Belief Rough Set Classifier Based on Generalization Distribution Table (D-BRSC-GDT)

In this Section, we propose two approaches of classification called Dynamic Belief Rough Set Classifier (D-BRSC) and Dynamic Belief Rough Set Classifier based on Generalization Distribution Table (D-BRSC-GDT) which are based on dynamic approach of feature selection. These classifiers are built from uncertain data under the belief function framework. The uncertainty appears only in decision attributes and is handled by the TBM. Before describing the main steps of the construction procedure of D-BRSC and D-BRSC-GDT especially the feature selection, we need to first describe the modified basic concepts of rough sets under uncertainty [23] such as decision table, indiscernibility relation (tolerance relation), set approximation, positive region, reduct and core.

4.1   Basic Concepts of Rough Sets under Uncertainty

In order to extend the rough sets under the belief function framework, we need to redefine the basic concepts of rough sets in this new context. This subsection describes the modified definitions of decision table, indiscernibility relation (tolerance relation), set approximations, positive region and dependency of attributes. These adaptations were originally proposed in [23].

Uncertain Decision Table (UDT). Our UDT is given by $A = (U, C \cup \{ud\})$, where $U = \{o_j : 1 \leq j \leq n\}$ is characterized by a set of certain condition attributes $C = \{c_1, c_2, ..., c_k\}$, and an uncertain decision attribute ud. The value set of ud, is called $\Theta = \{ud_1, ud_2, ..., ud_s\}$. We represent the uncertainty of each object o_j by a bba m_j expressing beliefs on decisions defined on the power set of Θ. These bba's are generally given by an expert (or several experts) to partial uncertainty, they can also present the two extreme cases of total knowledge and total ignorance.

Table 1. Uncertain decision table

U	Age	Property	Income	Unpaid Loan -credit	
o_1	Young	Middle	Average	No	$m_1(\{Yes\}) = 0.95$ $m_1(\{No\}) = 0.05$
o_2	Old	Middle	High	No	$m_2(\{Yes\}) = 1$
o_3	Old	Middle	Average	No	$m_3(\{Yes\}) = 0.7$ $m_3(\Theta) = 0.3$
o_4	Young	Less	High	Yes	$m_4(\{No\}) = 0.9$ $m_4(\Theta) = 0.1$
o_5	Young	Less	High	Yes	$m_5(\{No\}) = 1$
o_6	Young	Middle	Average	No	$m_6(\{No\}) = 0.9$ $m_6(\Theta) = 0.1$
o_7	Young	Middle	Low	No	$m_7(\{No\}) = 1$
o_8	Old	Less	High	No	$m_8(\{Yes\}) = 1$
o_9	Young	Greater	Average	No	$m_9(\{Yes\}) = 1$

Example: *Let us use Table 1 to describe our UDT. It contains nine objects (clients), four certain condition attributes $C = \{Age, Property, Income, Unpaid-credit\}$ and an uncertain decision attribute ud with two possible values $\{Yes, No\}$ representing Θ. The decision Yes means that a bank accepts to give the loan to the client and the decision No means that the bank refuses to give the loan. This table is useful for a bank to plan a loan policy. For example, for the object o_3, belief of 0.7 is exactly committed to the decision Yes, whereas belief of 0.3 is assigned to the whole of frame of discernment Θ (ignorance). Through a bba, we can also represent the certain case, like for the objects o_2, o_5 and o_7.*

Tolerance relation. The indiscernibility relation for the condition attributes U/C is the same as in the certain case because their values are certain. However, the indiscernibility relation for the decision attribute $U/\{ud\}$ is not the same as in the certain case. The decision value is represented by a bba. For this reason, the indiscernibility relation will be denoted by *tolerance relation*.

To redefine $U/\{ud\}$ in the new context, we need to assign each object o_j to the right tolerance class X_i representing the decision value ud_i. For each object o_j, we should check the distance between its bba m_j and the decision value ud_i. The idea is to use a distance measure between the bba m_j and a certain bba m (such that $m(\{ud_i\}) = 1$).

Many distance measures between two bba's were developed [1,6,7,11,21,31] which can be characterized into two kinds:

- Distance measures based on pignistic transformation [1,6,21,31]: For these distances, one unavoidable step is the pignistic transformation of the bba's. Since, there is no bijection between bba's and pignistic probabilities (transformations from the power set to the set). This kind of distance may lose information given by the initial bba's. Besides, we can obtain the same pignistic transformation on two different bba distributions. So, the distance between the two obtained results does not reflect the actual similarity between the starting bba distributions.

- Distance measures directly defined on bba's and on the power set Θ were developed in [7,11]. The first one developed by Fixen and Mahler [7] is a pseudo-metric, since the condition of non-degeneracy is not respected. In our case, we choose the distance measure proposed in [11] which satisfies more properties such as:

1. Non-negativity: $\text{dist}(m_1,m_2) \geq 0$
2. Non-degeneracy: $\text{dist}(m_1,m_2) = 0 \Longleftrightarrow m_1 = m_2$
3. Triangle inequality: $\text{dist}(m_1,m_2) \leq \text{dist}(m_1,m_3) + \text{dist}(m_3,m_2)$
4. Symmetry: $\text{dist}(m_1,m_2) = \text{dist}(m_2,m_1)$

For every ud_i, we define a tolerance class as follows:

$$X_i = \{o_j | dist(m, m_j) < 1 - threshold\}, \; such \; that \; m(\{ud_i\}) = 1 \tag{26}$$

Besides, we define a tolerance relation as follows:

$$IND_{\{ud\}} = U/\{ud\} = \{X_i | ud_i \in \Theta\} \tag{27}$$

Where $dist$ is a distance measure between two bba's [11]:

$$dist(m_1, m_2) = \sqrt{\frac{1}{2}(\| \vec{m_1} \|^2 + \| \vec{m_2} \|^2 - 2 < \vec{m_1}, \vec{m_2} >)} \tag{28}$$

$$0 \leq dist(m_1, m_2) \leq 1 \tag{29}$$

Where $< \vec{m_1}, \vec{m_2} >$ is the scalar product defined by:

$$< \vec{m_1}, \vec{m_2} > = \sum_{i=1}^{|2^\Theta|} \sum_{j=1}^{|2^\Theta|} m_1(A_i) m_2(A_j) \frac{|A_i \cap A_j|}{|A_i \cup A_j|} \tag{30}$$

with A_i, $A_j \in 2^\Theta$ for $i, j = 1, 2, \cdots, |2^\Theta|$. $\| \vec{m_1} \|^2$ is then the square norm of $\vec{m_1}$.

It should be noted here that we have replaced the term equivalence class from the certain decision attribute case by tolerance class for the uncertain decision attribute, because the resulting classes may overlap.

Example: *Let us continue with the same example to compute the equivalence classes based on condition attributes in the same manner as in the certain case: $U/C = \{\{o_1, o_6\}, \{o_2\}, \{o_3\}, \{o_4, o_5\}, \{o_7\}, \{o_8\}, \{o_9\} \}$ and to compute the tolerance classes based on the uncertain decision attribute $U/\{ud\}$. If the user fixed the value of the threshold at 0.1, we would obtain the following results.*

For the uncertain decision value Yes and $m(\{Yes\}) = 1$:

$dist(m, m_1) = 0.07 \; < 0.9$
$dist(m, m_2) = 0.00 \; < 0.9$

$dist(m, m_3) = 0.34 \; < 0.9$
$dist(m, m_4) = 0.93 \; > 0.9$
$dist(m, m_5) = 1.00 \; > 0.9$
$dist(m, m_6) = 0.93 \; > 0.9$
$dist(m, m_7) = 1.00 \; > 0.9$
$dist(m, m_8) = 0.00 \; < 0.9$
$dist(m, m_9) = 0.00 \; < 0.9$
So, $X_1 = \{o_1, o_2, o_3, o_8, o_9\}$

For the uncertain decision value No and $m(\{No\}) = 1$:

$dist(m, m_1) = 0.92 \; > 0.9$
$dist(m, m_2) = 1.00 \; > 0.9$
$dist(m, m_3) = 0.66 \; < 0.9$
$dist(m, m_4) = 0.06 \; < 0.9$
$dist(m, m_5) = 0.00 \; < 0.9$
$dist(m, m_6) = 0.06 \; < 0.9$
$dist(m, m_7) = 0.00 \; < 0.9$
$dist(m, m_8) = 1.00 \; > 0.9$
$dist(m, m_9) = 1.00 \; > 0.9$
So, $X_2 = \{o_3, o_4, o_5, o_6, o_7\}$

$$U/\{ud\} = \{\{o_1, o_2, o_3, o_8, o_9\}, \{o_3, o_4, o_5, o_6, o_7\}\}$$

Set approximation. Like the tolerance relation $U/\{ud\}$, the set approximation concept should be redefined in the new uncertain context. Hence, to compute the new *lower* and *upper* approximations for our UDT, we should follow two steps:

1. We combine the bba's for each equivalence class from U/C using the operator mean [12]. The latter is more suitable than the rule of combination in eqn. (21) which is proposed especially to combine different beliefs on decision for one object and not different bba's relative to different objects. The combination using operator mean [12] can reduce the conflict between pieces of evidence. The combined bba's for the equivalence class $[o_j]_C$ containing the object o_j is computed as follows:

$$\bar{m}_{[o_j]_C}(E) = \frac{1}{|[o_j]_C|} \sum_{o_i \in [o_j]_C} m_i(E), \; for \; all \; E \subseteq \Theta \tag{31}$$

2. We compute the new *lower* and *upper* approximations for each tolerance class X_i from $U/\{ud\}$ based on uncertain decision attribute ud_i as follows:

$$\underline{C}(X_i) = \{o_j | [o_j]_C \subseteq X_i \, and \, dist(m, \bar{m}_{[o_j]_C}) \leq threshold\} \tag{32}$$

such that $m(\{ud_i\}) = 1$

$$\bar{C}(X_i) = \{o_j | [o_j]_C \cap X_i \neq \emptyset\} \tag{33}$$

In the new *lower* approximation, all equivalence classes from U/C are included in X_i where the distance between the combined bba $\bar{m}_{[o_j]_C}$ and the certain bba m (such that $m(\{ud_i\}) = 1$) is less than a *threshold*. However, the *upper* approximation is computed in the same manner as in the certain case.

As in conventional rough set theory, the boundary region of X_i is defined as:

$$BN_C(X_i) = \bar{C}(X_i) - \underline{C}(X_i) \tag{34}$$

Note that in the case of uncertainty the *threshold value* gives more flexibility to the tolerance relation and the set approximations.

Example: *We continue with the same example to compute the new lower and upper approximations. After the first step, we obtain the combined bba for each equivalence class from U/C using the operator mean. Table 2 represents the combined bba for the equivalence classes $[o_1]_C$ and $[o_4]_C$. Note that to simplify the notation, we have used $m_{1,6}$ to mean $\bar{m}_{[o_1]_C}$.*

Table 2. The combined bba for the subsets $\{o_1, o_6\}$ and $\{o_4, o_5\}$

Object	$m(\{Yes\})$	$m(\{No\})$	$m(\Theta)$
o_1	0.95	0.05	0
o_6	0	0.9	0.1
$m_{1,6}$	0.47	0.47	0.05
o_4	0	0.9	0.1
o_5	0	1	0
$m_{4,5}$	0	0.95	0.05

Next, we compute the *lower* and *upper* approximations for each tolerance class $U/\{ud\}$. We will use *threshold* $= 0.1$.

For the uncertain decision value Yes, let $X_1 = \{o_1, o_2, o_3, o_8, o_9\}$. The subsets $\{o_2\}$, $\{o_3\}$, $\{o_8\}$ and $\{o_9\}$ are included in X_1. We should check the distance between their bba and the certain bba m (such that $m(\{Yes\}) = 1$).

$dist(m, m_2) = 0.00 < 0.1$
$dist(m, m_3) = 0.34 > 0.1$
$dist(m, m_8) = 0.00 < 0.1$
$dist(m, m_9) = 0.00 < 0.1$

$\underline{C}(X_1) = \{o_2, o_8, o_9\}$, $\bar{C}(X_1) = \{o_1, o_2, o_3, o_6, o_8, o_9\}$ and $BN_C(X_1) = \{o_1, o_3, o_6\}$

For uncertain decision value No, let $X_2 = \{o_3, o_4, o_5, o_6, o_7\}$. The subsets $\{o_3\}$, $\{o_4, o_5\}$ and $\{o_7\}$ are included in X_2. We should check the distance between their bba and the certain bba m (such that $m(\{No\}) = 1$).

$$dist(m, m_3) \quad = 0.66 > 0.1$$
$$dist(m, m_{4,5}) = 0.06 < 0.1$$
$$dist(m, m_7) \quad = 0.00 < 0.1$$

$\underline{C}(X_2)=\{o_4, o_5, o_7\}$, $\bar{C}(X_2)=\{o_1, o_3, o_4, o_5, o_6, o_7\}$ and $BN_C(X_2)=\{o_1, o_3, o_6\}$

Dependency of attributes and positive region. With this new *lower* approximation, we can define the new positive region denoted $UPos_C(\{ud\})$:

$$UPos_C(A, \{ud\}) = \bigcup_{X_i \in U/\{ud\}} \underline{C}(X_i) \tag{35}$$

The attribute ud depends on the set of attributes C to a degree k, if

$$k = \gamma_C(A, \{ud\}) = \frac{|UPos_C(A, \{ud\})|}{|U|} \tag{36}$$

Example: *Let us continue with the same example, to compute the positive region and dependency of attributes of A.*

$$UPos_C(A, \{ud\})=\{o_2, o_4, o_5, o_7, o_8, o_9\}, \gamma_C(A, \{ud\}) = \tfrac{6}{9}$$

Reduct and core. Using the new formalism of positive region, we can find the reduct of C as a minimal set of attributes $B \subseteq C$ such that:

$$UPos_B(A, \{ud\}) = UPos_C(A, \{ud\}) \tag{37}$$

The core is the most important subset of attributes. It is the intersection of all reducts.

$$UCore_C(A, \{ud\}) = \bigcap URed_C(A, \{ud\}) \tag{38}$$

Where $URed_C(A, \{ud\})$ is the set of all reducts of A relative to ud.

Example: *Let us continue with the same example in Table 1 to compute the possible reducts using the new formalism of positive region. We find that only the subset $\{Property, Income, Unpaid - credit\}$ has the same positive region as the whole set of condition attributes.*

$$UPos_{\{Property, Income, Unpaid-credit\}}(A, \{ud\})= UPos_C(A, \{ud\})$$

There is only one reduct: $B=\{Property, Income, Unpaid - credit\}$.

Belief decision rules. The decision rules induced from the uncertain decision table are called belief decision rules where the decision is represented by a bba.

Example: *The belief decision rule relative the object o_3 from the Table 1 is as follows: If $Age = Young$ and $Property = Middle$ and $Income = Average$ and $Unpaid - credit = No$ Then $m_3(\{Yes\}) = 0.7$ $m_3(\Theta) = 0.3$.*

4.2   Dynamic Belief Rough Set Classifier (D-BRSC)

Our first technique of classification under the belief function framework called the Dynamic Belief Rough Set Classifier (D-BRSC) was originally proposed in [26]. This technique is based only on the basic concepts of Rough Sets (RS). The construction procedure of the D-BRSC is described by means of the following steps:

Step 1. Feature selection based on dynamic reduct: It consists of removing the superfluous condition attributes that are not in reduct. This leaves us with a minimal set of attributes that preserve the ability to perform same classification as the original set of attributes. However, our uncertain decision table (UDT) shown in subsection 4.1 is characterized by a high level of uncertain and noisy data. One of the issues with such data is that the resulting reducts are not stable, and are sensitive to sampling. The belief decision rules generated are not suitable for classification. The solution to this problem is to redefine the concept of dynamic reduct [2] in the new context as we have done in [25]. The rules calculated by means of dynamic reducts are better predisposed to classify unseen objects, because they are the most frequently appearing reducts in sub-decision systems created by random samples of a given decision system.

According to the uncertain context, we can redefine the concept of dynamic reduct as follows:

$$UDR(A, F) = URed_C(A, \{ud\}) \cap \bigcap_{A' \in F} URed_C(A', \{ud\}) \qquad (39)$$

Where F is a family of subtables of A. Any element of $UDR(A, F)$ is called an F-dynamic reduct of A.

In some cases in the uncertain context, we need to apply the general notion called (F, ε)-dynamic reducts which is more flexible, where $0 \leq \varepsilon \leq 1$. The set $UDR_\varepsilon(A, F)$ of all (F, ε)-dynamic reducts is defined by:

$$UDR_\varepsilon(A, F) =$$

$$\left\{ R \in URed_C(A, \{ud\}) : \frac{|\{A' \in F : R \in URed_C(A', \{ud\})\}|}{|F|} \geq 1 - \varepsilon \right\} \qquad (40)$$

If $R \in URed(A', ud)$, then the number $\frac{|\{A' \in F : R \in URed_C(A', \{ud\})\}|}{|F|}$ is called the stability coefficient of R (relative to F).

Example: *Let us continue with the same example in Table 1. Let us remind that the set $B = \{Property, Income, Unpaid - credit\}$ is the only reduct of A. The two subtables $A' = \{o_1, o_2, o_3, o_4, o_9\}$ and $A'' = \{o_5, o_6, o_7, o_8, o_9\}$ form a family F of sub-decision tables. Both of them have the same relative reduct as the whole uncertain decision table A. So, the set $B = \{Property, Income, Unpaid - credit\}$ is also the dynamic reduct relative to the chosen family F which is described in Table 3.*

Table 3. The dynamic reduct

U	Property	Income	Unpaid-credit	Loan	
o_1	Middle	Average	No	$m_1(\{Yes\}) = 0.95$	$m_1(\{No\}) = 0.05$
o_2	Middle	High	No	$m_2(\{Yes\}) = 1$	
o_3	Middle	Average	No	$m_3(\{Yes\}) = 0.7$	$m_3(\Theta) = 0.3$
o_4	Less	High	Yes	$m_4(\{No\}) = 0.9$	$m_4(\Theta) = 0.1$
o_5	Less	High	Yes	$m_5(\{No\}) = 1$	
o_6	Middle	Average	No	$m_6(\{No\}) = 0.9$	$m_6(\Theta) = 0.1$
o_7	Middle	Low	No	$m_7(\{No\}) = 1$	
o_8	Less	High	No	$m_8(\{Yes\}) = 1$	
o_9	Greater	Average	No	$m_9(\{Yes\}) = 1$	

Step 2. Elimination of the redundant objects: After removing the superfluous condition attributes, we can find some redundant objects having the same condition attribute values. They may not have the same bba on decision attribute. So, we combine their bba's using the operator mean as a rule of combination as follows:

$$\bar{m}_{[o_j]_B}(E) = \frac{1}{|[o_j]_B|} \sum_{o_i \in [o_j]_B} m_i(E), \ \ for \ all \ E \subseteq \Theta \tag{41}$$

Where B is the relative reduct of C with respect to ud and $[o_j]_B$ is the equivalence class containing the object o_j with respect to the condition attribute subset B.

Please note that the operator mean [12] is more suitable to combine these bba's than the rule of combination in eqn. (21) which is proposed especially to combine different beliefs on decision for one object and not different bba's relative to different objects.

Example: *After removing the superfluous condition attributes and the redundant objects for the uncertain decision table, we obtain Table 4. Let us recall that we have used the notation $m_{1,3,6}$ to mean $\bar{m}_{[o_1]_B}$ with B is equal to $\{Property, Income, Unpaid - credit\}$.*

Step 3. Elimination of the superfluous condition attribute values: To further simplify the uncertain decision table (UDT), we can eliminate some attribute values. We also need to redefine the concept of value reduct and value core as follows:

For each belief decision rule of the form: **If** $C(o_j)$ **then** m_j.
For all $B \subset C$, Let $X = \{o_k | B(o_j) = B(o_k) \, and \, j \neq k\}$
If $X = \emptyset$ **then** $B(o_j)$ is a value reduct of o_j.
Else If $Max(dist(m_j, \bar{m}_{[o_k]_C})) \leq$ threshold **then** $B(o_j)$ is a value reduct of o_j.

Table 4. Combined bba's of redundant objects

U	Property	Income	Unpaid-credit	Loan
o_1, o_3, o_6	Middle	Average	No	$m_{1,3,6}(\{Yes\}) = 0.55$ $m_{1,3,6}(\{No\}) = 0.32$ $m_{1,3,6}(\Theta) = 0.13$
o_2	Middle	High	No	$m_2(\{Yes\}) = 1$
o_4, o_5	Less	High	Yes	$m_{4,5}(\{No\}) = 0.95$ $m_{4,5}(\Theta) = 0.05$
o_7	Middle	Low	No	$m_7(\{No\}) = 1$
o_8	Less	High	No	$m_8(\{Yes\}) = 1$
o_9	Greater	Average	No	$m_9(\{Yes\}) = 1$

The set of all indispensable attribute values from C for o_j with respect to ud will be called value core and will be denoted by $UCore_C^j(A, \{ud\})$.

$$UCore_C^j(A, \{ud\}) = \bigcap URed_C^j(A, \{ud\}) \tag{42}$$

Where $URed_C^j(A, \{ud\})$ is the family of all reduct value of C for o_j with respect to ud.

Example: *We compute the value reduct of each decision rule from Table 4 with threshold $= 0.1$. We take one decision rule from the table: If $Property = Middle$ and $Income = Low$ and $Unpaid - credit = No$ then m_7.*

Let us split the rule into:

1. *If $Property = Middle$ then $m_7 : X = \{o_1, o_2, o_3, o_6\}$ and $dist(m_7, m_{1,3,6}) > 0.1$. So, $Property = Middle$ is not a reduct value.*
2. *If $Income = Low$ then $m_7 : X = \emptyset$. So, $Income = Low$ a reduct value.*
3. *If $Unpaid - credit = No$ then $m_7 : X = \{o_1, o_2, o_3, o_6, o_8, o_9\}$ and $dist(m_7, m_{1,3,6}) > 0.1$. So, $Unpaid - credit = No$ is not a reduct value.*

If we compute the value reduct for all belief decision rules, we obtain Table 5. Note that the object o_2 has two reduct values.

Step 4. Generation of belief decision rules: After the simplification of the uncertain decision table, we can generate shorter and significant belief decision rules. With simplification, we can improve the time and the performance of classification of unseen objects. From one example of UDT, we can find many solutions of simplification which have many reducts and value reducts. The number of solutions, denoted nb, is computed as follows:

Table 5. Simplified uncertain decision table

U	Property	Income	Unpaid-credit	Loan
o_1, o_3, o_6	Middle	Average	No	$m_{1,3,6}(\{Yes\}) = 0.55$ $m_{1,3,6}(\{No\}) = 0.32$ $m_{1,3,6}(\Theta) = 0.13$
o_2	-	High	No	$m_2(\{Yes\}) = 1$
$o_{2'}$	Middle	High	-	$m_{2'}(\{Yes\}) = 1$
o_4, o_5	-	-	Yes	$m_{4,5}(\{No\}) = 0.95$ $m_{4,5}(\Theta) = 0.05$
o_7	-	Low	-	$m_7(\{No\}) = 1$
o_8	-	High	No	$m_8(\{Yes\}) = 1$
o_9	Greater	-	-	$m_9(\{Yes\}) = 1$

$$nb = |URed_C(A, \{ud\})| * \prod_{j=1}^{n} |URed_C^j(A, \{ud\})| \qquad (43)$$

Example: *Table 5 gives one solution of rule classification described as follows:*
If Property=Middle and Income=Average and Unpaid-credit=No then $m_{1,3,6}$
($\{Yes\}$) = 0.55 $m_{1,3,6}(\{No\}) = 0.32$ $m_{1,3,6}(\Theta) = 0.13$
If Income=High and Unpaid-credit=No then $m_{2,8}(\{Yes\}) = 1$
If Unpaid-credit=Yes then $m_{4,5}(\{No\}) = 0.95$ $m_{4,5}(\Theta) = 0.05$
If Income=Low then $m_7(\{No\}) = 1$
If Property=Greater then $m_9(\{Yes\}) = 1$

The generated belief decision rules are used to classify unseen objects. To simplify this task, we can compute the pignistic probability to m_j relative to each decision rule using eqn. (25).

4.3 Dynamic Belief Rough Set Classifier Based on Generalization Distribution Table (D-BRSC-GDT)

In this subsection, we propose our second classification approach called Dynamic Belief Rough Set Classifier based on Generalization Distribution Table (D-BRSC-GDT) which can generate a set of rules from uncertain databases with the minimal description length, having large strength and covering all instances. This method is based on the hybrid system GDT-RS [3,29]. In this subsection, we detail the main steps to construct our D-BRSC-GDT method allowing the discovery of classification rules from UDT under the belief function framework based on dynamic approach for feature selection.

Step 1. Creation of the GDT: This step can be omitted because the prior distribution of a generalization can be calculated using eqns. (13) and (14). Note that the standard GDT depends only on condition attributes and not on decision attribute values. In the uncertain context, our GDT still has the same structure as in [30].

Step 2. Feature selection based on dynamic reduct: It is similar to the first step in the construction procedure relative to the D-BRSC. Let us recall that this step consists of removing the superfluous condition attributes that are not in dynamic reducts. This leaves us with a minimal set of attributes that preserve the ability to perform the same classification as the original set of attributes.

Example: *Let us continue with Table 1 to compute the dynamic reduct of our uncertain decision table using eqn. (39). We find that the subset $B=\{Property, Income, Unpaid - credit\}$ is also the dynamic reduct using a family F formed by two other subtables $A'=\{o_1, o_2, o_3, o_8, o_9\}$ and $A''=\{o_4, o_5, o_6, o_7, o_9\}$. The two subtables have the same reduct as the whole decision table A. Table 3 shows the dynamic reduct relative to the chosen family F.*

Step 3. Definition of the compound object: Consider the indiscernibily classes with respect to the condition attribute subset B as one object, called the compound object o'_j and defined as follows:

$$o'_j = \{o_i | B(o_i) = B(o_j)\} \tag{44}$$

U' is the set of all compound objects obtained from U over the subset of condition attributes B. For objects that compose each compound object, we combine their bba's using the operator mean as rule of combination as follows:

$$m'_j(E) = \frac{1}{|o'_j|} \sum_{o_j \in o'_j} m_j(E), \forall E \subseteq \Theta \tag{45}$$

Example: *Let us continue with the same example. By applying the step 3, we obtain the following tables:*

Step 4. Elimination of the contradictory compound objects: For any compound object o'_j from U', compute $r_{ud_i}(o'_j)$ which represents a noise rate for each decision value ud_i. If there exists a ud_i such that $r_{ud_i}(o'_j) = min\{r_{ud_{i'}}(o'_j) | ud_{i'} \in \Theta\} < T_{noise}$ (threshold value), then we assign the decision value corresponding to ud_i to the object o'_j. If there is no $ud_i \in \Theta$ such that $r_{ud_i}(o'_j) < T_{noise}$, we treat the compound object o'_j as a contradictory one, and set the decision value of o'_j to $\perp$(uncertain).

The noise rate is originally calculated using eqn. (17). This is not appropriate in our uncertain context since the decision value is represented by a bba. In this new situation, the noise rate will be computed as follows:

$$r_{ud_i}(o'_j) = dist(m'_j, m), \text{such that } m(\{ud_i\}) = 1 \tag{46}$$

Table 6. The compound objects

U	U'	Property	Income	Unpaid-credit
o_1, o_3, o_6	o'_1	Middle	Average	No
o_2	o'_2	Middle	High	No
o_4, o_5	o'_4	Less	High	Yes
o_7	o'_7	Middle	Low	No
o_8	o'_8	Less	High	No
o_9	o'_9	Greater	Average	No

Table 7. The combined bba's of each compound object

U'	Property	Income	Unpaid-credit	Loan
o'_1	Middle	Average	No	$m'_1(\{Yes\}) = 0.55$ $m'_1(\{No\}) = 0.32$ $m'_1(\Theta) = 0.13$
o'_2	Middle	High	No	$m'_2(\{Yes\}) = 1$
o'_4	Less	High	Yes	$m'_4(\{No\}) = 0.95$ $m'_4(\Theta) = 0.05$
o'_7	Middle	Low	No	$m'_7(\{No\}) = 1$
o'_8	Less	High	No	$m'_8(\{Yes\}) = 1$
o'_9	Greater	Average	No	$m'_9(\{Yes\}) = 1$

The idea is to use the distance between two bba's m'_j and a certain bba m (such that $m(\{ud_i\}) = 1$). In this manner, we can check that the decisions of all instances belonging to the compound object are close to a certain case. So, it is considered as a non-contradictory compound object.

Example: *By applying the step 4 to Table 7, we obtain Table 8.*

Step 5. Minimal description length of decision rule: Let U'' be the set of all non-contradictory compound objects. From U'', select one compound object o'_j. Create a discernibility vector (the row or the column with respect to o'_j in the discernibility matrix) for o'_j.

The discernibility matrix [16] of A is a symmetric $n \times n$ matrix. Each entries a_{ij} in A consists of the set of attributes upon which objects o_i and o_j differ.

$$a_{ij} = \{c \in C | c(o_i) \neq c(o_j)\} \text{ and } ud(o_i) \neq ud(o_i) \text{ for } i, j = 1, ..., n \qquad (47)$$

Table 8. Contradictory and non-contradictory compound objects

U'	Property	Income	Unpaid-credit	Loan
o'_1	Middle	Average	No	$\perp$
o'_2	Middle	High	No	Yes
o'_4	Less	High	Yes	No
o'_7	Middle	Low	No	No
o'_8	Less	High	No	Yes
o'_9	Greater	Average	No	Yes

Next, we compute all the so-called local relative reducts for the compound object o'_j by using the discernibility function $f_A(o'_j)$. It is a boolean function of k boolean variables corresponding to the k condition attributes defined as below:

$$f_A(o'_j) = \bigwedge \{\bigvee a_{ij} | 1 \le i \le n, a_{ij} \ne \emptyset\} \tag{48}$$

The set of all prime implicants of $f_A(o'_j)$ determines the sets of all reduct values of the compound object o'_j.

Example: *According to Table 8, the discernibility vector for the compound object o'_2 is as follows.*

Table 9. Discernibility vector for o'_2

U'	$o'_1(\perp)$	$o'_2(Yes)$	$o'_4(No)$	$o'_7(No)$	$o'_8(Yes)$	$o'_9(Yes)$
$o'_2(Yes)$	Income	$\emptyset$	Property,Unpaid-credit	Income	$\emptyset$	$\emptyset$

We obtain two reducts: $\{Property, Income\}$ and $\{Income, Unpaid\text{-}credit\}$ by applying the indiscernibility function : $f_A(o'_2){=}(Income) \wedge (Property \vee Unpaid{-}credit) \wedge (Income){=} (Property \wedge Income) \vee (Income \wedge Unpaid - credit)$

Step 6. Selection of the best rules: We start by constructing the decision rules obtained from the local reducts for the object o'_j and revising their strength using eqn. (15). Then, we select only the best rule for o'_j having the best strength.

Example: *According to the same example, the following rules are acquired for object o'_2:*

1. *If $Property = Middle$ and $Income = High$ then Yes with $S = (\frac{1}{2}{*}1){*}(1){=}$ 0.5*
2. *If $Income = High$ and $Unpaid - credit = No$ then Yes with $S = (\frac{1}{3}*2)*(1){=}$ 0.66*

The rule If Income = High and Unpaid − credit = No then Yes is selected for the compound object o_2' due to its strength.

Step 7. Stopping criterion: Let $U'' = U'' - \{o_j'\}$. $If\ U'' \neq \emptyset$, then go back to *Step* 5. Otherwise, STOP.

Example: *As a result, we obtain a set of decision rules that are able to classify unseen objects shown in the Table 10.*

Table 10. Decision rules

U	rules	strengths
o_2', o_8'	If *Income = High* and *Unpaid − credit = No* then *Yes*	0.66
o_4'	If *Unpaid − credit = Yes* then *No*	0.11
o_7'	If *Income = Low* then *No*	0.16
o_9'	If *Property = Greater* then *Yes*	0.16

4.4 Computational Complexity

Our two classification systems D-BRSC and D-BRSC-GDT are based on the reduct set computation. The time requirement of the decision rule generation can be too expensive when the uncertain decision table has too many attributes or different values of attributes or objects. According to the size of the UDT, the reduct set computation can be exponential and the problem of calculating a minimal reduct is NP-hard [16]. For these reasons, it would be better to apply some heuristic algorithms to avoid the costly computation of a minimal reduct [24].

Note that the time complexity of the D-BRSC and the D-BRSC-GDT is $O(n \times m^2 \times Nr_{max})$, where n is the number of instances in a given database, m stands for the number of attributes, Nr_{max} is the maximal number of reducts.

5   Experimentations

In our experiments, we have performed several tests on real databases obtained from the U.C.I. repository[1] to evaluate our two classifiers: D-BRSC and D-BRSC-GDT. A brief description of these databases is presented in Table 11.

5.1   Constructing Uncertainty in Databases

These databases were artificially modified in order to include uncertainty in the decision attribute. We took different degrees of uncertainty (Low, Middle and High) based on increasing values of probabilities P used to transform the actual decision value d_i of each object o_j to a bba $m_j(\{d_i\}) = 1 - P$ and $m_j(\Theta) = P$.

---

[1] http://www.ics.uci.edu/ mlearn/MLRepository.html

Table 11. Description of databases

Database	#instances	#attributes	#decision values
W. Breast Cancer	690	8	2
Balance Scale	625	4	3
C. Voting records	497	16	2
Zoo	101	17	7
Nursery	12960	8	3
Solar Flares	1389	10	2
Lung Cancer	32	56	3
Hyes-Roth	160	5	3
Car Evaluation	1728	6	4
Lymphography	148	18	4
Spect Heart	267	22	2
Tic-Tac-Toe Endgame	958	9	2

- Low degree of uncertainty: we take $0 < P <= 0.3$
- Middle degree of uncertainty: we take $0.3 < P <= 0.6$
- High degree of uncertainty: we take $0.6 < P <= 1$

A larger P gives a larger degree of uncertainty.

5.2   Evaluation Criteria and Technique of Sampling

The relevant criteria used to judge the performance and the quality of our proposed classifiers: D-BRSC and D-BRSC-GDT are as follows:

1. *The classification accuracy* represents the percent of correct classification (PCC) of the objects belonging to testing set. For each testing instance, we determine its corresponding rule and we look for the most probable decision corresponding to this rule using the pignistic probability computed from its bba.
2. *The time requirement* represents the number of seconds needed for the construction procedure.
3. *The size* represents the number of belief decision rules generated from the classifier.

Each database is divided into ten parts. Nine parts are used as the training set, the last is used as the testing set. The procedure is repeated ten times, each time another part is chosen as the testing set. This method, called a cross-validation, permits an efficient estimation of the evaluation criterion. In this paper, we perform ten-fold cross-validation tests with different data splits and we report the average of the evaluation criteria.

The technique for sampling of uncertain decision table and dynamic reduct computation in the uncertain context is similar to the technique employed in the certain case [2] which consists of the following steps.

- In the first step, some of samples of the uncertain decision table are computed. Next, reducts for all these subtables are calculated.
- In the second step, reducts with the stability coefficients higher than a fixed threshold are extracted. They are considered as dynamic reducts.

5.3   Experimental Results

In this subsection, we detail different results obtained from our two classifiers (D-BRSC and D-BRSC-GDT) relative to the three evaluation criteria. To evaluate them, we compare the experimental results with those obtained from Static Belief Rough Set Classifier (S-BRSC) [27] and Static Belief Rough Set Classifier based on Generalization Distribution Table (S-BRSC-GDT) [28]. The latter two are based on static reduct in the feature selection step.

To further evaluate the performance of our two proposed dynamic classifiers (D-BRSC, D-BRSC-GDT) with their static versions (S-BRSC, S-BRSC-GDT), we compare the results with those given by another classifier not based on rough set methodology. We choose the Belief Decision Tree (BDT) approach developed in [4]. The latter is a classification technique in an uncertain environment where the uncertainty is represented through the TBM and appears in the decision attribute of training objects. The model generated by the BDT also consists of belief decision rules. Hence, the comparison between all the classifiers is possible. From the procedures of building BDT, we mention the averaging and the conjunctive approaches.

Note that the pruning step is indispensable to improve the size and the accuracy of the BDT. In [22], we have concluded that post-pruning is better than pre-pruning [5] in term of size and accuracy. For this reason, we focus in our experimentations only on the post-pruned BDT in averaging and conjunctive approaches.

The classification accuracy criterion. We start with the more important evaluation criterion, i.e. the classification accuracy. Tables 12, 13 and 14 summarize the experimental results for the PCC relative to the proposed classifiers for all databases and for different degrees of uncertainty. Table 12 reports the results relative to the D-BRSC and the S-BRSC. On the other hand, Table 13 shows the results relative to the D-BRSC-GDT and the S-BRSC-GDT. Table 14 summarizes the results relative to the averaging and the conjunctive pruned BDT.

From these Tables, we conclude that reducing uncertain and noisy databases using dynamic feature selection approach is more suitable for classification of the unseen cases than the static approach. The mean PCC (low, middle, high) obtained from D-BRSC and D-BRSC-GDT is better than the mean PCC obtained from S-BRSC and S-BRSC-GDT. It is true for all databases and for all degrees of uncertainty. For example, the mean PCC for *Balance Scale* database is 87.33% with D-BRSC and 83.57% with S-BRSC. The mean PCC for *Lung Cancer* database is 80.21% with D-BRSC-GDT and 75.49% with S-BRSC-GDT.

Table 12. The PCC relative to D-BRSC and S-BRSC

Database	D-BRSC PCC (%)					S-BRSC PCC (%)				
	Low	Middle	High	Mean	S. deviation	Low	Middle	High	Mean	S. deviation
W. Breast Cancer	89.82	89.63	89.21	89.55	0.31	84.43	84.15	83.65	84.07	0.39
Balance Scale	87.59	87.33	87.07	87.33	0.26	82.72	82.58	82.41	82.57	0.15
C. Voting records	98.91	98.77	98.48	98.72	0.21	98.54	98.31	98.09	98.31	0.22
Zoo	96.56	96.42	95.85	96.27	0.37	94.16	94.03	93.54	93.91	0.32
Nursery	97.64	97.18	97.03	97.28	0.31	96.37	95.94	95.71	96.03	0.33
Solar Flares	92.66	92.58	92.42	92.55	0.12	89.69	89.61	89.57	89.62	0.06
Lung Cancer	81.79	81.54	81.19	81.50	0.30	76.61	76.32	76.18	76.37	0.21
Hyes-Roth	98.83	98.32	97.62	98.27	0.60	97.58	97.24	97.08	97.28	0.25
Car Evaluation	85.71	85.22	85.19	85.37	0.29	83.94	83.68	83.14	83.58	0.40
Lymphography	86.49	86.31	85.75	86.18	0.38	84.73	84.51	84.02	84.45	0.36
Spect Heart	89.47	89.25	89.11	89.27	0.18	87.44	87.33	87.14	87.30	0.15
Tic-Tac-Toe	88.72	88.45	88.23	88.46	0.24	86.87	86.71	86.33	86.63	0.27

Additionally, we conclude that generating uncertain decision rules using D-BRSC yields better PCC than using D-BRSC-GDT. For example, the mean PCC for *W. Breast Cancer* database is 89.55% with D-BRSC and 86.49% with D-BRSC-GDT. Besides, we also conclude that our classification systems based on rough sets are more accurate than the pruned BDT in averaging and conjunctive approaches. We further note that the PCC slightly increases when the uncertainty decreases for all the approaches.

Finally, we also conclude that the standard deviation of the PCC (low, middle, high) obtained from our classification systems based on rough sets: D-BRSC, S-BRSC, D-BRSC-GDT and S-BRSC-GDT is less than the standard deviation obtained from the pruned BDT (averaging and conjunctive). For example, the standard deviation of the PCC for *Spect Heart* database is 0.65 with conjunctive BDT and 0.18 with D-BRSC.

The time requirement criterion. We now move to the second evaluation criterion representing the time requirement needed for the construction procedure to our classification systems. Tables 15, 16 and 17 give the different results obtained from all databases for the time requirement relative to the BRSC (static and dynamic), the BRSC-GDT (static and dynamic) and the pruned BDT (averaging and conjunctive) respectively. Note that the time requirement is almost the same for the different degrees of uncertainty. Note also that from the ten-fold cross-validation tests, we report the minimum, the maximum and the average of the time requirement. These tables show that the two approaches of classification based on static rough sets namely the S-BRSC and the S-BRSC-GDT are faster than the two approaches of classification based on dynamic rough sets: D-BRSC and D-BRSC-GDT. Especially, with large databases like the *Nursery* database where the mean time requirement goes from 380 seconds with S-BRSC

Table 13. The PCC relative to D-BRSC-GDT and S-BRSC-GDT

Database	D-BRSC-GDT PCC (%)					S-BRSC-GDT PCC (%)				
	Low	Middle	High	Mean	S. deviation	Low	Middle	High	Mean	S. deviation
W. Breast Cancer	86.79	86.52	86.18	86.49	0.30	83.76	83.55	83.17	83.48	0.29
Balance Scale	83.66	83.34	83.16	83.39	0.25	80.34	80.21	80.03	80.19	0.15
C. Voting records	98.83	98.55	98.26	98.54	0.28	98.46	98.22	98.03	98.23	0.21
Zoo	96.58	96.31	96.08	96.32	0.25	93.61	93.46	93.02	93.36	0.30
Nursery	97.08	97.01	96.98	97.02	0.05	96.84	96.27	95.87	96.32	0.48
Solar Flares	89.86	89.75	89.59	89.73	0.13	89.67	89.61	89.56	89.61	0.05
Lung Cancer	80.39	80.15	80.08	80.21	0.16	75.78	75.45	75.24	75.49	0.27
Hyes-Roth	98.70	98.50	97.31	98.17	0.75	97.77	97.43	97.13	97.44	0.32
Car Evaluation	84.69	84.31	84.25	84.41	0.23	81.59	81.23	81.20	81.34	0.21
Lymphography	86.16	86.11	85.96	86.08	0.10	84.41	84.17	84.05	84.21	0.18
Spect Heart	88.59	88.36	88.19	88.38	0.20	87.34	87.28	87.07	87.23	0.14
Tic-Tac-Toe	87.54	87.41	87.19	87.38	0.17	86.49	86.36	86.21	86.35	0.14

Table 14. The PCC relative to the averaging and conjunctive BDT

Database	Averaging pruned BDT PCC (%)					Conjunctive pruned BDT PCC (%)				
	Low	Middle	High	Mean	S. deviation	Low	Middle	High	Mean	S. deviation
W. Breast Cancer	83.41	83.01	82.07	82.83	0.68	83.46	83.01	82.17	82.88	0.65
Balance Scale	78.05	77.72	77.67	77.82	0.20	78.15	77.83	77.76	77.91	0.21
C. Voting records	98.21	97.70	97.74	97.84	0.28	98.28	97.76	97.71	97.91	0.32
Zoo	91.90	93.32	91.36	91.53	0.32	91.94	93.36	91.41	91.57	0.32
Nursery	95.79	95.08	95.07	95.31	0.41	95.84	95.13	95.11	95.36	0.42
Solar Flares	85.69	85.49	85.35	85.52	0.19	85.78	85.61	85.46	85.61	0.16
Lung Cancer	74.91	74.70	74.38	74.65	0.21	74.63	74.36	74.07	74.35	0.28
Hyes-Roth	83.64	83.17	82.30	83.02	0.67	83.66	83.31	82.14	83.03	0.80
Car Evaluation	73.38	73.01	72.87	73.11	0.26	73.49	73.11	72.97	73.19	0.27
Lymphography	79.21	78.92	78.91	79.01	0.17	79.25	78.97	78.94	79.05	0.17
Spect Heart	83.25	82.79	82.01	82.67	0.62	83.46	83.01	82.17	82.88	0.65
Tic-Tac-Toe	84.83	84.68	84.35	84.60	0.24	83.91	83.75	83.42	83.69	0.25

to 1106 seconds with D-BRSC. These tables also show that the S-BRSC and the S-BRSC-GDT are slightly faster than the pruned BDT in both averaging and conjunctive approaches. Besides, we also conclude from these tables that S-BRSC-GDT and D-BRSC-GDT approaches, which can avoid many iterations, are slightly faster than S-BRSC and D-BRSC, respectively. For example, the mean time requirement for *Balance Scale* database is 297 seconds with D-BRSC and 282 seconds with D-BRSC-GDT.

Table 15. The time requirement (seconds) relative to D-BRSC and S-BRSC

| Database | D-BRSC | | | S-BRSC | | |
	Min	Max	Mean	Min	Max	Mean
W. Breast Cancer	279	421	393	149	159	154
Balance Scale	282	312	297	121	134	129
C. Voting records	217	267	235	104	115	110
Zoo	204	256	221	96	107	101
Nursery	1079	1128	1106	271	387	380
Solar Flares	398	447	412	152	163	157
Lung Cancer	181	207	187	48	48	48
Hyes-Roth	199	246	205	85	101	91
Car Evaluation	402	487	426	171	181	178
Lymphography	208	274	217	100	109	102
Spect Heart	229	268	233	106	111	109
Tic-Tac-Toe	303	356	312	137	144	139

Table 16. The time requirement (seconds) relative to D-BRSC-GDT and S-BRSC-GDT

| Database | D-BRSC-GDT | | | S-BRSC-GDT | | |
	Min	Max	Mean	Min	Max	Mean
W. Breast Cancer	367	423	381	133	139	136
Balance Scale	268	312	282	122	128	126
C. Voting records	210	256	224	102	110	106
Zoo	203	236	207	95	96	96
Nursery	859	895	867	349	378	356
Solar Flares	384	412	391	143	151	146
Lung Cancer	168	195	174	42	42	42
Hyes-Roth	185	221	198	79	87	82
Car Evaluation	395	437	418	155	156	156
Lymphography	202	231	209	93	103	97
Spect Heart	211	244	226	86	96	89
Tic-Tac-Toe	385	405	293	126	129	127

The size criterion. Our third evaluation criterion is the number of the learned uncertain decision rules generated from our classifiers: BRSC (static and dynamic), BRSC-GDT (static and dynamic), and pruned BDT (averaging and conjunctive). Tables 18, 19 and 20 summarize the different results for the size relative to the chosen databases relative to our classification systems. Note that the size is almost the same for the different degrees of uncertainty. Note also that from the ten-fold cross-validation tests, we report the minimum, the maximum and the average of the size. From these tables, we conclude that that the classification based on static rough sets namely S-BRSC and S-BRSC-GDT give a few

Table 17. The time requirement (seconds) relative to averaging and conjunctive BDT

Database	Averaging BDT			Conjunctive BDT		
	Min	Max	Mean	Min	Max	Mean
W. Breast Cancer	151	159	156	152	163	158
Balance Scale	136	143	139	139	143	141
C. Voting records	114	123	117	118	129	122
Zoo	100	107	103	102	109	107
Nursery	378	406	386	383	411	397
Solar Flares	158	166	160	159	166	161
Lung Cancer	54	56	56	54	56	56
Hyes-Roth	91	97	93	91	97	94
Car Evaluation	183	197	189	189	197	193
Lymphography	105	110	108	108	113	112
Spect Heart	109	116	111	112	119	115
Tic-Tac-Toe	146	155	149	148	155	152

Table 18. The size relative to D-BRSC and S-BRSC

Database	D-BRSC			S-BRSC		
	Min	Max	Mean	Min	Max	Mean
W. Breast Cancer	50	55	52	48	53	50
Balance Scale	59	66	63	58	61	59
C. Voting records	52	57	54	49	53	51
Zoo	40	49	44	36	40	38
Nursery	223	239	233	221	232	227
Solar Flares	125	132	129	118	125	121
Lung Cancer	28	28	28	27	27	27
Hyes-Roth	41	47	44	32	41	39
Car Evaluation	174	183	179	171	181	174
Lymphography	56	61	59	53	59	57
Spect Heart	52	55	54	50	54	53
Tic-Tac-Toe	120	131	125	117	127	123

more combined decision rules than the classification based on dynamic rough sets, i.e. D-BRSC and D-BRSC-GDT. For example, the number of the uncertain decision rules induced from *Balance Scale* database goes from 59 using S-BRSC to 63 using D-BRSC. These tables also show that BRSC-GDT (static and dynamic) gives smaller models than BRSC (static and dynamic) as well as pruned BDT (averaging and conjunctive) for all the databases and for all the degrees of uncertainty. For example, the number of the uncertain decision rules induced from *W. Breast Cancer* database goes from 38 using S-BRSC-GDT to 50 using S-BRSC and to 45 using averaging BDT. Finally, we conclude that pruned BDT (averaging and conjunctive) gives a few more combined decision rules for all the databases than BRSC (static and dynamic).

Table 19. The size relative to D-BRSC-GDT and S-BRSC-GDT

Database	D-BRSC-GDT			S-BRSC-GDT		
	Min	Max	Mean	Min	Max	Mean
W. Breast Cancer	40	41	41	36	39	38
Balance Scale	49	52	51	47	51	48
C. Voting records	42	45	40	39	40	39
Zoo	30	31	30	26	26	26
Nursery	191	203	194	186	192	188
Solar Flares	108	117	110	99	106	101
Lung Cancer	27	27	27	21	21	21
Hyes-Roth	30	34	32	29	30	29
Car Evaluation	158	163	161	151	156	154
Lymphography	43	45	44	41	42	41
Spect Heart	40	44	42	39	42	40
Tic-Tac-Toe	108	115	110	102	110	106

Table 20. The size relative to averaging and conjunctive BDT

Database	Averaging BDT			Conjunctive BDT		
	Min	Max	Mean	Min	Max	Mean
W. Breast Cancer	43	47	45	45	49	47
Balance Scale	49	52	51	50	53	52
C. Voting records	40	45	42	42	46	44
Zoo	32	32	32	35	36	35
Nursery	199	203	202	205	210	207
Solar Flares	109	118	112	115	120	117
Lung Cancer	24	24	24	25	26	25
Hyes-Roth	33	34	33	34	35	35
Car Evaluation	167	173	170	169	176	173
Lymphography	47	51	49	50	52	51
Spect Heart	44	45	45	47	48	47
Tic-Tac-Toe	112	117	115	114	120	118

6   Conclusion and Future Work

In this paper, we have presented two new approaches of classification called
Dynamic Belief Rough Set Classifier (D-BRSC) and Dynamic Belief Rough Set
Classifier based on Generalization Distribution Table (D-BRSC-GDT). Both are
based on rough sets and induced from uncertain data under the belief function
framework. The feature selection step relative to their construction procedure is
based on dynamic approach.

We have done experimentations on real databases to evaluate our proposed classifiers based on three evaluation criteria. To evaluate our approaches, we compare
the results with those obtained from Static Belief Rough Set Classifier(S-BRSC)

and Static Belief Rough Set Classifier based on Generalization Distribution Table (S-BRSC-GDT). To further evaluate the performance of our two proposed dynamic classifiers (D-BRSC, D-BRSC-GDT) with their static versions (S-BRSC, S-BRSC-GDT), we compare the results with those given by another similar classifier called the Belief Decision Tree (BDT), which is not based on rough set methodology.

According to the experimental results, we find that generating decision rules based on dynamic approach of feature selection is more suitable for classification process than the static one. In addition, the experimental results show that all our classification systems based on rough sets (dynamic and static versions) are more accurate than the pruned BDT in averaging and conjunctive approaches. As a future work, we plan to work with uncertainty in condition attribute values.

Acknowledgements

The authors would like to thank the anonymous reviewers for their detailed comments, criticisms, and suggestions that have helped greatly in improving the quality of this paper.

References

1. Bauer, M.: Approximation algorithms and decision making in the Dempster-Shafer theory of evidence - An empirical study. IJAR 17(2-3), 217–237 (1997)
2. Bazan, J., Skowron, A., Synak, P.: Dynamic reducts as a tool for extracting laws from decision tables. In: Raś, Z.W., Zemankova, M. (eds.) ISMIS 1994. LNCS (LNAI), vol. 869, pp. 346–355. Springer, Heidelberg (1994)
3. Dong, J.Z., Zhong, N., Ohsuga, S.: Probabilistic rough induction: The GDT-RS methodology and algorithms. In: Raś, Z.W., Skowron, A. (eds.) ISMIS 1999. LNCS (LNAI), vol. 1609, pp. 621–629. Springer, Heidelberg (1999)
4. Elouedi, Z., Mellouli, K., Smets, P.: Belief decision trees: Theoretical foundations. International Journal of Approximate Reasoning 28(2-3), 91–124 (2001)
5. Elouedi, Z., Mellouli, K., Smets, P.: A Pre-Pruning Method in Belief Decision Trees. In: The Ninth International Conference on Information Processing and Management of uncertainty in Knowledge-Based Systems, IPMU 2002, Annecy, France, vol. 1, pp. 579–586 (2002)
6. Elouedi, Z., Mellouli, K., Smets, P.: Assessing sensor reliability for multisensor data fusion within the transferable belief model. IEEE Trans. Syst. Man Cyben. 34(1), 782–787 (2004)
7. Fixen, D., Mahler, R.P.S.: The modified Dempster-Shafer approach to classification. IEEE Trans Syst Man Cybern. 27(1), 96–104 (1997)
8. Grzymala-Busse, J.W.: Rough set strategies to data with missing attribute values. In: Workshop Notes, Foundations and New Directions of Data Mining, the 3rd International Conference on Data Mining, Melbourne, Florida, pp. 56–63 (2003)
9. Grzymala-Busse, J.W., Siddhaye, S.: Rough Set Approaches to Rule Induction from Incomplete Data. In: Proceedings of the IPMU 2004, Perugia, Itally, July 4-9, vol. 2, pp. 923–930 (2004)

10. Hong, T.P., Tseng, L.H., Chien, B.C.: Learning fuzzy rules from incomplete quantitative data by rough sets, pp. 1438–1442. IEEE, Los Alamitos (2002)
11. Jousseleme, A.L., Grenier, D., Bosse, E.: A new distance between two bodies of evidence. Information Fusion 2(2), 91–101 (2001)
12. Murphy, C.K.: Combining belief functions when evidence conflicts. Decision Support Systems 29, 1–9 (2000)
13. Pawlak, Z.: Rough Sets. International Journal of Computer and Information Sciences 11(5), 341–356 (1982)
14. Pawlak, Z.: Rough Sets: Theoretical Aspects of Reasoning About Data. Kluwer Academic Publishing, Dordrecht (1991)
15. Shafer, G.: A mathematical theory of evidence. Princeton University Press, Princeton (1976)
16. Skowron, A., Rauszer, C.: The discernibility matrices and functions in information systems. In: Slowinski, R. (ed.) Intelligent Decision Support, pp. 331–362. Kluwer Academic Publishers, Boston (1992)
17. Skowron, A., Grzymala-Busse, J.W.: From rough set theory to evidence theory. In: Advances in the Dempster-Shafer Theory of Evidence, New York, pp. 193–236 (1994)
18. Smets, P., Kennes, R.: The transferable belief model. Artificial Intelligence 66(2), 191–234 (1994)
19. Smets, P.: The transferable belief model for quantified belief representation. In: Gabbay, D.M., Smets, P. (eds.) Handbook of Defeasible Reasoning and Uncertainty Management Systems, vol. 1, pp. 267–301. Kluwer, Doordrecht (1998)
20. Smets, P.: Application of the transferable belief model to diagnostic problems. International Journal of Intelligent Systems 13(2-3), 127–157 (1998)
21. Tessem, B.: Approximations for efficient computation in the theory of evidence. Artif. Intell. 61(2), 315–329 (1993)
22. Trabelsi, S., Elouedi, Z., Mellouli, K.: Pruning belief decision tree methods in averaging and conjunctive approaches. International Journal of Approximate Reasoning 46, 91–124 (2007)
23. Trabelsi, S., Elouedi, Z.: Learning decision rules from uncertain data using rough sets. In: The 8th International FLINS Conference on Computational Intelligence in Decision and Control, Madrid, Spain, September 21-24, pp. 114–119. World Scientific, Singapore (2008)
24. Trabelsi, S., Elouedi, Z.: Heuristic method for attribute selection from partially uncertain data using rough sets. International Journal of General Systems 39(3), 271–290 (2010)
25. Trabelsi, S., Elouedi, Z., Lingras, P.: Dynamic reduct from partially uncertain data using rough sets. In: Sakai, H., Chakraborty, M.K., Hassanien, A.E., Ślęzak, D., Zhu, W. (eds.) RSFDGrC 2009. LNCS (LNAI), vol. 5908, pp. 160–167. Springer, Heidelberg (2009)
26. Trabelsi, S., Elouedi, Z., Lingras, P.: A comparison of dynamic and static belief rough set classifier. In: Szczuka, M., Kryszkiewicz, M., Ramanna, S., Jensen, R., Hu, Q. (eds.) RSCTC 2010. LNCS (LNAI), vol. 6086, pp. 366–375. Springer, Heidelberg (2010)
27. Trabelsi, S., Elouedi, Z., Lingras, P.: Belief rough set classifier. In: Gao, Y., Japkowicz, N. (eds.) AI 2009. LNCS (LNAI), vol. 5549, pp. 257–261. Springer, Heidelberg (2009)

28. Trabelsi, S., Elouedi, Z., Lingras, P.: Rule discovery process based on rough sets under the belief function framework. In: Hüllermeier, E., Kruse, R., Hoffmann, F. (eds.) IPMU 2010. LNCS (LNAI), vol. 6178, pp. 726–736. Springer, Heidelberg (2010)
29. Zhong, N., Dong, J.Z., Ohsuga, S.: Data mining: A probabilistic rough set approach. In: Rough Sets in Knowledge Discovery, vol. 2, pp. 127–146 (1998)
30. Zhong, N., Ohsuga, S.: Using generalization distribution tables as a hypotheses search space for generalization. In: Proceedings of Fourth International Workshop on Rough Sets, Fuzzy Sets, and Machine Discovery, RSFD 1996, pp. 396–403 (1996)
31. Zouhal, L.M., Denoeux, T.: An evidence-theory k-NN rule with parameter optimization. IEEE Trans. Syst. Man Cybern. 28(2), 263–271 (1998)

Author Index

GPSR Compliance

The European Union's (EU) General Product Safety Regulation (GPSR) is a set of rules that requires consumer products to be safe and our obligations to ensure this.

If you have any concerns about our products, you can contact us on ProductSafety@springernature.com

In case Publisher is established outside the EU, the EU authorized representative is:

Springer Nature Customer Service Center GmbH
Europaplatz 3
69115 Heidelberg, Germany

Batch number: 09490872

Printed by Printforce, the Netherlands